Rating Valuation

Rating Valuation: Principles and Practice has long been the standard go-to guide for both students studying rating valuation and practitioners needing a comprehensive reference book covering rating law, valuation and, importantly, practice. This fifth edition brings the reader up to date with the changes for the 2023 Rating Revaluation and developments in case law, as well as highlighting the differences between the law in England and Wales. A comprehensive chapter covers rates in Northern Ireland.

Starting with the basics, the book goes on to provide more in-depth detail for advanced readers, using clear, accessible and engaging analysis and example valuations throughout to break down what many see as a complex subject.

Whether you are studying to pass your APC, or just want an overview of the changes following the latest revaluation, *Rating Valuation: Principles and Practice* will give you all you need to understand rating valuation.

Patrick H. Bond is former Head of Rating Valuation at the Valuation Office Agency's head office, a past president of the Rating Surveyors' Association and was chairman of the RICS Rating Diploma Holders' Section in its 2022 Centenary year. He is a visiting lecturer in Rating at Bayes (formerly Cass) Business School, City University, and an assessor on the RICS Rating Diploma Study Course as well as being an IRRV tutor on both Rating and Valuation courses. He is a well-known conference speaker including at various international conferences and contributes articles in the professional press.

Peter K. Brown is a former Professor of Property Taxation at Liverpool John Moores University, UK, and a David C. Lincoln Research Fellow, Lincoln Land Institute, Cambridge, Massachusetts, USA. He was a member of the second Wood Committee looking into the rateability of plant and machinery. He was a regular speaker at conferences and contributed to articles in the professional press.

Rating Valuation

Principles and Practice

Fifth Edition

Patrick H. Bond BSc (Est. Man.)
**FRICS Dip.Rating IRRV (Hons) and
Peter K. Brown** MRICS Dip.Rating
IRRV (Hons)

 Routledge
Taylor & Francis Group

LONDON AND NEW YORK

Designed cover image: Patrick H. Bond

Fifth edition published 2024
by Routledge
4 Park Square, Milton Park, Abingdon, Oxon, OX14 4RN

and by Routledge
605 Third Avenue, New York, NY 10158

Routledge is an imprint of the Taylor & Francis Group, an Informa business

First edition published by Estates Gazette 2002
Second edition published by Estates Gazette Books 2006
Third edition published by Routledge 2011
Fourth edition published by Routledge 2017

British Library Cataloguing-in-Publication Data
A catalogue record for this book is available from the British Library

ISBN: 978-1-032-28552-8 (hbk)
ISBN: 978-1-032-27828-5 (pbk)
ISBN: 978-1-003-29736-9 (ebk)

DOI: 10.1201/ 9781003297369

Typeset in Times New Roman
by Newgen Publishing UK

Contents

Cases

Statutes

England

Wales

Northern Ireland

Acknowledgements

The authors would like to thank all those people who have assisted in the preparation of the fifth edition of this book, particularly Sim Biring, Dario Costantin, Wayne Cox, Paul Dalling, Mandy Franklin, Dave Hughes, Heather King, Ian Cameron McIntosh, Billy Joss, Paul May, Adrian Mills, Steve Nettleton, Anthony Robinson, Chris Royle and Henry Spence.

1 Introduction

1.1 Introduction to rating

Rating is the oldest of the United Kingdom's taxes with the possible exception of stamp duty. Its origins can be traced back to before 1601. The reasons for its venerable origins lie in its particular advantages as a tax.

Today it is comparatively easy to forecast a tax's yield, though Chancellors of the Exchequer do not always find that the levels of receipts match the predictions they have been given. In the past the usual need was to raise a particular sum for a known purpose or at least to raise a target amount. The raising of the sum needed to be spread among the local population in a suitable way. A poll tax, in theory, would raise the target sum on the basis of dividing the sum to be raised by the number of people and getting each person to pay the resulting figure. However, it is usually fairer and certainly more successful to tax people who have money rather than to try and tax everyone including the poor. The early measure of wealth was the amount of land and buildings a person occupied. Persons with wealth or a substantial income not derived from land were extremely rare. It was found that a practical and acceptable way to raise a sum of money was to apportion the required sum 'rateably' among occupiers in proportion to the annual rental value of their occupation of land and buildings. A local authority, or in the past a body known as the Guardians of the Poor, could raise a known sum to pay its annual expenses by dividing the sum by the total of the rateable value in its area. Looking at a simple example, if a local authority's area had 200 properties in it, 100 with a rateable value (RV) of £250 each, 50 with an RV of £500 each and 50 with an RV of £1,000 each, the total value in its area would be:

100 @ £250 RV = £25,000
50 @ £500 RV = £25,000
50 @ £1,000 RV = £50,000
Total RV = £100,000

DOI: 10.1201/9781003297369-1

If the local authority needed to raise a sum of £50,000, the rate would have to be set at half the RV or 50p in the £, i.e.

£100,000 @ 50p = £50,000

In addition to giving a great deal of certainty in their yield, rates have another great advantage in being a tax on something that cannot be hidden: at least not easily. There are stories nevertheless of people escaping paying rates on new properties for many years because the local authority had not noticed them! The idea is made even easier by rates being a tax on occupation, rather than ownership, which may be more difficult to identify. All the local authority has to do, in simple terms, is go to a building, knock on the door and ask the occupier to pay the rates for the property. When looking at other taxes by way of comparison, income can be hidden, VAT may not be paid where the builder is paid cash or, as was found with the unsuccessful 1990 Community Charge or Poll Tax experiment, people can simply disappear from local authority registers to evade a poll tax.

Rating has developed over the years and since 1990, with the Local Government Finance Act 1988 coming into effect, ceased to be a local tax set by the local authority. Instead, the level of the rate, or 'poundage', was to be set nationally by the Government and Senedd or Welsh Parliament though local authorities, as billing authorities, remained responsible for collecting rates. Additionally in 1990, the rating of houses and flats was ended and rating was made into a tax on non-domestic property only. Since 1993, a new tax on domestic property has been raised, the Council Tax, based on capital rather than rental values.

The 'nationalisation' of local rates in 1990 has been a little ameliorated by the recent 'Localism' agenda of the Government and local 'Rate Retention'. This is explained a little in section 1.8.

A popular, but misleading, term for rates is 'business rates'. As Professor Graham Zellick, formerly President of the Valuation Tribunal for England (VTE), said in *Coad and others* v *Gidman (VO)*, unreported: 'It is true that non-domestic rates are frequently described as "business rates", even by government, but there is no statutory basis for this. It is a colloquialism – and a misleading one at that.'

Misleading indeed, as rates are payable on properties which are neither domestic nor exempt from rating, and there is no requirement at all for a business to be run in the property for it to be assessed for rates. It is best not used and that is the approach taken in this book. Nonetheless it is popularly adopted as a term for non-domestic rates. Until comparatively recently, the phrase did not appear in statute, but the Business Rates Supplement Act 2009 does use the term in providing for county councils, unitary authorities and the Greater London Authority (GLA) to have the power to levy a supplement on the national non-domestic rate. This power is principally intended to fund the London 'Crossrail' scheme.

1.2 Historical

Rating is normally regarded as having commenced with the Poor Relief Act 1601, one of the last Acts of Queen Elizabeth I's reign. However, this Act largely repeated an earlier enactment of 1597, which itself consolidated earlier acts. The 1601 Act remained the basic Rating Act, though supplemented by other legislation, up until its consolidation into the General Rate Act 1967.

The Poor Relief Act 1601 required the Overseers of the Poor to raise sums of money for the relief of the 'poor by taxation of every inhabitant, parson, vicar and other, and of every occupier of lands, houses, tithes inappropriate or propriations of tithes, coal mines or saleable underwood in the said parish ...'.

The Act did not actually specify what the basis of the taxation was to be. It may well have been that the intention was to tax visible wealth and, certainly, the inclusion of 'inhabitant' as well as 'occupier' suggests that the intention was to tax not just the annual value of the occupation of land but also any other wealth an inhabitant possessed. The idea seems to have been to tax according to the ability of the person to contribute to the relief of the poor, and this ability would not have meant just the value of occupied landed property. In 1635, Dalton's *Country Justice*, commenting on the rating provisions of the 1601 Act, makes it clear that rating was about a fair distribution of the rate and both the value of stock in trade as well as landed property should be taken into account:

> *In these taxations there must consideration be had, first to equality, and then to estates. Equality, that men be equally rated with their neighbours, and according to equal proportion. Estates, that men be rated according to their estates of goods known, or according to the known yearly value of their lands, farms, or occupyings.*

The eventual removal of liability from the 'inhabitant' by the Poor Rate Exemption Act 1840 left the occupier as the only rateable person. Rental value had, by then, become regarded as indicating what a property would yield and thereby giving the measure of the 'ability' of the occupier. Confirmation of this appeared in statute with the Parochial Assessments Act 1836 defining 'net annual value' as: 'the rent at which the hereditament might reasonably be expected to let from year to year ...'

Setting a standard basis for assessing the occupier across the country was an early attempt at establishing uniformity of assessment and was followed by the Union Assessment Act 1862. This Act required assessment committees to be set up to supervise valuations within unions of parishes and valuation lists to be prepared of all rateable hereditaments in a parish, and persons aggrieved by a valuation list could object on the grounds of unfairness or incorrectness in any valuation.

The Union Assessment Act 1862 also required a valuation list to show both a gross value and a rateable value for a hereditament. The gross figure represented the rental value assuming a landlord was responsible for repairs and insurance. The rateable value represented the rental value after an amount had been deducted for these expenses. This requirement lasted for most hereditaments until 1990 and now all rateable properties only have a rateable value.

The Valuation (Metropolis) Act 1869 further improved the system by requiring new valuation lists to be prepared every five years. This pattern of quinquennial revaluations remains part of the rating system today. This Act only applied to London and it was not until the Rating and Valuation Act 1925 that the requirement for quinquennial revaluations was extended outside the capital. The Rating and Valuation Act 1925 applied outside London and made further improvements. It abolished the overseers and transferred their functions to local authorities. County valuation committees were appointed to promote uniformity, and a Central Valuation Committee was appointed to make representations which became general guidance for all the local authority rating valuers. Instead of ratepayers objecting to valuation lists, a new system of appeal by making a proposal was introduced.

The system of various committees and each local authority preparing its own valuation list was found unsatisfactory, with different local authorities undertaking the preparation and maintenance of lists in different ways and conflicts of interest arising because local authorities both assessed and were the beneficiaries of the rates. The Local Government Act 1948 sought to remedy this by separating the function of assessing rateable values and preparing valuation lists away from the function of collecting rates. It required the Commissioners of Inland Revenue to appoint valuation officers to prepare and amend valuation lists. It abolished the assessment committees and, in their place, created independent local valuation courts.

To encourage uniformity, the system of representations of the Central Valuation Committee, which was abolished, was replaced by guidance from the valuation officers' head office under the direction of a chief valuer. This guidance is now generally available and can be accessed on Gov.uk at the Valuation Office pages, www.gov.uk/government/organisations/valuation-office-agency.

The Local Government Act 1948 was consolidated into the General Rate Act 1967. Under this Act, in general terms, properties that were mainly land were assessed direct to Rateable Value as were industrial properties. Rateable Value assumed a letting on full repairing and insuring (FRI) terms. All other properties were assessed to Gross Value, which assumed a letting on the basis that the tenant was responsible for internal repairs and insurance.

The date of valuation for the General Rate Act 1967 was the day the valuation list came into force.

Under the General Rate Act 1967 the rental value was a ceiling and a reduction in assessment could be obtained where the rental value had fallen to a lower value than that in the list. Consequently, the valuer had to ensure that the rental value of the property was always in excess of the tone value.

The General Rate Act 1967 remained the principal rating act until the changes to the rating system in 1990, when domestic rating was ended with the Local Government Finance Act 1988 coming into force. This retained the separate responsibility of valuation officers to maintain lists of values, now called rating lists, for the local authorities. Valuation courts were renamed valuation tribunals, and rating legislation was rewritten to tie in with the Community Charge legislation, with an extensive use of enabling powers in the primary legislation giving ministers the power to make regulations covering much of the workings of the rating system. Particularly with the regulations covering the rights of appeal, these have been subject to frequent revision. The relevant law is, therefore, found not in the main act, but in a multitude of statutory instruments often altered by amending statutory instruments.

1.3 Other approaches to local property taxation

The British system of local property taxation or rates can be found in many Commonwealth countries. However, even among these there is a divergence of approach to property taxation.

There are three main approaches to property taxes involving lists of property values, as opposed to more ad-hoc taxes such as those on capital gains or on realised development value which only follow specific acts by taxpayers.

The first approach, as adopted in the United Kingdom, is to tax occupiers on the rental value of property. This approach is also adopted in Ireland, Italy, France and Belgium.

The second approach is based on the capital or selling value rather than the rental value. Where this occurs, it is more common to tax the owner rather than the occupier. The Czech Republic, Denmark, Austria and Latvia all adopt this approach.

A third approach is to assess the value of the land only, either as bare land or including the value of infrastructure and site works but not building works. This is known as Land Value Taxation or sometimes Site Value Rating. This was advocated by the economist Henry George in the nineteenth century and has been proposed for adoption in the United Kingdom by a range of organisations. In most cases the tax is levied on the owner of the land, rather than the occupier.

The majority of European countries use some form of land tax based on one of the three methods. Only Malta does not have a system of local property taxation. It is favoured and encouraged by the International Monetary Fund, the World Bank and others as a visible and reliable way of raising tax. In Eastern Europe, former communist countries have looked to develop this form of tax to give local authorities their own tax base. Without their own tax base local authorities, as in the past, are merely agents of the central government and this central control does not encourage the desired development of local democracy and a measure of autonomy.

Table 1.1 Number and value of hereditaments by type by rateable value 2017–2023

Generic property type	No. 2017	RV 2017	No. 2023	RV 2023
Advertising	36,085	114,607,729	36,915	95,220,928
Camping	8,727	315,268,633	8,620	278,982,858
Car parks	80,437	717,602,417	78,095	712,941,104
Hotels	102,459	1,650,907,542	97,282	2,248,457,014
Markets	964	35,876,118	1,002	34,398,846
Miscellaneous commercial	41,482	1,119,724,835	40,193	865,663,767
Offices	436,517	16,404,015,100	437,141	14,900,650,207
Public houses, clubs	52,990	1,565,798,676	53,224	1,811,959,989
Restaurants, cafés	50,318	1,735,190,421	49,142	1,729,762,266
Shops	514,173	14,756,178,476	514,161	16,524,741,400
Vehicle showrooms and repair	39,568	1,497,075,270	39,408	1,292,161,881
Warehouses	257,803	9,607,342,375	256,067	7,270,277,313
Education	48,414	4,832,038,875	48,172	4,287,629,391
Docks & power	4,690	692,831,900	4,760	643,724,605
Factories, workshops	270,032	8,242,370,729	267,803	6,531,491,514
Community, leisure	96,068	2,163,245,191	95,750	2,023,641,029
Miscellaneous	83,094	3,878,625,128	83,098	3,781,413,784
Transport	17,272	591,839,466	16,283	457,439,423
Defence related	1,127	415,915,785	1,377	350,395,715
Total	2,142,220	70,336,454,666	2,128,493	65,840,953,034

Note: the above compares draft rating lists and is for illustration purposes only.

1.4 Rateable properties

Correctly, properties are not rateable, but people who occupy properties are rateable in respect of their occupation. Rating is a tax on the occupation of land including buildings. Rateable properties have a technical name given in the legislation. A rateable property is known as a hereditament.

The data above are derived from the draft rating lists figures published by the Valuation Office Agency (VOA) November 2023.

The number of hereditaments appearing in rating lists in England and Wales is approximately 2,140,000.

Table 1.2 shows how many local list hereditaments fall into different rateable value bands.

Table 1.3 shows the number of hereditaments and their values broken down by region.

1.5 Assessment

While local authorities remain the collectors of rates, they have not, since the early 1950s, been responsible for determining rateable values. This responsibility passed to statutory officers, valuation officers, appointed by the Board of HM Revenue and Customs for each local authority area. These valuation

Table 1.2 Number of hereditaments by rateable value

<£12,000)	1,342,721
£12,000–£25,000	3665,72
£25,000–£50,000	201,285
£50,000–£75,000	76,739
£75,000–£100,000	39,732
£100,000–£500,000	98,078
£500,000–£1,000,000	10,327
£1,000,000–£5,000,000	6,308
£10,000,000–£20,000,000	92
>£20,000,000	28
Grand total	2,141,882

officers are responsible for preparing and maintaining lists of rateable values for each billing authority area called 'Rating Lists'.

Up until 1990, domestic properties were rateable, but since 1993 they have been subject to the Council Tax.

1.6 Revaluation

The intention from the 1950s was that properties would be revalued by valuation officers preparing new rating lists every five years. In *Dawkins (VO)* v *Ash Brothers and Heaton Ltd* [1969] 2 AC 336, Lord Pearce said:

> *Rating seeks a standard by which every hereditament in this country can be measured in relation to every other hereditament. It is not seeking to establish the true value of any particular hereditament but rather its value in comparison with the respective value of the rest.*

The object of revaluation is to ensure this standard is maintained. Over time the relative values of properties do change. One shopping centre may become more popular and valuable than another; one type of property may become substantially outmoded, whereas another may enjoy a rise in demand. If revaluations do not occur, the relativities between properties and localities become fixed. Revaluation enables the relativities to be looked at afresh and corrected using the test of rental value to give relative worth. Regular revaluations are essential to ensure fairness between rate or taxpayers.

The early hope for five-yearly – quinquennial – revaluations was not well achieved early on, with the gaps from 1956 to 1963, 1963 to 1973 and 1972 to 1990 being well in excess of that objective. More recently the aim has been achieved, though there was a short postponement of the 2015 Revaluation to 2017. Recent revaluations have, therefore, taken place in England and Wales in 1990, 1995, 2000, 2005, 2010 and 2017 with the most recent being the revaluation in 2023. The Government has signalled its intention to move to a three

Table 1.3 Number of hereditaments and their values broken down by region

RV bands	East Midlands	East of England	London	North East	North West	South East	South West	West Midlands	Yorkshire and The Humber
<£12,000)	106,066	130,589	140,770	6,3421	199,247	170,616	158,021	132,759	144,974
£12,000–£25,000	26,288	38,948	72,211	12,489	41,296	56,484	36,934	35,367	31,117
£25,000–£50,000	13,514	2,0745	45,031	6,487	22,205	32,144	19,524	18,058	15,818
£50,000–£75,000	5,029	7,890	18,213	2,553	8,237	12,495	6,947	6,699	5,902
£75,000–£100,000	2,404	4,125	10,188	1,241	4,095	6,462	3,494	3,358	3,024
£100,000–£500,000	5,646	9,548	28,135	3,139	10,071	15,316	7,761	8,045	7,085
£500,000–£1,000,000	586	978	3,655	263	896	1612	690	769	613
£1,000,000–£5,000,000	431	563	2,180	187	601	868	368	517	418
£10,000,000–£20,000,000	1	5	63	2	4	5	2	4	2
>£20,000,000	–	4	15	–	3	4	–	–	2
Grand total	159,965	213,395	320,461	89,782	286,655	296,006	233,741	205,576	208,955

yearly revaluation cycle with the next revaluation being on 1 April 2026, but this will require primary legislation.

Council Tax was introduced in 1993 and, whilst there was a revaluation of Council Tax bands in Wales in 2005, the Council Tax revaluation planned for 2007 in England was cancelled after much work had been undertaken, and no date has yet been announced for a new revaluation. This leaves the perhaps surprising position at the time of publication of the fifth edition of this book that the valuation basis in England is almost a third of a century out of date! The Welsh Government undertook a revaluation based on values at 1 April 2003 and having effect from 2005, and is considering a further revaluation coming into effect in 2025.

1.7 Collection

Rating in England, Wales and Scotland can now be seen as a national and not a local tax.

In introducing the new post-1990 rating system, the Government sought to achieve uniformity so that a business in one local authority area would be paying rates at the same level as its competitor in the adjoining area. It was not seeking an identity of bills, as one business might enjoy advantages flowing from location or quality and size of premises that another business in a different location did not have. These differences would be reflected in rents paid and therefore in the rateable values based on the rents. However, the Government considered the previous practice of individual authorities setting their own percentages of rateable value to pay as rates, known as poundages, to be unfair.

In recent years consideration has been given to allowing billing authorities to set or retain an amount of business rates collected in their area.

A single poundage for England and a separate one for Wales was introduced and known as the Non-Domestic Multiplier. Its rise, year on year, was set so as not to exceed the rate of inflation rather than being based on the local need to raise a particular sum of money or, as might be expected of a national tax, on the Chancellor's requirements for taxation income each year. The change to the multiplier is calculated using the annual rise in the Retail Price Index (RPI) to the September before the start of the new rate year (properly 'Chargeable Financial Year') which runs from 1 April to 31 March.

The original multipliers for England and Wales In 1990/91 were 34.8p and 36.8p respectively.

From 2005/06 onwards, a small business multiplier reduces the multiplier slightly for small hereditaments in England but not Wales. Up until 2017 this was for properties in Greater London with a rateable value of less than £26,000 and for properties elsewhere with a rateable value of less than £18,000. From 1 April 2017 it is £51,000 for the whole of England.

Table 1.4 Uniform Business Rate (UBR) 2010–2024

Year	England	Wales	England small hereditament
2010/11	41.4p	40.9p	Reduced by 0.7p for small hereditaments
2011/12	43.3p	42.8p	Reduced by 0.7p for small hereditaments
2012/13	45.8p	45.2p	Reduced by 0.8p for small hereditaments
2013/14	47.1p	46.4p	Reduced by 0.9p for small hereditaments
2014/15	48.2p	47.3p	Reduced by 1.1p for small hereditaments
2015/16	49.3p	48.2p	Reduced by 1.3p for small hereditaments
2016/17	49.7p	48.6p	Reduced by 1.3p for small hereditaments
2017/18	47.9p	49.9p	Reduced by 1.3p for small hereditaments
2018/19	49.3p	51.4p	Reduced by 1.3p for small hereditaments
2019/20	50.4p	52.6p	Reduced by 1.3p for small hereditaments
2020/21	51.2p	53.5p	Reduced by 1.3p for small hereditaments
2021/22	51.2p	53.5p	Reduced by 1.3p for small hereditaments
2022/23	51.2p	53.5p	Reduced by 1.3p for small hereditaments
2023/24	51.2p	53.5p	Reduced by 1.3p for small hereditaments

The City of London is able to set a different multiplier from the rest of England, and for 2023/24 this was 52.6p with a reduction of 1.3p for small hereditaments.

A Business Rate Supplement (BRS) can be levied by certain councils, but only the GLA has adopted the procedure and had it approved. This is to fund the London 'Crossrail' project. The GLA decided to set an additional multiplier of 2p per pound of rateable value from April 2010 on the occupiers (or owners if vacant) of properties with a rateable value of over £55,000 in all 32 London boroughs and the Corporation of London with a rateable value of over £55,000. A property with a rateable value of £100,000 therefore incurs an annual BRS contribution of £2,000 (i.e. 2% of £100,000). The term 'Business Rate Supplement' is, because of the rateable value threshold, something of a misnomer as some 83% of properties do not incur a liability as their rateable values are £55,000 or less. A more accurate term might be Large Business Rate Supplement.

Despite the rate being a national tax, collection remains the responsibility of the district, borough and unitary local authorities (properly called billing authorities in this role). There are 336 billing authorities in England and Wales.

1.8 Developments in the system

In 1990 when the rating system was substantially reformed the idea of a locally set rate was ended and instead a national multiplier was set separately for England and Wales. There had been considerable concern at the disparity in 'poundages' levied by different councils and having a single national multiplier was seen as ensuring a fairer system.

From 1990 until 2013 this approach continued with local government simply collecting the rates and passing it, net of allowed collection costs, to central government. The total rates income was pooled and redistributed to local government on a broadly per capita needs basis. This changed in 2013 under the Local Government Finance Act 2012, which allowed councils to retain up to 50% of the rates collected from their area. The remaining 50% was still pooled and redistributed to local authorities as revenue support grant as before. The amount retained by each local authority is up to 50% because some have more rates income than others. 'Rates rich' councils paid a 'tariff' used to support 'top-up' payments made to those whose rate income would be insufficient for their needs.

The Rates Retention Scheme, therefore, provides for the redistribution of some rates revenue between local authorities, based on an assessment of the difference between their level of rate revenue and the funding they need to provide services. Every local authority either pays funds into a central pot via a 'tariff' or receives money from the central pot via a 'top-up'.

The spring 2016 budget proposed that, by 2020, councils would retain not only 50% but up to 100% of rates income; they would be allowed to apply discounts to the national multiplier and for those with an elected mayor they would be able to levy supplements to the multiplier to fund infrastructure projects. A bill was introduced in 2017 but did not complete its legislative stages before the 2017 general election and was not later reintroduced.

Another substantial change from April 2017 was that small businesses that occupy properties with a rateable value of £12,000 or less would pay no rates and that this would be a permanent relief. This has been criticised by some commentators as whilst small businesses may individually need some support for a time it is obvious that not all need support; indeed many will be very profitable and able to contribute. This took a large number of occupations out of rating altogether: according to the Government some 600,000 ratepayers. It should be noted that it applies to small business occupiers and not simply any occupation of properties with modest rateable values.

Various reliefs and exemptions have been introduced over the years with laudable purposes of providing relief to certain categories of ratepayers.

A 2016 Northern Ireland consultation document 'Reforming the Rating System – Spurring Economic Growth' appeared to recognise the indiscriminate nature of a blanket relief. In paragraph 33 it described the existing tapering small business relief in Northern Ireland as including 'significant deadweight (many firms do not need the subsidy) and a total absence of any outcome based measures'.

In paragraph 36, the consultation document said:

As this policy is intended to 'stimulate' investment in cities, towns and villages it will be important to ensure that it is not regarded as an ongoing entitlement and eventually become capitalised into higher rents. Consequently it is proposed that the scheme will be operated for a period of three years before being phased out over a further two year period. An evaluation will then take place after this five-year period to judge the effectiveness of the policy.

This highlighted the real problem when trying to use rates to achieve a policy objective in that reliefs become translated into an ability to offer higher rents so the benefit of the relief ends up in the freeholder's pocket. If two or more people are competing for premises and know they will no longer need to pay rates they are able, and do, offer more in rent in order to secure the premises. This propensity is known as 'equation theory'. The idea of this is a tenant is not particularly concerned with the amount of rent or the amount of rates paid but is concerned by the total outgoings of occupation. If the amount of rates is depressed by a relief then this leaves more money available to pay rent, and a prospective tenant enjoying such relief may outbid another possible tenant who does not enjoy the relief.

There are therefore likely benefits in having targeted reliefs in the short term but reliefs should, perhaps, not be long term or permanent. It is also arguably a good principle of taxation that everyone pays something.

In November 2021 the Government published a consultation document 'Business Rates Review: Technical Consultation', which sought views on a number of issues for England including:

- more frequent revaluations;
- more responsibility placed on the ratepayer to supply information relating to the property, including its use, tenancy agreements, trade and account information, cost information and changes to the property;
- reform of the appeals system;
- relief for improvements;
- green measures;
- a range of administrative issues including the simplification of the system of discretionary relief.

A Non-Domestic Rating bill had its first reading in the House of Commons on 28 March 2023 to implement the above reforms. As drafted the bill proposes:

- more frequent revaluations, to ensure rate demands keep up with economic trends;
- administrative reforms to deliver a sustainable shorter revaluation cycle, intended to enable more accurate rating lists and enable the disclosure of more information to ratepayers about their valuations;
- measures to support decarbonisation and investment worth over £700 million, including a relief for low carbon heat networks and a new Improvement Relief so that, from April 2024, ratepayers will not see an increase in rates payable from qualifying improvements made to their properties for 12 months;
- removal of the statutory requirement for revenue neutrality within Transitional Relief;
- enabling the Digitalising NDR project, which will modernise the system, improve the targeting of rates relief, generate better data for central and local government and help to improve compliance;

• improvements to the administration of NDR, including replacing RPI with CPI as the measure of inflation used in the annual indexation of the multiplier.

1.9 Transitional arrangements

The relative simplicity of the concept of Rateable Value × Multiplier = Rates Payable has been very much complicated by schemes of transitional relief or phasing first introduced by the Government from 1990.

The Government realised that the twin effects of a revaluation in 1990 and the introduction of the single multipliers would mean some businesses being faced with very large increases in rates. These would be in areas that had seen out of the ordinary rental growth since 1973 and/or were in areas where the previous poundages set by the local authority had been comparatively low. The Government considered it unfair and likely to cause major criticism if these increases in rates came in straightaway, so a system to phase them in slowly was devised. The Government decided the scheme should be self-financing by also phasing in decreases in rates liabilities caused by the revaluation and the change to national multipliers.

The paradox of this scheme was soon noted by commentators. It resulted in businesses which had seen low rental growth and were perhaps in declining industries paying more than their 'true' liability, whereas businesses in successful areas where values had sharply increased would be paying less than their 'true' liability. The difference in many cases seemed to be between the northern industrial properties that became subject to downward phasing and the southern offices and shops which had enjoyed exceptional growth. The two classes of business were rather euphemistically described as 'gainers' and 'losers': 'gainers' being those whose properties were subject to downwards phasing and had had poor rental growth; 'losers' being those in upward transition who had enjoyed good rental growth and general business success!

However, for most properties, the effects of phasing would have worked themselves through by the end of the five-yearly revaluation cycle, had not the effects of the scheme for businesses subject to upwards phasing been modified on the grounds of helping business generally. Transitional relief remains a feature of the rating system notwithstanding the distortion in payments that it caused. There were still occupiers of some properties paying rates which have been subject to phasing throughout the 1990, 1995, 2000 and 2005 revaluations right up to the start of the 2009/10 year when the 2005 scheme finished. Their occupiers had, therefore, been paying rates not based on current property values but on the basis of their rates liability in 1989/90 calculated using the 1973 valuation list rateable value, with year-on-year phased increases according to the transitional relief formulae.

For the 2010 Revaluation that came into effect on 1 April 2010, the transitional relief scheme in England reverted to being over the life of the rating lists rather than ending after four years as the 2010 scheme had done but was only a scheme based on the 2009/10 liability and not any years before. This continued for the 2017 rating lists for England and a new scheme introduced for Wales. For 2023 the approach adopted in England is that there will be no downward transition, i.e. properties whose rate liability has decreased due to the 2023 revaluation will get the full benefit of the reduction immediately. For properties where the liability has increased, subject to provisions, the increase will be phased in over effectively three years. For full details see section 16.11.

In Wales the scheme will limit increases in rate bills as a result of the revaluation on 1 April 2023. Eligible ratepayers will pay 33% of their additional liability in the first year (2023–24) and 66% in the second year (2024–25), before reaching their full liability in the third year (2025–26). See section 16.13.

1.10 Wales

The introduction of devolved government for Wales has permitted the Welsh Government to choose to make different regulations from those applying to England. This allows innovative ideas to be tried out and solutions particularly suited to Wales to be adopted, but it does add an unwelcome complexity for the practitioner and student.

With effect from the 1 April 2015, non-domestic rates were fully devolved to the Welsh Government. 'The Devolution of Business Rates to Wales', Report to the Minister for Economy, Science and Transport from the Business Rates Panel (February 2015), outlined the implications of devolution and some of the issues facing the Welsh Government in the future.

The Non-Domestic Rating (Miscellaneous Provisions) (Wales) Regulations 2017 (2017/327) W.82 makes a number of provisions required as a result of the devolutions of rates to the Welsh Assembly Government.

Unless otherwise stated, this book refers to the English legislation. Any difference in the Welsh legislation will be identified at the end of each chapter.

On 12 July 2022 the Welsh Government published a consultation document, 'A Fairer Council Tax'. Responses were required by 22 October 2022.

The Welsh Government is considering:

- Conducting a revaluation of all 1.5 million domestic properties across Wales because the council tax bands are almost 20 years out-of-date. It said, 'The evidence suggests we are unable to achieve our goal of more progressive tax bands without doing this. We would not seek to raise additional revenue overall through this exercise.'
- Implementing an updated property list with effect from 1 April 2025, based on valuations from 1 April 2023. It said, 'This is the earliest timeframe we could implement an updated property list.'
- Adding more council tax bands and changing the tax rates charged for each band, to make the system more progressive, shifting more of the tax burden to the highest end of property values.

- Putting in place a transitional relief scheme to smooth the transition for households affected by changes.
- Ensuring local council revenues are not impacted overall, by redistributing the Welsh Government's Revenue Support Grant according to the latest property information.

On 21 September 2022 the Welsh Government published a consultation document 'Reforming Non-Domestic Rates in Wales'. Responses were required by 14 December 2022.

The document sets out number of proposals on which it sought views, which are summarised below:

- more frequent revaluations;
- improving the flow of information between the Welsh government and ratepayers;
- providing the Welsh Government with more flexible legislation to amend reliefs and exemptions;
- reviewing the reliefs and exemptions;
- greater scope to vary the NNDR multiplier;
- improving the administration of the valuation functions and rating lists;
- tackling avoidance.

1.11 Scotland

Rates are also levied in Scotland but due to the differences in the legal system in Scotland the legislation is quite different from that in England and Wales, and falls outside the scope of this book. Council Tax also applies in Scotland.

1.12 Northern Ireland

Northern Ireland continues to levy rates on both domestic and non-domestic property. While the system is broadly based on the English and Welsh system, the legislation did not develop in parallel with the English and Welsh system during the Troubles and was somewhat frozen in time. Now the legislation has been updated and a further revaluation occurred in 2020. A description of the system is contained in Chapter 19.

1.13 The Irish Republic

The Irish Republic has a similar system to Northern Ireland, certainly as regards its basis given the common root of the pre-partition rating system and the Griffith's Valuation. The system stayed very much the same for many years with net annual values, whilst amended for changes, not updated by revaluations. Innovative methods have been utilised to undertake revaluations

Table 1.5 Illustrative rateable values

	2010 rating lists	*2017 rating lists*	*2023 rating lists*
Airports			
Heathrow Airport	£216,480,000	£215,000,000	£210,000,000
Liverpool Airport	£5,640,000	£4,580,000	£1,400,000
Birmingham Airport	£12,500,000	£12,550,000	£10,080,000
Manchester Airport	£32,100,000	£27,630,000	£25,080,000
Conference centres			
NEC, Birmingham	£9,400,000	£7,250,000	£8,710,000
Harrogate International Centre	£1,120,000	£1,170,000	£1,250,000
Football grounds			
Aston Villa	£3,140,000	£2,870,000	£2,350,000
Everton	£2,579,000	£2,240,000	£3,170,000
Liverpool	£4,000,000	£3,170,000	£5,770,000
Arsenal	£7,250,000	£6,130,000	£8,650,000
Manchester United	£7,620,000	£6,090,000	£8,680,000
Newcastle United	£4,310,000	£4,000,000	£4,540,000
Hotels			
Hilton Park Lane, London	£7,200,000	£8,340,000	£3,495,000
The Dorchester, London	£7,000,000	£11,565,000	£3,600,000
The Waldorf, London	£3,160,000	£3,735,000	£1,920,000
The Adelphi, Liverpool	£710,000	£655,000	£362,000
Miscellaneous			
Bank of England	£6,810,000	£8,140,000	£7,290,000
British Library	£12,700,000	£12,570,000	£16,410,000
National Gallery	£3,790,000	£4,410,000	£6,270,000
BBC Broadcasting House		£25,910,000	£22,250,000
Houses of Parliament	£14,700,000	£12,910,000	£17,240,000
Old Bailey	£3,000,000	£3,230,000	£3,120,000
Tower of London	£1,790,000	£3,400,000	£2,750,000
Wormwood Scrubs	£1,480,000	£1,090,000	£1,430,000
Channel Tunnel			£40,000,000
Industrial			
BNFL, Sellafield	£67,500,000	£40,630,000	£49,190,000
Jaguar Cars, Birmingham	£5,290,000	£6,380,000	£8,740,000
Jaguar Cars, Halewood	£6,160,000	£6,040,000	£7,170,000
Leisure			
O2 Arena	£4,500,000	£10,265,000	£11,100,000
Blackpool Tower	£520,000	£545,000	£482,500
Alton Towers	£6,320,000	£8,380,000	£7,000,000
Wimbledon Tennis	£6,000,000	£6,940,000	£16,250,000
Retail			
Selfridges, Oxford Street	£22,370,000	£31,060,000	£16,820,000
Harrods, Knightsbridge	£22,700,000	£25,700,000	£18,000,000
John Lewis, Oxford Street	£12,950,000	£19,910,000	£6,570,000
Networks – central list			
Network Rail	£227,000,000	£384,000,000	£450,000,000
BT	£244,460,000	£714,870,000	£586,540,000
United Utilities Water	£122,020,000	£136,930,000	£96,590,000

and modernise the system with ratepayer assisted valuations, use of private sector valuers and, importantly, a rolling revaluation.

The Revaluation Programme provides for the ongoing revaluation of all rateable property in Ireland on a rolling basis rating area by rating area rather than in a single national revaluation. Deferral of a number of revaluations were made due to the impact of the COVID-19 pandemic, but in 2022 the Commissioner of Valuation made fresh Valuation Orders for the rating authority areas of Clare, Donegal, Dún Laoghaire-Rathdown, Galway, Kerry and Mayo County Councils and Galway City Council. Some of these are second revaluations.

The revaluations will assess the rental value of each rateable property at the valuation date of 1 February 2022. It is expected the new valuations will be published on 22 September 2023 and will become effective for rating purposes from 1 January 2024.

1.14 Illustrative rateable values

Before moving on to examine the subject in more detail, it is worth just noting some of the rateable values that have been placed on well-known properties in the draft 2023 rating lists. While these values may well change before the list becomes effective and as ratepayers challenge their assessments, they do indicate the significance of the tax to both ratepayers and the Government. They also highlight how values change between revaluations.

1.15 Current issues and forthcoming changes

Whilst every effort has been made to ensure this book is as up to date as possible as of August 2023, inevitably in a fast-moving subject area the authors are aware of issues, draft legislation and cases that have not been settled, enacted or decided at the time the book goes to press. It is important to be aware of changes that may have occurred.

Appendix 2 provides a short summary of late breaking issues, cases, etc., that they were not able to include within the main body of the book.

2   The rateable occupier

2.1 Introduction

Rating is a tax on occupation not ownership. Occupiers are rateable in respect of the land and buildings they occupy unless the properties are exempt or domestic. As Lord Russell of Killowen said in the leading case of *Westminster City Council* v *Southern Railway Co*, *The Railway Assessment Authority and WH Smith and Son Ltd* [1936] AC 511: 'The occupier, not the land, is rateable; but, the occupier is rateable in respect of the land which he occupies.'

This chapter is concerned with the concept of the Rateable Occupier, the person who is going to have to pay the tax. The rateable property, or 'hereditament', will be considered in the next chapter.

Rates have always been a tax on occupation. This idea is central to the rating system. An immediate objection to this idea can be raised, 'but aren't rates also paid on empty property?' They are indeed, but this has only been the case since 1966 and rates on empty property are best seen as separate and not part of the 'main' rating system. This approach of treating empty property rating as an 'add on' is in line with both the historical development of rating and the statutory scheme which provides separately for the rating of empty property. Chapter 17 deals with the rating of empty property.

Rates are levied on the basis of the value of the occupation of the hereditament to the occupier. In *Poplar Assessment Committee* v *Roberts* [1922] 2 AC 93, Lord Buckmaster said:

> *It is in respect of his occupation that the rate is levied, and the standard in the Act is nothing more but a means of finding out what the value of that occupation is for the purposes of the assessment.*

The standard, or measure, of the value of the benefit of occupation to the occupier is set out in the Local Government Finance Act 1988 and is the rental value of the property. This measure of rental value is covered in Chapter 6.

The liability of persons in occupation of rateable property to pay rates is set out in section 43 of the Local Government Finance Act 1988:

DOI: 10.1201/9781003297369-2

43-(1) A person (the ratepayer) shall as regards a hereditament be subject to a non-domestic rate in respect of a chargeable financial year if the following conditions are fulfilled in respect of any day in the year –

(a) on the day the ratepayer is in occupation of all or part of the hereditament, and

(b) the hereditament is shown for the day in a local non-domestic rating list in force for the year.

The Act does not define occupier (though it does define owner!). In section 65(2) it refers to the General Rate Act 1967, which was the principal rating act prior to the changes brought about by the Local Government Finance Act 1988, which came into effect on 1 April 1990:

Whether a hereditament or land is occupied, and who is the occupier, shall be determined by reference to the rules which would have applied for the purposes of the 1967 Act had this Act not been passed (ignoring any express statutory rules such as those in sections 24 and 46A of that Act).

Consequently, the 'old' rules still apply. However, if the General Rate Act 1967 is examined, while section 16 establishes an occupier's liability, it does not define or give these rules that enable an occupier to be identified.

The rules have evolved over the centuries of rating in case law. One of the best summaries of what constitutes rateable occupation can be found in the judgement of Lush J in *R* v *Assessment Committee of St Pancras* [1877] 2 QB 581:

> *It is not easy to give an accurate and exhaustive definition of the word 'occupier.' Occupation includes possession as its primary element, but it also includes something more. Legal possession does not of itself constitute an occupation. The owner of a vacant house is in possession, and may maintain trespass against anyone who invades it, but as long as he leaves it vacant he is not rateable for it as an occupier. If, however, he furnishes it, and keeps it ready for habitation whenever he pleases to go to it, he is an occupier, though he may not reside in it one day in a year.*
>
> *On the other hand, a person who, without having any title, takes actual possession of a house or piece of land, whether by leave of the owner or against his will, is the occupier of it. Another element, however, besides actual possession of the land, is necessary to constitute the kind of occupation which the Act contemplates, and that is permanence. An itinerant showman who erects a temporary structure for his performances, may be in exclusive actual possession, and may, with strict grammatical propriety, be said to occupy the ground on which his structure is placed, but it is clear that he is not such an occupier as the statute intends. As the poor-rate is not made day by day or week by week, but for months in advance, it would be absurd to hold that a*

person, who comes into a parish with the intention to remain there a few days or a week only, incurs a liability to maintain the poor for the next six months. Thus a transient, temporary holding of land is not enough to make the holding rateable.

This explanation was further developed in the decision of the Court of Appeal in *Laing (J) & Son* v *Kingswood AC* [1949] 1 KB 344. In this case the Court laid down four essential ingredients for determining whether an occupier was in rateable occupation. Those ingredients were:

- actual occupation;
- exclusive occupation;
- beneficial occupation;
- occupation for not too transient a period.

If any one of the four essential ingredients is missing then the occupier cannot be in rateable occupation. In such cases it may be necessary to examine whether someone else may be in rateable occupation or whether a different rateable hereditament can be identified (see Chapter 3).

2.2 Actual occupation

Actual occupation in its simplest form can mean the physical presence of the occupier on the land, for example the occupier of a shop, office or factory. However, it can also mean the acts by which the occupier makes use of or even controls the land.

We can see from the decision in *R* v *Assessment Committee of St Pancras* [1877] 2 QB 581 that:

'Legal possession does not of itself constitute an occupation.' and that*: 'On the other hand, a person who, without having any title, takes actual possession of a house or piece of land, whether by leave of the owner or against his will, is the occupier of it.'*

Thus, legal possession, the legal right to occupy, is not enough to satisfy the rating test of actual occupation. If you purchase the freehold or take a lease of a shop you have legal possession – you are entitled to occupy it, but you are not in rateable occupation until you, in simple terms, physically occupy it. Merely entering to check all is well and carrying out redecoration works are not sufficient to indicate actual occupation.

Legal title is in fact largely irrelevant as rating is concerned with occupation and not ownership. Indeed trespassers and licensees may find themselves in rateable occupation and liable for rates although they hold no legal title.

In *Crowther Smith* v *New Forest Union* (1890) 54 JP 324, the issue of legal possession was examined. The appellants were the owners of the building land that they were trying unsuccessfully to sell. They used the land in no other way than to post a notice saying the land was for sale. The Court of Appeal held that the owners were not rateable because they were not in actual occupation, although they had legal possession of the property. Legal possession was not sufficient to make the appellant an occupier and therefore rateable. If the owners had used the land, however slightly, then they would have been rateable. The owner of a vacant property which is left vacant is not rateable.

In *Briant Colour Printing Co. Ltd* v *London Borough of Southwark* [1977] RA 101 33, workers who staged a sit in at a factory and excluded the management were held to be liable for rates during their period of occupation. Similarly, squatters in *Tomlin* v *Westminster City Council* [1990] RA 79, occupying the former Cambodian Embassy, were also found to be in occupation of the property and thus liable for rates.

It is well established that whether or not a person is an occupier is a matter of fact, not of law.

Legal title may occasionally need to be looked at but only to enlarge and not to cut down the apparent character of the use of the land.

In the majority of cases, actual occupation will involve a physical presence on the property and an actual use. In such cases, few problems arise. However, there have been cases where the physical presence of a person on the land or premises has not been held necessary to establish actual possession.

In *R* v *Melladew* [1907] 1 KB 192, the owner of warehouse premises, carrying on the trade of storage as a warehouseman for the nearby docks, found that his storage accommodation was not fully occupied. Melladew moved some goods so that one of the warehouse buildings was vacated and entirely unused. He intended to reuse this building when an improvement in the demand for storage accommodation made it worthwhile. He was held rateable, even though the premises were vacant, as a warehouseman would normally expect his premises to be empty from time to time and even in this state the warehouse was still being 'used' for the purpose of the business as it was ready to receive goods when business picked up.

Farwell LJ:

has the person to be rated such use of the tenement as the nature of the tenement and of the business connected with it render it reasonable to infer was fairly within his contemplation in taking or retaining it?

Collins MR:

It is important to remember ... that occupation, which is the basis of liability (to pay rates) necessarily varies with the nature of the subject matter. It is clear ... that the intention of the occupier in respect of the hereditament is a governing factor in determining ... whether rateable occupancy has been established.

In *R* v *Melladew* [1907] 1 KB 192, the Court looked at the intention of the rate-payer to decide whether there was rateable occupation. Melladew's intention was to make use of the warehouse when there was sufficient demand to make it worthwhile. This can be contrasted with *Overseers of Bootle* v *Liverpool Warehousing Co. Ltd* [1901] 65 JP 740, where the circumstances were similar to those in *R* v *Melladew* [1907] 1 KB 192, except that the owner had no intention of reoccupying the warehouse during the rate period and had withdrawn the premises entirely from his trade. Again, the Court looked at the intention of the occupier and found that there was no actual rateable occupation.

A similar situation arose in *SSE plc* v *Moore (VO)* [2023] UKUT 24 (LC) where a gas power station was mothballed pending an improvement in economic activity (which happened within about a year). Whilst the parallels were not exact as significant work had to be undertaken to bring the power station back into use, in both Melladew and this case the essential position was the same. The property remained in occupation for the purposes of the relevant business, notwithstanding an interruption in actual business activity, respectively actual storage and actual electricity generation, caused by market conditions.

While regard can be had to the intention of the occupier, there must be something more than a mere intention in order to attract rateability. For example, in *Associated Cinema Properties Ltd* v *Hampstead Borough Council* [1944] 1 KB 412, the company leased two empty dwelling houses in order to have accommodation available for use as offices if their normal premises were bombed and destroyed in the war. The houses were not prepared in any way and no furniture or fixtures were placed on the premises. In fact, the premises were never used as the company was not bombed out.

The Court of Appeal held:

> *A mere intention to occupy premises on the happening of a future uncertain event cannot without more, be regarded as evidence of occupation. The company was not in rateable occupation because it had never been in actual possession.*

In *Calmain Properties Ltd (Formerly NMT Properties Ltd)* v *Rotherham Metropolitan Borough Council* [1989] 2 EGLR 165, the nature and the intention of the occupier was considered. NMT was a property development and owning company that had constructed a building comprising four warehousing bays. Upon completion of the building, three of the bays were occupied, under a lease, by a warehousing company which was an associated company of NMT. The fourth bay was vacant. NMT instructed agents to market the vacant unit and sales particulars were drawn up advertising the unit as 'available as a building lease or alternatively with full on site facilities provided by one of the leaders in the field of storage, materials handling and forwarding'.

Rotherham Council saw parallels with *R* v *Melladew* [1907] 1 KB 192 and believed NMT was offering warehousing services and therefore considered the warehouse bay was in rateable occupation. The Court found this was not the case:

- NMT was not in rateable occupation.
- NMT had never acted as warehousemen, and there was nothing to suggest that it would do so.
- NMT had never occupied or used the premises as a warehouse.
- NMT did not hold itself out as warehousemen.

As per Dillon LJ, in agreement:

> *of the four ingredients of rateable occupation, the one in issue in the present case is actual occupation ... they had never actually occupied it in any ordinary sense for any business purposes ... You could only say that they were occupying as warehousemen if they had indeed decided to carry on that business rather than to grant a building lease.*

In *Southend-on-Sea Corp.* v *White* (1900) 65 JP 7, a seaside shop was occupied during the season and then at the end of the season stock and furniture were removed leaving only a few fixtures and chattels. There was a definite intention to return as soon as the next season began. The tenants were held rateable for the whole year as they had been in occupation in the past and had every intention of returning in the future. Intention went to show 'present occupation and user'. An interesting comparison was made in this case to a fruit tree; one could not suggest the tree was only occupied when fruit was on the branches!

In some cases, property may be kept empty in order to satisfy specific requirements. In *Liverpool Corporation* v *Chorley Union* [1913] All ER 194, the Corporation, which was the owner and occupier of a reservoir and waterworks system, had acquired some 11,000 acres of land, which formed the gathering ground for its waterworks. Part of the land acquired was agricultural and part was moorland and, in order to reduce the risk of pollution, the Corporation had demolished or left vacant some of the farm houses and had limited the use of that land to sporting purposes and afforestation. It was held by the House of Lords that the Corporation was in fact in occupation:

> *an owner in possession is prima facie occupier unless occupation is shown to be by someone else ... The intention of the alleged occupier is a governing factor in determining whether rateable occupancy is established ... The fact that the alleged occupier is not physically on the property either in person or in works or chattels placed on the land is not necessarily the test. There are many cases where the occupier does not have to go on the land ...*
>
> *Where the alleged occupier is a recent purchaser one test is whether, being the owner and having put no other person in possession, he has such use of the land as is reasonable to infer that he intended to obtain when he bought it,*

that use of being one which constitutes a benefit to him not in the sense that he makes a profit by it but in the sense that the occupation is of value to him …

> *The Council … bought the land for the purpose of excluding those dangers. It was worth their while to pay a large sum for it to ensure the absence of population which may contaminate or consume. They have put no other in occupation. They are enjoying the benefits for which they bought the land.*

In *Arbuckle Smith and Co. Ltd* v *Greenock Corporation* [1960] AC 813, a warehouse was being converted into a bonded store and was in the hands of the builders. The Court said:

> *an owner of land must have made some actual use of the premises in the relevant year before he can be called on to pay rates.*
>
> *I broadly accept the argument for the appellants which can be put in this way. The sole purpose of the appellants in acquiring the premises was to use them as a bonded store in connection with their business as warehousemen. The alterations were necessary in order that this purpose might, if alterations were approved by the Customs and Excise, receive effect. Yet activity carried on in relation to premises, the sole objective of which is to make the premises fit for the only use which is contemplated, does not amount to the kind of actual user as is essential to rateable occupation. So long as the activities were confined to the making the premises fit for a contemplated use, the premises were not serving the appellant's purpose as warehousemen. The premises were not being applied to the purpose for which they existed but were in an antecedent stage.*

2.3 Exclusive or paramount occupation

The second ingredient of rateable occupation is exclusive occupation. The occupation of the property need not necessarily be exclusive in the sense that it is taken to mean that it excludes all others, but rather it is exclusive for a particular purpose. A rateable occupier should have the right to carry out the purpose of his or her occupation without anyone else on the premises doing the same thing. If there is a competing occupier for the same purpose then one or other but not both will be the rateable occupier. This requirement can give rise to difficulties when there are a number of persons who seemingly have rights of occupation, e.g. a landlord and lodgers in a house.

The leading case on exclusivity is the House of Lords case, *Westminster City Council* v *Southern Railway Co, The Railway Assessment Authority, and WH Smith and Son Ltd* [1936] AC 511, The case concerned a bookstall on Victoria Station where WH Smith sold newspapers and magazines. WH Smith was held under a licence which bound it to observe the railway company's bylaws. These contained restrictions on the sort of publications that could be sold. The railway company could also close the station at will and, as the only access to the bookstall was through the station, this would prevent customers reaching the bookstall. The railway company did in fact lock the station's main gates at certain times.

The House of Lords found the restrictions imposed were no more than restrictive covenants in a lease and that WH Smith was in occupation of the bookstall since its position in relation to the bookstall was paramount for the purpose for which the bookstall was occupied. Lord Russell said:

> *The question … is whether the premises in question have been so carved out of the railway hereditament, to which they or their sites belonged, as to be capable of a separate assessment, or whether they have, though let out, been so let out as still to leave them in the occupation of the Railway Company.*

Lord Russell's analysis discussed the question of control and whether the railway company or WH Smith's occupation was paramount:

> *Where there is no rival claimant to the occupancy, no difficulty can arise; but in certain cases there may be a rival occupancy in some person who, to some extent, may have occupancy rights over the premises. The question in every such case must be one of fact – namely, whose position in relation to occupation is paramount, and whose position in relation to occupation is subordinate; but, in my opinion, the question must be considered and answered in regard to the position and rights of the parties in respect of the premises in question, and in regard to the purpose of the occupation of those premises. In other words, in the present case, the question must be, not who is in paramount occupation of the station, within whose confines the premises in question are situate, but who is in paramount occupation of the particular premises in question.*

The above wording has been long regarded as requiring an examination not of who was in paramount control of the station but rather who was in paramount control of the part that was let out. The understanding being that an examination of purpose was required and WH Smith were clearly in occupation since its position in relation to the bookstall, the particular part under consideration, was paramount for the purpose for which the bookstall itself was occupied. This understanding was found to be mistaken. In *Cardtronics UK Ltd* v *Sykes (VO)* [2020] UKSC 21 the Supreme Court, in considering the sites of automatic teller machines (ATMs) built into supermarket walls and operated by bank companies rather than the supermarket chain, took great interest in an earlier case, *Holywell Union Assessment Committee* v *Halkyn District Mines Drainage Co.* [1885] AC 117 where Lord Herschell had said:

> *Where a person already in possession has given to another possession of a part of his premises, if that possession be not exclusive he does not cease to be liable to the rate, nor does the other become so. A familiar illustration of this occurs in the case of a landlord and his lodger. Both are, in a sense, in occupation, but the occupation of the landlord is paramount, that of the lodger subordinate.*

The Supreme Court considered that, unlike Southern Railway and WH Smith, there had been a joint economic benefit in having the ATMs in the store, Indeed the ATM, unlike the selling of newspapers in Southern Railway, was not a distinct business activity but an integral part of the business activity of the store. It saw the ATM as a 'lodger' rather than a separate occupier.

In considering paramount control it is not now correct to start from an examination of the purpose of the occupation of the smaller part but to examine its occupation in relation to the 'host' premises of which it forms a part. Does general control lie with the host occupier?

The following will provide a framework for the evaluation of the facts in any particular case:

1. Start from the larger premises that contain the particular part under consideration.
2. Does the host continue to use the premises for its own purposes? – if yes, this suggests the part is not in separate occupation.
3. Is the situation akin to landlord and lodger? – if yes, this suggests the part is not in separate occupation.
4. Are the host and smaller part operating distinct business activities? – if yes, this suggests the part *is* in separate occupation.
5. Do the parties share the economic fruits of the specific activity for which the space is used? – if yes, this suggests the part is not in separate occupation.

Cases where there are larger premises containing smaller parts that may or may not be in separate rateable occupation are not uncommon. Quite how the comparatively new decision in Cardtronics will play out with regard to serviced offices, self-storage and food courts has yet to become clear.

A recent Valuation Tribunal for England case illustrates the approach in practice: *Kevin Prosser KC (Pump Court Tax Chambers)* v *Ricketts (VO) 2023 Appeal No. CHG100352561*. It concerned whether nine barristers' rooms and support staff rooms occupied by an unincorporated association set up by a group of self-employed barristers was one hereditament or many? The question revolved around paramount occupation and the Southern Railway and Cardtronics cases.

The VTE President considered there was no rivalry between the individual barristers and chambers as a whole in the use of the rooms. Even though each individual barrister operated as a separate business entity, none of them truly operated independently from chambers as a whole. Each of the individual rooms occupied by individual barristers was clearly complementary to the business of chambers and of mutual benefit to both and therefore in reality the occupation of the rooms by barristers was subordinate to chambers whose occupation was paramount. The reality was that individual members were lodgers whose entitlement to a room ceased when they ceased to be members of chambers. The appeal was dismissed (it has, however, been appealed).

While domestic property is no longer rateable, the cases of *Helman* v *Horsham and Worthing Assessment Committee* [1949] 1 All ER 776 and *Field Place Caravan Park Ltd* v *Harding (VO)* [1965] RA 521 illustrate the principle of paramount control.

In *Helman* v *Horsham and Worthing Assessment Committee* [1949] 1 All ER 776, the resident owner of a house let rooms on the first floor to a lodger and shared the bathroom and WC, landings and hall, etc., with himself. It was held that as the owner continued to live in the property there was no rateable occupation by the lodger as there was no clear evidence of any intention by the owner to abandon control of the lodger's rooms. As per Lord Denning:

> *On these facts the test of rateable occupation was whose occupation was paramount. So long as the householder was living in the house he was prima facie regarded ... as being in occupation of the whole, whether his contract with the 'incomer' was demise or licence.*

In the *Helman* case it was decided that the owner had retained such a sufficient degree of control over the property to be the rateable occupier. In *Field Place Caravan Park Ltd* v *Harding (VO)* [1965] RA 521, a caravan site contained a number of residential caravans which had been in position for more than a year. Each had a small garden and was connected in a temporary way to electricity and drainage but still had wheels attached and could be moved quickly in the event of fire. It was held by the Court of Appeal that each occupier of a caravan on the site was in rateable occupation of his or her own caravan.

Lord Denning said: '*you have to look at the enjoyment by the occupier of the premises for which he occupies, and the extent to which the site operator can interfere with that enjoyment.*'

In *Andrews* v *Hereford Rural District Council* [1963] RVR 168, the appellant owned a farm which included an open gravel pit fenced off from the agricultural land. He agreed with a gravel company that it could work the pit under licence. The company had to obtain the appellant's approval when it wished to work a new part of the pit. The company worked the pit between 1959 and 1961 and extracted some 40,000 tons of gravel. During this time three other persons were given oral permission by the appellant to remove 200/300 tons of gravel. It was held that the owner was in rateable occupation because the company was not in occupation as it did not have a substantially exclusive licence.

In *Re the Appeal of Heilbuth (VO)* [1999] RA 109, the Lands Tribunal considered whether 55 starter workshop units in Enfield, converted from a 1930s factory, were in the occupation of the individual licensees or whether the whole centre was in the landlord's occupation. The Tribunal considered that the licensees could reasonably expect from their agreements to be able to carry out their business during normal business hours without interference or interruption by the landlords. It did not find this exclusivity affected by the landlords retaining control over the main gate or having a duplicate set of keys to the units in order to enter for a variety of purposes. This did not give the landlords a right to occupy or conduct business in the units or give them paramount control.

In *Bradford (VO)* v *Vtesse Networks Ltd* [2006] RA 57, the Lands Tribunal decided that a telecommunications network, constructed of pairs of glass fibres little thicker than human hair, extending between Henley-on-Thames and London, was in the paramount occupation of the ratepayer company. The ratepayers argued they were part of the landlord's larger telecommunications system. The fibre pairs were wound together with many other fibres used by the landlord in a cable in ducts in a trench and the ratepayers had never seen them and did not know exactly where the fibres were. Nonetheless, the Tribunal found the ratepayer company in occupation because this did not affect its ability to enjoy the use of the fibres for its purpose of transmitting data. The Tribunal considered the ratepayers had exclusive use of the fibre pairs; no one else could send signals through the fibres and it was their own equipment that provided the laser light pulse for signal transmission. The Court of Appeal found no error of law in the Tribunal's judgement.

It can often be a fine line in deciding who is the rateable occupier and of what. Each case has to be decided on its own facts. In *Esso Petroleum Co.* v *Walker (VO)* [2013] RA 355, this was very strongly illustrated. The valuation officer thought Roadchef was in occupation of the petrol filling element of a motorway service station, just as it was of the main part of the motorway service station, because it very much operated the petrol delivery function. The ratepayers thought otherwise and argued paramountcy of occupation fell to Esso. The tribunal found that Roadchef arranged most of the day-to-day running of the site but considered this 'interfered hardly at all' with the occupation of the premises by Esso for the purpose for which Esso occupied them. It said the lack of interference in the day-to-day running by Esso did not equate to a lack of control but showed a stable and compliant system to achieve Esso's purposes as set out in the agency agreement and Esso's handbook.

Exclusivity means there can only be one rateable occupier for a particular purpose. If there is more than one occupier for a particular purpose then one or other must be in paramount occupation as illustrated by the *Westminster City Council* v *Southern Railway Co, The Railway Assessment Authority and WH Smith and Son Ltd* [1936] AC 511 case. This is, however, for one particular purpose. It is possible, though unusual, to have two or more rateable occupiers of the same piece of land but occupying it for different purposes.

Pimlico Tramway Co. v *Greenwich Union* (1893) LR 9 QB concerned the rateability of tram lines laid in a public highway, over which vehicles and pedestrians could pass. It was held that, notwithstanding the fact that the tramway company could not exclude the public from driving over the surface of the rails, the company was rateable for having exclusive use of the rails by trams with flanged wheels.

In *Holywell Union Assessment Committee* v *Halkyn District Mines Drainage Co.* [1885] AC 117, the owner of land granted the Halkyn District Mines Drainage Co. exclusive rights of drainage through tunnels and water courses.

The owner reserved to himself the right to work the minerals. It was held that the company was in rateable occupation for the purpose for which it was occupying. It was implied the owners could also be regarded as in rateable occupation for a different purpose if they were to mine the minerals and the occupiers of a tramway in the tunnels might also be separately rateable for that other use.

Lord Herschell LC said:

> *It was strongly contended, on behalf of the respondents, that they could not be liable to be rated, inasmuch as they were not in exclusive occupation. There are many cases where two persons may, without impropriety, be said to occupy the same land, and the question has sometimes arisen which of them is rateable. Where a person already in possession has given to another possession of a part of his premises, if that possession be not exclusive he does not cease to be liable to the rate, nor does the other become so. A familiar illustration of this occurs in the case of a landlord and his lodger. Both are, in a sense, in occupation, but the occupation of the landlord is paramount, that of the lodger subordinate.*
>
> *In the present case, in my opinion, on the true construction of the deed, the possession of the respondents is paramount, and any rights which the Duke has are subordinate. The respondents alone have the right of using the tunnels for the primary purpose for which they have been constructed. The Duke has no such right, and, in my opinion, the respondents are in occupation of the tunnels and works.*
>
> *A question was raised with regard to a tramway which has been laid down along a part of the tunnel for the purpose of carrying minerals and other materials. It is not necessary to consider whether the occupiers of this tramway could be separately rated in respect of it. The fact that its construction and use are permitted does not, in my judgement, prevent the respondents being in occupation of the land.*

In *Bartlett (VO)* v *Reservoir Aggregates Ltd* [1985] 2 EGLR 171, the rateability of a mineral working situated in a reservoir owned by Thames Water was considered. Thames Water wanted its reservoir deepened to increase its capacity and reached an agreement with a minerals company, Reservoir Aggregates, that the company could extract sand and gravel from a stratum below the reservoir, thereby deepening it.

Reservoir Aggregates and Thames Water argued that Thames Water was in sole rateable occupation. The valuation officer argued that Reservoir Aggregates occupied the mineral workings within the stratum and Thames Water separately occupied the reservoir. The Court of Appeal decided Reservoir Aggregates' occupation was exclusive and its purpose of extracting gravel was seen as separate from Thames Water's alternative purpose of having the reservoir deepened.

These cases illustrate the rating concept of dual occupation, where it is possible to have two separate rateable occupiers of the same piece of land but only if each occupation satisfies the four tests of rateable occupation and, in particular, each occupation is paramount or exclusive for the purpose for which the particular occupier occupies the land.

If there are no competing occupations then the occupier is in occupation for the whole of the benefit of the land and not just for the occupier's own particular purpose. This is summed up by the rating maxim of 'occupation of part is occupation of the whole'. Where there are two or more persons having similar use of the land and each cannot exclude the other from using the land in the same way, the occupation which is rateable will be the person who is in 'paramount' occupation, i.e. the person who has control over the use to which the others may make of the property.

2.4 Beneficial occupation

The occupation of the property must be of some benefit or value to the rateable occupier. This does not mean that it must be profitable, though clearly this will be a prime motive for property occupation. Other reasons may include fulfilling a statutory requirement or some other special need of the occupier. The question really is whether someone would be prepared to pay a rent for the property. This would indicate that the person considered the occupation to be of value.

Thus, sewage farms, town halls, public libraries and local authority schools are all rateable because, where a statutory authority has a duty to perform and must occupy land to carry it out, then such occupation is beneficial even though it is unprofitable in commercial terms. This view has been extended to cover cases where public authorities do not have a duty but are empowered to carry out tasks which require the occupation of land. While under no compulsion, it is reasonable to suppose that if the authority wishes to carry out the task then it will be prepared to pay rent to secure the land/buildings.

When someone is in apparent occupation and is unable to derive any benefit, then beneficial occupation does not exist.

In *Hare* v *Overseers of Putney* [1881] 7 QBD 223, Putney Bridge was a toll bridge purchased by the Metropolitan Board of Works and then opened to the public free of toll.

It was held that the Metropolitan Board of Works was not a rateable occupier of the bridge.

As per Brett LJ:

> *They are no more occupiers of the bridge than the owner of a street is the occupier of a street after it has been dedicated to the public. It seems to me that the MBW have no power to do anything to or on the bridge except keep it up for the benefit of the public. The public have a right to use the whole of the bridge and have not merely a right of way across it. The whole bridge is to be kept up for the use and benefit of the public.*

As per Bramwell LJ:

> *No doubt the property is in them … An action for trespass would have to be brought by them, but that does not make them occupiers so as to be liable to be rated.*

In *Lambeth Overseers* v *London County Council* [1897] AC 625 (The Brockwell Park case), the London County Council purchased a park under a special act and maintained it for recreational purposes. It was held that the London County Council was in the position of trustee for the public, and as the public had the right to use the park in perpetuity, the London County Council was not in occupation. The land was 'struck with sterility' as no one could derive a greater benefit from it than anyone else. The public at large was not, of course, a rateable occupier.

It should not be thought these old cases are simply that – old. The issues and outcomes remain very relevant today. In *Milton Keynes* v *Alexander (VO)* [2018] UKUT 0366 the valuation officer considered a car park set out as 18 marked parking bays for use by coaches and buses was in beneficial occupation notwithstanding the parking spaces were provided free of charge. The area was distinct from the main highway, though pedestrians and bicyclists could traverse the area. In the Upper Tribunal's view, the property was clearly part of the highway and the parking of buses on designated parts of a highway was a normal arrangement. The Tribunal considered the council was not in rateable occupation of the property.

The case of *R* v *School Board for London* (1886) 17 QBD 738 also concerned whether there was beneficial occupation of a school. The assessment had been arrived at on the contractor's basis by taking a percentage of the cost of construction to arrive at the assessment. It was held:

> *The school board which could make no profit as a tenant, had rightly been considered as a tenant and the gross and rateable values calculated by the rent which the board might reasonably be expected to pay for the premises for the use as a school.*
>
> *It is said that the school board ought to be excluded because it can never obtain any beneficial interest from its tenancy; but it can be a tenant; it has a duty to perform which may induce or force it to be a tenant; it follows therefore that it would be wrong to exclude the board from the list of possible tenants whether it is in a position of owner or as occupier.*
>
> *It can be said that the school board is not to be considered as a possible tenant, because, though it occupies the premises, it can make no profit out of them; but a man who occupies a house for his own comfort may well be excluded. The term 'sterility' has been introduced into this case because as a general rule a profit is produced, but it does not by any means follow that because there is no profit there is no value … The only question is whether the*

person to be considered as a tenant could reasonably be expected to take the premises from any motive.

Consequently, the School Board was held to be in rateable occupation because it had a statutory duty to fulfil and thus would be prepared to pay a rent for the property, even though it could make no profit.

Similarly, in the case of *London County Council* v *Erith and West Ham (Church Wardens and Overseers)* [1893] AC 552, the London County Council was the owner of a pumping station and sewage works and an outfall of sewers constructed above ground, and while put to this use the land and premises were incapable of yielding a profit and the Council was, in reality, the only tenant. It was held that the true test of beneficial occupation was not whether a profit could be made, but whether occupation was of value.

> *I think the learned judge here points to the true test: whether the occupation is such to be of value. The possibility of making a pecuniary profit is not, in my opinion, the test of whether the occupation is of value.*

2.5 Transience

Casual or occasional occupations such as the itinerant showman in Lush J's statement quoted earlier are not rateable because they lack the necessary degree of permanence. The itinerant showman is not in rateable occupation of his booth, even though he may be in actual and exclusive possession and deriving a benefit for which he would be prepared to pay, or indeed is paying, a rent, because his occupation is not permanent.

Permanent means continuous possession as opposed to a mere temporary holding of land. Title is again not relevant, and the fact that the occupier may be on a weekly tenancy subject to, say, a week's notice will not mean his occupation lacks a sufficient degree of permanence provided it has existed for sufficient time. Thus, regard has to be had to the occupation which has occurred and is likely to occur.

In *Cory* v *Bristow* (1877) 2 Apps Cas 262, moorings in the bed of the River Thames that were subject to removal at one week's notice were found to be in rateable occupation.

To be rateable, the occupation of the land comprising the hereditament must exist for not too transient a period. However, when considering this aspect the courts will consider not only the length of occupation but also its character and nature.

This aspect of the essential requirements of rateable occupation has caused a great deal of litigation, especially in relation to builders' huts and motorway construction sites' borrow pits.

In general terms, the occupation, to be rateable, must have been in existence for a period in excess of 12 months. There have been some instances where occupation for a lesser period has been found to be rateable. These cases are mainly where the use of the property is so extensive in a short period that it attracts rateability.

In *London County Council* v *Wilkins (VO)* [1956] 49 R&IT 495 [1957] AC 362, builders, under contract to erect a school on land belonging to the London County Council, erected temporary huts. Three huts remained in position for 18 months and one remained for 21 months. The House of Lords considered the Lands Tribunal had not misdirected itself in holding the builders' huts rateable, since the occupation of the huts had sufficient permanence and were not too transient in nature.

Again, in *McAlpine (Sir Robert)* v *Payne (VO)* [1969] RA 368, the Lands Tribunal held that five separately assessed sheds which had remained in situ for between six and seven months were not rateable. While the Tribunal expressed the opinion that, in the generality of cases, a life of less than 12 months was too transient to establish rateability, it accepted the agreed view of the parties that the question should be considered not on the facts known at the date of proposal, but on those at the date of the appeal hearing.

It may be that 'permanence' signifies no more than continuous, as opposed to intermittent, physical possession of the soil.

Further cases on builders' huts led to a working rule on permanence, namely that if a hut or caravan has been (or is likely to be) on a site for more than 12 months then the occupation is rateable and if it has been on a site for less than 12 months it is not.

The above cases have concerned what are generally termed 'temporary structures', which by their very nature are removed from the site once construction has been completed. In the case of *Dick Hampton (Earth Moving) Ltd* v *Lewis (VO) and United Gravel Co. Ltd* v *Sellick (VO)* [1975] RA 269, the valuation officer sought to assess gravel pits which had been occupied for a period of less than 12 months. During the construction of a motorway the contractors had entered into agreements for the extraction of soil and gravel from adjoining land for the purpose of the construction of embankments. These pits were outside the line of the motorway. In one pit on a seven acre site some 310,891 tons of fill were extracted in six months and from the other over 700,000 tons were taken in nine months from ten acres.

The ratepayers argued that one of the essential ingredients of rateable occupation is that 'possession must not be for too transient a period'.

The Court noted the working rule of 12 months on site for a builder's hut to become rateable but considered that it was a mistake to elevate this ingredient to a principle of law or to construct out of it a working rule. They pointed out that there is no authority for such an interpretation and that the House of Lords in *London County Council* v *Wilkins (VO)* [1956] 49 R&IT 495 [1957] AC 362 had only held that there must be a sufficient degree of 'permanence'.

As per Lord Denning:

If some degree of 'permanence' is necessary, these borrow pits are as permanent as anything could be. The landscape has been changed forever. Huge slices have been dug out of the hillside, leaving gaping voids which will never be filled up. They are far more permanent than the most massive of buildings. It has all been done in a few months, but I cannot believe that it is grounds for exemption from rates.

As per Lord Russell:

In the present class of case I regard the period of occupation as of vastly less importance than its quality and consequences.

The Court decided that the occupation of the gravel pits was sufficiently permanent for rateable occupation to be established.

The cases heard have tended to concern the placing of temporary buildings or caravans on sites or, as in *Dick Hampton (Earth Moving) Ltd* v *Lewis (VO) and United Gravel Co. Ltd* v *Sellick (VO)* [1975] RA 269, the use of land without buildings. However, the four ingredients of rateable occupation apply equally to hereditaments that comprise permanent buildings on sites. A reasonable question to ask is whether the occupation of a shop or an office floor for one, three or six months gives rise to rateable occupation because it might be said the occupation seems to be rather transient. The situation is, however, different when dealing with permanent buildings rather than contractors' huts and the like. Builders' huts are clearly analogous to the wayfarer or itinerant showman envisaged by Lush J, where the occupation may or may not prove to be sufficiently permanent depending on the facts of the case. However, with a permanent building, the occupation, whether by the present occupier or some successor in occupation, is clearly permanent as the building will be there for many years to come. As the Lands Tribunal said in *McAlpine (Sir Robert)* v *Payne (VO)* [1969] RA 368:

There remains the question 'Is there a minimum period for establishing that an occupation is not too transient?' and if there is 'What is it?' It is I think clear that for permanent buildings there is no fixed period: theoretically at any rate occupation of a house or a shop for even a day could attract rateability. But the authorities I think show clearly that different considerations apply to temporary structures. These are less likely to be 'settlers' and more likely to be 'wayfarers' …

2.6 Rateable chattels

The cases of *Field Place Caravan Park Ltd* v *Harding (VO)* [1965] RA 521 and *London County Council* v *Wilkins (VO)* [1957] AC 362 both involve what has become known as rateable chattels.

It is now well established that certain objects can, if enjoyed with land and with a sufficient degree of permanence, become rateable with it, e.g. caravans, builders' huts, floating clubhouses and floating restaurants. In *Rudd (VO)* v

Cinderella Rockerfellas Ltd [2003] EWCA Civ 529, the Lands Tribunal found that a former ferry which had been converted into a nightclub and moored on the River Tyne was rateable with the land and should be assessed as such.

Thus, the permanence test applies to two situations:

- the permanence needed before a person can become the rateable occupier of land; and
- the permanence needed before a chattel can become rateable with land and buildings.

2.7 Summary

In order to incur rate liability there must be a rateable hereditament and a rateable occupier.

There are four essential ingredients for rateable occupation – actual occupation, exclusive occupation, beneficial occupation and transience or permanence. If any one of the essential ingredients is missing, then there cannot be rateable occupation.

The concept of paramount occupation arises where there may, on the face of it, be more than one potential occupier of a property. The paramount occupier is the one who exercises control over the use of the property.

2.8 Wales

The rules relating to rateable occupation are the same for both England and Wales.

3   The hereditament

3.1 Introduction

For any valuation, the valuer needs to know what the property to be valued comprises and what its boundaries are. The term used for a single unit of rateable property is a hereditament.

The previous chapter established there can be only one rateable occupier of a piece of land and if there are two competing occupations one needs to be identified as the rateable occupier by using the tests of actual, beneficial, exclusive or paramount occupation and that the occupation cannot be too transient. But it is difficult to establish who is the occupier of a piece of land until you know what the piece of land is in the first place – the hereditament.

As Lord Carnwath said in *Cardtronics UK Ltd* v *Sykes (VO)* [2020] UKSC 21, 'although it may be convenient for the purpose of analysis to separate the issues of hereditament and occupation, they are in truth linked.'

There is indeed an element of 'chicken and egg' about the two concepts. The tests of rateable occupation and the hereditament may sometimes have to be considered in parallel.

The Local Government Finance Act 1988 does not actually go so far as to define hereditament but, in section 64(1), it 'helpfully' states:

64.- (1) A hereditament is anything which, by virtue of the definition of hereditament in section 115(1) of the 1967 Act, would have been a hereditament for the purposes of that Act had this Act not been passed.

This at least indicates that a definition exists, but it is contained in the General Rate Act 1967, which was repealed by the Local Government Finance Act 1988. It defines hereditament as a:

property which is or may become liable to a rate, being a unit of such property which is, or would fall to be, shown as a separate item in the valuation list.

DOI: 10.1201/9781003297369-3

This does not actually help a great deal! Indeed, Sir Alan Ward in *Reeves (LO) v Northrop* [2013] RA 295 in the Court of Appeal said that 'If prizes are to be offered for legislative gobbledegook then the foregoing would surely qualify'!

And as Lord Sumption said in *Woolway (VO) v Mazars* [2012] UKUT 165 (LC); (CA); [2015] UKSC 53: 'Notwithstanding more than four centuries of experience, the question how a hereditament is to be identified for rating purposes remains in important respects unclear.'

However, the definition does indicate a hereditament is a single unit of property which will have a separate entry in a list. Case law has amplified this definition and set down various rules for defining what constitutes a hereditament. In practice, identifying the hereditament in most cases is straightforward, but some circumstances do present difficulties. For many years the leading case on identifying the hereditament was *Gilbert (VO) v Hickinbottom (S) & Sons Ltd* [1956] 2 QB 240 in which Parker LJ summed up the requirements as:

> *Whether or not premises in one occupation fall to be entered in the valuation list as one or more hereditaments depends upon a number of considerations. Without attempting an exhaustive list, the following considerations can be mentioned: (1) Whether the premises are in more than one rating area. If so, they must be divided into at least the same number of hereditaments as the rating areas in which the premises are situated. (2) Whether two or more parts of the premises are capable of being separately let. If not, then the premises must be entered as a single hereditament. (3) Whether the premises form a single geographical unit. (4) Whether though forming a single geographical unit the premises by their structure and layout consist of two or more separate parts. (5) Whether the occupier finds it necessary or convenient to use the premises as a whole for one purpose, or whether he uses different parts of the premises for different purposes.*

This case has been very substantially superseded by the Supreme Court's decision in *Woolway (VO) v Mazars* [2012] UKUT 165 (LC); (CA); [2015] UKSC 53. Nonetheless, Parker LJ's summary does provide a helpful list of rules for determining what constitutes a single hereditament. Summarising the rules for the present day, to be a hereditament a property must:

- be within one or more billing authorities' areas;
- have a single rateable occupier;
- be capable of separate occupation;
- be a single geographical unit;
- be put to a single purpose;
- have a single definable position.

If a property does not satisfy any of these rules then it will not be a single hereditament. Let us consider each of these rules in turn in the following sections.

3.2 Within one or more billing authorities' areas

Parker LJ mentions in *Gilbert (VO)* v *Hickinbottom (S) & Sons Ltd* [1956] 2 QB 240 that a hereditament cannot straddle a rating area boundary. Prior to 1990, if it did it became two or more hereditaments depending upon the number of local authority areas the property was situated in. This rule was appropriate when each local authority levied its own rate and might adopt different multipliers between different areas within its boundaries.

With the advent of the National Non-Domestic Rate, providing for a uniform multiplier over the whole country, the Government decided the creation of such artificial hereditaments was unnecessary. The Non-Domestic Rating (Miscellaneous Provisions) Regulations 1989 (SI 1989/1060) states:

Cross-boundary property

6.-(1) This regulation applies to any unit of property ('relevant property') which by virtue of section 64(1) of the Act comprises separate hereditaments solely by reason of being divided by a boundary between (billing) authorities.

(2) Relevant property shall be treated as one hereditament and … as situated throughout any relevant period in the area of the (billing) authority in whose area is situated that part of the property which would but for this regulation be the hereditament appearing to the relevant valuation officer or officers to have, on the relevant day, the greater or (as the case may be) the greatest rateable value.

The new rule basically provides that the whole property, which would have been treated as a single hereditament had it not straddled a boundary, is to be treated as if it was in only one billing authority's area and will only appear in its rating list. The decision as to which list it appears in is decided by determining which area has the part with the greatest rateable value. Hereditaments straddling the boundary between England and Wales or Scotland will still divide into artificial hereditaments.

The old rule was often expressed as requiring a hereditament to be within a single parish or rating area. This also covered where a property was either outside or partly outside any local government area. The boundary of local authority areas with tidal boundaries is the low watermark. This aspect of the old rule still applies. If the whole or part of a property is beyond the low watermark and outside the local authority's boundary then it, or the part beyond the boundary, cannot form, or be part of, a hereditament. Examples of such exparochial property include the part of a mine or timber piers extending beyond the low watermark. The part of such a mine or pier which does come within a billing authority's boundaries will be rateable and constitute a hereditament.

Care needs to be taken in just using the low watermark as a guide, because an Act of Parliament may have made the part of a mine, pier, etc., beyond the low watermark part of the local authority's area. In addition, such things as solid concrete piers or breakwaters which are an 'accretion' from the sea are deemed part of the adjoining parish by the Local Government Act 1972 and will not therefore be outside the local authority's area.

For cross boundary property in Wales see section 3.12.

3.3 A single rateable occupier

There can be only one rateable occupier for a hereditament. If there are two rateable occupiers then there will be two hereditaments.

Since *Allchurch* v *Hendon Union* [1891] 2 QB 436, it is no longer considered necessary for properties to be structurally severed before two hereditaments can exist, providing the different pieces of property in different occupations can be clearly identified.

3.4 Capable of separate occupation

Before a property can be identified as more than one hereditament, it must be shown that each part is capable of separate occupation. If two parts of a property are so bound together that they cannot, or one part cannot, be separately let, then two separate hereditaments cannot exist. Clearly, if parts are in fact separately occupied the test is satisfied; the difficulty is when there are not separate occupations.

The general rule in rating is that 'occupation of part is occupation of the whole'. In *R* v *Aberystwyth* (1808) 10 ESAT 353, it was said:

> *There has been no instance where a man has been permitted to carve out the occupation of his house in the manner now attempted, locking up one room and then another, but using as much of the house as he found convenient. This would make a new system of occupation by sub-divisions.*

It is not therefore possible to avoid rates on, for example, some of the rooms on a floor of a block of offices held under a lease by merely emptying them of furniture and locking their doors. Occupation of part of the floor is occupation of the whole, as 'vacant and to let' the property would let as a single unit. The extent to which any part is used is a personal matter for the tenant. This means that where an occupier has previously occupied a building or part of a building but part of it is now vacant, the occupier will always be treated as occupying the whole area. This will, as in the *R* v *Aberystwyth* (1808) 10 ESAT 353 case, depend on whether the part is capable of being separately occupied or let. The decision in *Moffatt (VO)* v *Venus Packaging* [1977] 243 EG 391 illustrates this. In this case there were adjoining and inter-connecting factory buildings, one in use by Venus Packaging, one empty but previously used by Venus Packaging and which they still had the right to occupy. The Lands Tribunal held they constituted two separate hereditaments because the two parts were capable of being separately let and were used for different uses. The Tribunal commented that 'use' and 'non use' (i.e. it being vacant) were different uses.

3.5 A single geographical unit

The basic rule is if land in one occupation is neither within the same curtilage nor contiguous (meaning touching) but is fragmented then it will be more than one hereditament.

In *University of Glasgow* v *Assessor for Glasgow* (1952) SC 504, the university was in occupation of a number of buildings outside the main body of the University. The question was whether the University should in total comprise one unit of assessment or whether these other buildings should each form their own unit of assessment. It was held that each building was a separate unit of assessment because there was geographical separation and a capacity of being separately assessed.

This requirement to be contiguous was the basic rule under *Gilbert (VO)* v *Hickinbottom (S) & Sons Ltd* [1956] 2 QB 240. The simplest way of considering the rule was to ask whether the land could be enclosed by a ring fence. If all the land could be bounded by an imaginary fence it was likely to be a single hereditament. However, the Supreme Court in *Woolway (VO)* v *Mazars* [2012] UKUT 165 (LC); (CA); [2015] UKSC 53] decided contiguity was not enough by itself and considered the parts of the hereditament needed to directly inter-communicate as well.

The Supreme Court set out a Geographic Test based on visual or cartographic unity to establish whether there was a single geographic unit. It regarded this as the primary test to determine what constituted a hereditament.

3.6 Geographical Test

First, this Geographic Test looks at whether the land and buildings under consideration constitute a single unit on a plan. This might be a single unit horizontally or vertically – it is not a plan in two dimensions only. Normally this will be a self-contained piece of property, i.e. all parts of which are physically accessible from all other parts, without having to go onto other property. This might be, for example, by walking down a corridor that joins two buildings together where both corridor and buildings are in sole occupation, or by leaving one building and walking across a yard between or across grounds surrounding buildings where the buildings, yard and grounds are all in sole occupation.

Connection by common parts such as stairs in an office building is not sufficient to make, for example, the second and sixth floors of an office building occupied by a company into a single hereditament. This was the subject of the *Woolway* case. The common parts are no different from a private road used by a number of different occupiers and do not give contiguity or give a single unit on a plan.

Unity on a plan is not, as mentioned, just a question of contiguity: it is not simply a question of whether two pieces of property touch. There also needs to be 'direct communication' between the two parts. Normally, if it is necessary to go out of one part and into the other part via the street or common parts (for example, from one floor to another in an office building) this will mean the two parts are separate hereditaments notwithstanding there being a single occupier. Having just a wall or ceiling between the two parts is not good enough.

It might be thought that floors next to each other in a building (e.g. the second and third floors), or two buildings (such as shops) side by side, occupied by a single occupier might constitute one hereditament because they are contiguous. However, usually there will be no direct communication between them and to reach each of the floors it will be necessary to go out into the shared common parts which are not in the occupation of the occupier, or with next door buildings go out into the street or shopping mall.

It might be thought that floors next to each other in a building, (e.g. the second and third floors), or two buildings (such as shops) side by side, occupied by a single occupier might constitute one hereditament because they are contiguous. However, usually there will be no direct communication between them and to reach each of the floors it will be necessary to go out into the shared common parts which are not in the occupation of the occupier, or with next door buildings go out into the street or shopping mall.

If direct communication were to be established, by piercing a staircase through the floor slab or a doorway through the wall then the occupier would usually be said to have created a new and larger hereditament in place of the two which previously existed.

Prima facie, where an occupier has two or more continuous and inter-communicating parts these will form a single hereditament because of the Geographic Test; conversely where an occupier has two or more discontinuous or non-intercommunicating occupations these will not, because of the Geographic Test, form a single hereditament. There are, however, exceptional circumstances where this basic position may need to be modified because of the other tests and in particular the Functional Test.

Connection by cables, wires, pipes and railway lines have generally not been held to create geographic unity. Therefore, premises separated by property in other occupations, including public highways, private roads (where the occupier does not have exclusive rights over the road), railway and canals are not geographically one merely because of such connections.

In *Harding and Clements* v *The Secretary of State for Transport* [2017] UKUT 135 (LC), [2017] RA 271 [2017] UKUT 135 (LC), [2017] RA 271, which, rather surprisingly was a blight notice rather than rating case, the Upper Tribunal examined whether two fields constituted a single hereditament despite being separated by a public road. Three ducts ran between them under the road, one carrying a water supply. The Tribunal noted in the Woolway case the discussion about 'direct communication' had referred to connection by a doorway or staircase, but it was not persuaded the Supreme Court necessarily intended to delimit the type of link which might satisfy the Geographic Test. Nonetheless the Tribunal did not accept the ducts amounted to a sufficient connection.

So, broadly speaking, to be one hereditament two parcels of land or parts of a building must not only be in single occupation but touch and intercommunicate.

In most cases it is fairly easy to determine whether land or buildings are contiguous or not, and whether they intercommunicate. Two shops side by side in single occupation with a doorway from one to the other or a passageway at the rear are contiguous and intercommunicating; two industrial units side by side in single occupation with a sole occupation parking or loading area to their front will be contiguous and intercommunicating; or three office floors in single occupation with a private stairway between them will be contiguous and intercommunicating.

Premises separated by a public road, railway line, canal, or buildings or land occupied by another are not contiguous and therefore cannot usually form a single hereditament.

Premises connected by a private road may be contiguous if the road is in the exclusive occupation of the occupier of the premises. The existence of a private estate road used by various occupiers on an estate will not be sufficient to make two geographically separate but singly occupied factories into one hereditament because the road will not be in the exclusive occupation of the occupier of the two factories. If two parts of a factory are separated by a railway line but there is a tunnel or bridge in the sole occupation of the factory occupier between the two parts, then it is likely the two parts will form a single hereditament. In *Newbold* v *Bibby and Baron Ltd* (1959) 4 RRC 345, connection by underground electric cables and a steam main were held insufficient to make two properties into a single hereditament. Whereas in *FC Brown Steel Equipment Ltd* v *Hopkins (VO)* [2022] UKUT 51 (LC) the Upper Tribunal was content that a factory with buildings either side of a public road but connected by a box section bridge 4 m × 3 m, approximately 87 m in length and 8 m off the ground did constitute a single geographical unit. The bridge was used for transport by conveyor of constructed office furniture (made in the factory) rather than for people. The VO had considered that as the bridge was not for pedestrians this was insufficient for it to constitute one hereditament, but the Tribunal noted there was nothing in *Woolway* to suggest the passage of people was necessary to satisfy the test; indeed in the context of industrial property such a limitation would not make sense. It said it was important not to set rigid tests or qualifying conditions that would automatically cause properties to be or not be regarded as a single hereditament.

It has long been accepted that there is an exception to the general rule about contiguity. In the *Gilbert* case the exception was described as being where the two parts were 'functionally essential' to each other. This test has been refined as the 'Functional Test' in *Woolway (VO)* v *Mazars* [2012] UKUT 165 (LC); (CA); [2015] UKSC 53 and can override the 'Geographic Test' in limited circumstances.

3.6.1 *The PICO test (Property in Common Occupation)*

To add further to the complexity of identifying the hereditament, Parliament decided it should reverse the decision in the Woolway case – to an extent. Requiring adjacent floors in multi-occupied buildings and adjoining buildings in the same occupation not only to touch but also to intercommunicate had not been the understanding of the law prior to Woolway. The Rating (Property in Common Occupation) and Council Tax (Empty Dwellings) Act 2018 sought to change this. However, it neither actually reversed Woolway nor codified what a hereditament comprised, rather it added an additional stage to identifying the hereditament.

First, the hereditament should initially be identified using the tests from the Woolway decision, primarily the Geographic Test. Second, the PICO Act test is applied. The aim is to largely remove the requirement properties should intercommunicate.

It is a two-stage approach. The rules from the Woolway case being applied first and then the result modified if the PICO test applies.

The PICO test applies only to those hereditaments that satisfy the 'contiguity condition.' This statutory contiguity exists between two hereditaments where they have at least part of a wall or fence in common between them. The wall could be a party wall between two buildings or a boundary wall or fence separating pieces of land, or the yards of buildings. The test does not make hereditaments that are not contiguous one hereditament.

It also exists between two hereditaments where they are separated by a floor rather than a wall, as might be the case in an office building. The wording is precise, 'the hereditaments are on consecutive storeys of a building and some or all of the floor of one hereditament lies directly above all or part of the ceiling of the other hereditament'. It is interesting to note the requirement that at least part of the upper hereditament has to be directly above the other. It is not permissible for the hereditaments to be merely on consecutive floors: they must, at least in part, be one above the other.

The wording for walls does not, at first sight, appear to have the feature requiring the two hereditaments to be at least in part adjacent as with ceilings. However, careful examination of the wording shows this is indeed a requirement. Simplifying it, 'some ... of a wall ... of one hereditament forms ... part of a wall ... of the other hereditament'. In both cases the piece of wall has to be an actual adjacent boundary between them. The wall cannot happen to merely touch both hereditaments in the course of its perhaps rather long length: it has to separate them in part.

Further wording does not disqualify a wall or ceiling from being a boundary, and therefore the hereditaments being statutorily contiguous, merely because there might be some void in the ceiling or wall occupied or owned by someone else (most likely the landlord). It is easy to envisage this applying to a ceiling and a floor but more difficult with a wall because the existence of a space may well mean there are two walls between the hereditaments not one. It will depend on construction. Certainly, it would seem if there are two adjacent buildings each with its own wall, perhaps inches apart, this will not be covered as there is not 'a wall' but two walls.

In *Roberts (VO)* v *Backhouse Jones Ltd [2020] UKUT 0038 (LC)* the Upper Tribunal considered what a wall meant in this context and what constituted a void disqualifying the PICO contiguity requirement. In the case there was a communal fire escape corridor separating two office suites. The Tribunal found no part of any wall of either suite formed any part of a wall of the other. It was 'not the presence of the corridor between the two suites which prevents them being contiguous …, it is the fact that they are not separated by a common wall'. The fire corridor between the two hereditaments was not a space within a wall, it was a space between two walls, neither of which enclosed both hereditaments.

If statutory contiguity applies then the PICO Act requires that the two hereditaments should be treated as one and merged, providing, of course, they have the same occupier. And unless they are used for wholly different purposes.

The legislation provides a rather complicated set of words to ensure where there is more than one hereditament in single occupation, it is not a requirement that each hereditament is statutorily contiguous with all the others so long as there is a continuity to the connection – a string of hereditaments thereby becoming one! So, for example:

- The first, second and third floors of an office block where each floor is occupied by XYZ Co. will be treated as one hereditament because they are statutorily contiguous.
- Two detached industrial units side by side but with their yards separated by a single fence will be treated as one hereditament, if in the same occupation, because, whilst the units do not share a common wall, a part of each hereditament shares a common fence.
- A set of ten rooms off one side of a common corridor will be one hereditament (but they will not be one with any rooms across the corridor) because the rooms on one side of the corridor are statutorily contiguous (whereas the rooms on the other side are separated by a corridor from the ten rooms).

Lastly, the legislation is extended to cover certain vacant hereditaments. It has the effect of permitting the continuation of hereditaments which are still treated as one because they had been occupied by a single occupier and met the conditions for statutory contiguity, notwithstanding there is no longer an occupier. It does not apply merely because vacant hereditaments are or have become owned by the same person. They do, indeed, need to be owned by the same person, but must also have ceased to be occupied on the same day, have remained unoccupied and deemed to be one hereditament under the new legislation when they were occupied.

The actual wording of the main legislation is:

(3ZA) In relation to England, where–

(a) two or more hereditaments (whether in the same building or otherwise) are occupied by the same person,
(b) the hereditaments meet the contiguity condition (see subsection (3ZC)), and
(c) none of the hereditaments is used for a purpose which is wholly different from the purpose for which any of the other hereditaments is used,

the hereditaments shall be treated as one hereditament.

(3ZC) The hereditaments meet the contiguity condition if–

(a) at least two of the hereditaments are contiguous, and
(b) where not all of the hereditaments are contiguous with each other–
   (i) one or more of the other hereditaments is contiguous with one or more of the hereditaments falling within paragraph (a), and
   (ii) each of the remaining hereditaments (if any) is contiguous with at least one hereditament that falls within sub-paragraph (i) or this sub-paragraph.

(3ZD) For the purposes of subsection (3ZC) two hereditaments are contiguous if–

(a) some or all of a wall, fence or other means of enclosure of one hereditament forms all or part of a wall, fence or other means of enclosure of the other hereditament, or
(b) the hereditaments are on consecutive storeys of a building and some or all of the floor of one hereditament lies directly above all or part of the ceiling of the other hereditament,

and hereditaments occupied or owned by the same person are not prevented from being contiguous under paragraph (a) or (b) merely because there is a space between them that is not occupied or owned by that person.

3.6.2 The Functional Test

Where two parcels of land or two buildings are geographically distinct, a Functional Test may, despite the separation and lack of direct intercommunication, enable them to be treated as a single hereditament but this will be so only where the use of one is necessary to the effectual enjoyment of the other. It is an exceptional situation. It might need to be considered where two parts are divided by a public road, e.g. a factory complex with buildings either side of a highway, or where two parts are divided by land in the occupation of a third party.

As mentioned, before the Supreme Court's decision in the *Woolway* case this question was examined on the basis of whether the two parts were 'functionally essential' to each other. Sometimes this was considered with regard to the particular way the occupier chose to use the two parts. Following the *Woolway* decision no account should be taken of the way the actual occupier chooses to use the premises. The Functional Test is an objective test, not one based on the way the ratepayer uses premises. The Functional Test is a question not of examining the particular needs or way of operating the two parts by the actual occupier but of looking at the nature of the actual premises and how any occupier would use them in the existing mode or category of use. As Lord Gill said:

> *It cannot be right that geographically separate premises should be valued as one hereditament simply because the ratepayer chooses to link his use of one with his use of the other. To modify the geographical test with considerations of functionality, in this sense of the word, is to add to a clear and objective test the uncertainty of a test that is dependent on whatever happens to be the ratepayer's choice of use.*
>
> *functionality … is not a reference to the use that the ratepayer chooses to make of the premises. It is a reference to a necessary interdependence of the separate parts that is objectively ascertainable.*

Or Lord Sumption:

> *the question whether the use of one section is necessary to the effectual enjoyment of the other depends not on the business needs of the ratepayer but on the objectively ascertainable character of the subjects.*

Great care needs to be taken in basing decisions on cases decided prior to *Woolway* as these will have been decided on the old 'functionally essential' test under *Gilbert (VO)* v *Hickinbottom (S) & Sons Ltd* [1956] 2 QB 240 and not the current understanding of the law from *Woolway (VO)* v *Mazars* [2012] UKUT 165 (LC); (CA); [2015] UKSC 53.

Whether the use of one part is necessary to the effectual enjoyment of the other was described in *Woolway* by Lord Sumption as something that 'could commonly be tested by asking whether the two sections could reasonably be let separately?'

To a large extent, of course, any two parts 'could' be let separately but the question is whether they could 'reasonably' be let separately. This appears to invoke a test of whether one part could be separately let and therefore separately occupied without significant detriment to the other part.

Lord Gill said: 'Properties that are discontiguous but nonetheless geographically linked, may constitute one hereditament if the occupation of one part would be pointless without the occupation of the other.'

Lord Neuberger echoed this in saying that 'strict necessity is not the test' but if one property 'could not sensibly be occupied or let other than with the other property' then the two parts should be treated as one hereditament.

So, it is a question of looking at the two parts and deciding if they could sensibly be occupied separately. It is not a test requiring both parts to be reasonably occupiable without significant detriment to the other but that the occupation of at least one part would be pointless without the other. In *Harding and Clements* v *The Secretary of State for Transport* [2017] UKUT 135 the Upper Tribunal said that mutual dependency was not essential, and it was sufficient that while one part could reasonably be separately occupied on its own the test was satisfied if the other part could not. It is, therefore, not a requirement that both parts have to be incapable of reasonable separate occupation.

It is important to note that treating two separate properties as one hereditament is very much the unusual situation. As Lord Neuberger said, 'Where premises consist of two self-contained pieces of property it would, in my view, require relatively exceptional facts before they could be treated as a single hereditament.' Lord Sumption also described the circumstance as 'rare'.

In cases before the *Woolway* case was decided it was very clearly established that the two parts needed to be close to each other – normally simply the distance of a road's width away.

In *Edwards (VO)* v *BP (Llandarcy) Ltd* (1974) LT RA 1, the Tribunal (Mr Stuart Daniels QC and Mr Emlyn Jones FRICS) stated the following, which included what has been seen as a very useful and often quoted analogy:

> *Consideration of these two cases leads us to the conclusion that the two separate properties which are not directly and physically contiguous could not properly be regarded as a single hereditament unless firstly, there is an essential functional link between the two parts and secondly, there is a substantial degree of propinquity. One might perhaps consider the analogy of a sparking plug where the gap between the parts is so small that it can physically be traversed in the course of the functioning of the whole. It might also be true to say that the stronger the spark the greater the gap that can be traversed.*

Note: The word 'propinquity', in the context of the above quotation, can be taken to mean 'nearness'.

In *Re the appeals of Evans (VO)* (2003) LT RA 173, it was held that the functional connection was not sufficiently strong to overcome a geographical separation of 380 m (in which seven intervening buildings were situated) between the two buildings.

In *Woolway (VO)* v *Mazars* [2012] UKUT 165 (LC); (CA); [2015] UKSC 53, the judges did not mention the degree of separation. However, the test of whether one part could be separately let, and therefore separately occupied without significant detriment to the other part, is very much more likely to be met where the two parts are geographically very close and the geographical separation is small. In the rare cases to which the Functional Test applies to aggregate two geographically distinct parts it is likely that if the two parts are not close they will reasonably be capable of separate letting and therefore remain two hereditaments.

3.7 Single purpose

If part of a property is used for a wholly different purpose from the rest and that part is capable of separate occupation, then it will form a separate hereditament. This simple sounding rule can cause difficulties, as in practice premises apparently used for quite different purposes have been found to be a single hereditament. The reason in these cases seems to be that the purposes were not substantially different or that one was ancillary to the other. A shop with separately lettable living accommodation above, if in one occupation, was usually assessed as one hereditament when domestic rating was in force, despite retail and living accommodation seeming to be completely different purposes. Regard is given to the primary use to which a property is put and associated uses are treated as part of the primary overall purpose. For example, within a factory complex it is likely that the buildings will have many uses: factory floor, offices, canteen and stores, and even, perhaps, for old factory complexes, a works playing field and pavilion. Usually all these different uses of the factory go to make up the essential characteristics of a factory purpose and would be treated as a single hereditament.

The leading case is *North Eastern Railway Co.* v *York Union* [1900] 1 QB 773. York station included locomotive sheds, a hotel, a pumping station and yards which were assessed as one hereditament. It was held that the hotel and refreshment room should be separately assessed but the rest should form only one hereditament. It was found, as a matter of fact, that the other parts could be separately occupied but, as at present laid out, they were adapted for use by the railway company itself and therefore fell to be assessed as one. The hotel and refreshment rooms were regarded as the only parts actually used for a wholly different purpose.

In *Butterley Co. Ltd* v *Tasker (VO)* [1961] 1 WLR 300, the Court of Appeal considered whether offices used by a company for the administration of a group of companies should be treated as a separate hereditament from a factory occupied by the ratepayers and situated 150 m away from the offices along a private road. Both parts were on a two square mile site used by the ratepayers for various purposes including agriculture. The Court of Appeal found them not to be contiguous because there was no evidence that the private road, which was part of a system of roads on the site, was incorporated with the factory or the offices. Instead it appeared the private road system, which the public could use on payment of a toll, should be a hereditament in its own right. The Court also considered the two parts could not in any case be a single hereditament because they were not dedicated to a single purpose: 'one is a factory area the other is devoted entirely to executive purposes'.

In *Moffatt (VO)* v *Venus Packaging* [1977] 243 EG 391, a company which had occupied a factory complex vacated part of it and left it unused. The occupied part included a covered loading bay, a portal frame factory building and car parking. It was separated from the unoccupied part by a brick wall with an interconnecting door. It was agreed between the parties that the unoccupied part was capable of being separately let. The valuation officer argued that 'occupation of part is occupation of the whole'. The ratepayers' view was that the use of the unoccupied part was wholly different from the occupied part as there was no use at all. The Lands Tribunal accepted that 'occupation of part of an hereditament is to be taken as occupation of the whole occupation' was 'still good law', but this did not help in determining whether there were one or two hereditaments. The Tribunal accepted the two parts were in wholly different uses and could be separately let and were therefore two hereditaments.

In *Poor Sisters Nazareth* v *Gilbert (VO)* [1984] RA 10, the Poor Sisters occupied two residential homes adjoining each other and fronting onto the same road. One home was a converted Victorian residence for the care of 32 children. The other was a purpose-built modern unit for the care of 60 elderly people. It might have been thought that, all other qualifications having been met, the single purpose of caring for those in need would be sufficient to assure combined assessment. The Lands Tribunal, however, thought not when weighed against the fact the homes were administered separately, they were not interdependent, one of them could revert to use as a private dwelling and either could be separately let.

In *Trafford MBC* v *Pollard (VO)* [2007] RA 49, the Lands Tribunal considered whether a property owned and occupied by a local authority comprising a school and sports centre on the same site should be assessed as a single hereditament or as two. The sports centre was used both by the school and as a public facility out of school hours. The Tribunal found this case difficult to decide but made its decision 'on balance'. It regarded the difference in use of the two parts of the site as an important consideration, but it was not the only one. While accepting that the two parts of the site were used differently, it noted a significant degree of interaction and functional connection between the two parts of the site and in particular was influenced by the original conception and actual use of the sports centre operation as a dual-use facility with the school. This showed that the local authority had a single purpose in using the whole site. It seems from this case that what is important is not so much the use of two parts of a contiguous site being different, but whether the purpose of the two parts is entirely different. Where wholly different uses are being undertaken then this will not be sufficient to justify separate assessment unless the purposes are different.

It is interesting that the (Rating (Property in Common Occupation) and Council Tax (Empty Dwellings) Act 2018 makes as one of the requirements for its 'PICO' provisions to apply that 'none of the hereditaments is used for a purpose which is wholly different from the purpose for which any of the other hereditaments is used'. This maintains the rule in the PICO situation that if a property is used for a wholly different purpose from another, they cannot form a single hereditament.

3.8 Single definable position

The requirement that a hereditament must have a single definable position usually causes no difficulty. What is occupied can usually be easily seen or established.

In *Spear* v *Bodmin Union* (1880) 49 LJMC 69, a stall-holder, who was entitled to reserve two sites within a market, was found not to have a rateable hereditament because he had a right to two stalls but not the right to place them on a definite portion of ground.

In *Peak (VO)* v *Burley Golf Club* [1960] 1 WLR 568, a golf club which had a licence to use part of the New Forest as a golf course was held not rateable for the course by the Court of Appeal. This was because there was no exclusive occupation for the purpose for which the land was used by the club as it appeared anyone could come and play golf on the course and also the extent of the hereditament could not be clearly defined.

As per Harman LJ: *'I cannot find any rateable hereditament defined; I should have thought that before levying a rate you must be able to say what the hereditament was on which you were to levy.'*

The requirement is for the location to be fixed and to have a definable boundary. There is no requirement for the boundary to be a physical one on the ground and painted lines defining the boundary were held sufficient in *Coxhead (VO)* v *Brentwood Urban District Council* [1972] RA 12. This case concerned individual car parking spaces for blocks of flats which the Lands Tribunal found could be identified on the ground either by painted lines or by expansion joints in the concrete surface. It also appears that delineation of the boundaries on a lease plan will be sufficient.

3.9 When does a hereditament come into existence?

Defining what constitutes a single hereditament can be difficult, but determining when the hereditament comes into existence can also be a problem. Normally this should be straightforward. For a new building the hereditament or hereditaments come into existence when it is ready for occupation, and for a hereditament created by division or merger it is when the change happens. It is important to know the date the hereditament comes into being in order to determine the effective date, i.e. the date from which rate charging can begin. In *Baker (VO)* v *Citibank* [2007] RA 93, the Lands Tribunal determined that the significant extension of the curtilage of a hereditament created a new hereditament. The case involved successive additional floors being erected and occupied alongside, and as an extension to, an existing office building.

The Lands Tribunal considered the inclusion of each additional floor was properly to be regarded as the creation of a new hereditament, replacing the one defined by the previous area of occupation. It rejected:

• The idea that the existing hereditament already included parts of the building that were incomplete or unoccupied.

- The respondent's submission that each extension did not result in a new hereditament because it was only a small percentage increase in area. It noted the area in actual terms was large (about 30,000 sq ft) and of considerable value.
- The proposition that the extent of the hereditament was defined by the boundary of the occupier's ownership and that physical changes within that boundary did not constitute the creation of a new hereditament. The Tribunal considered it 'unreal' to treat as part of the ratepayers' hereditament those three-dimensional spaces which, when filled out as parts of a building and made capable of occupation, would become part of the occupation of the ratepayer, simply because, in advance of its occupation of those areas, it had acquired rights in respect of them.

3.10 Relevant non-domestic hereditaments

The Local Government Finance Act 1988, as explained earlier, retained the old General Rate Act 1967 definition of a hereditament, and treats a hereditament as anything that would have been a hereditament under the pre-1990 rules. The Act, in section 64(4), defines a relevant hereditament as a hereditament consisting of lands, coal mines, other mines or advertising rights.

The mention of advertising rights is needed because rights are not rateable unless specifically mentioned in the Act. Rights on their own do not involve the occupation of land: they allow someone to do something on, or take something from, land occupied by another person. As rating is a tax on the occupation of land, such *incorporeal hereditaments* are not rateable unless they are specifically made rateable by the Act. The only rights now rateable on their own are advertising rights and the right to operate electricity and gas meters. Sporting rights were removed from rateability in 1997. In the case of meters the Central Rating List (England) Regulations 2005 (SI 2005/551) and the Central Rating List (Wales) Regulations 2005 (SI 2005/422) require electricity and gas meters not to be treated as one hereditament and to appear in the central lists rather than as thousands of separate entries in local lists. Rights can, however, be included in the value of a hereditament if they are an adjunct to the enjoyment of the occupation of that hereditament, e.g. a right of way. To remove houses and flats from being rateable, the Act in section 64(8) defines a non-domestic hereditament as one which either consists entirely of property which is not domestic, or is a composite hereditament. A composite hereditament is one which is both domestic and non-domestic, e.g. a shop with a flat over, where both are occupied together.

Section 42 of the Local Government Finance Act 1988 provides for relevant non-domestic hereditaments to appear in local non-domestic rating lists unless they are exempt, wholly domestic or should appear in a central rating list.

Section 43 of the Local Government Finance Act 1988 provides for ratepayers to be subject to a non-domestic rate for the hereditaments they occupy providing they are shown in a local non-domestic rating list which is in force.

3.11 Summary

In order to incur rate liability there must be a rateable hereditament and a rateable occupier.

There are six requirements for a property to become a hereditament – it must be within one or more billing authority's area; have a single rateable occupier; be capable of separate occupation; be a single geographical unit; be put to a single purpose; and have a single definable position.

The time when an actual hereditament comes into existence can cause problems.

3.12 Wales

Regulation 4 of the Non-Domestic Rating (Miscellaneous Provisions) (Wales) Regulations 2017 (2017/327) regulates how cross boundary property should be treated in Wales. The rules in respect of the hereditament are essentially the same in both England and Wales.

The PICO act does not apply to Wales, instead The Non-Domestic Rating (Property in Common Occupation) (Wales) Regulations 2022 SI 1025 (W. 218) brings the Welsh practice into line with the existing PICO provisions in England albeit only from 1 April 2023. It requires separate hereditaments to be merged if in the same occupation and satisfying the contiguity condition. The wording is almost exactly the same as in England … but not quite. The effect is the same.

4 The domestic/non-domestic borderline

4.1 Introduction

The replacement of the old pre-1990 rating system by a non-domestic rating system and, from 1993, a Council Tax based on capital values, created a requirement to identify what properties are domestic, what are non-domestic and what are both domestic and non-domestic at the same time. Prior to 1990 all properties were either rateable or exempt. Now, only wholly non-domestic properties or properties which are both domestic and non-domestic are rateable. For properties which are both domestic and non-domestic, only the non-domestic element is valued for rating.

Usually it is obvious whether a property is domestic or non-domestic, but this is not always the case. A factory or an office block may appear clearly non-domestic and a house may seem clearly domestic. However, even these properties can have a domestic or non-domestic use within them such as a caretaker's flat in the office block or a business run from home. For other property types the division between domestic and non-domestic may be even less clear. For example, a decision had to be made in drafting the legislation about whether particular types of property such as holiday cottages or guest houses should be treated as domestic or non-domestic. Further, the borderline between domestic and non-domestic usage can fluctuate during the course of a year.

4.2 Definition of domestic property

Section 66(1) of the Local Government Finance Act 1988, as amended, provides a basic definition of what constitutes domestic property:

Property is domestic if:-

(a) **it is used wholly for the purposes of living accommodation,**
(b) **it is a yard, garden, outhouse or other appurtenance belonging to or enjoyed with property falling within paragraph a) above,**

DOI: 10.1201/9781003297369-4

(c) **it is a private garage which either has a floor area of 25 square metres or less or is used wholly or mainly for the accommodation of a private motor vehicle, or**
(d) **it is private storage premises used wholly or mainly for the storage of articles of domestic use.**

The first part of the definition treats property as domestic if it is used wholly for the purposes of living accommodation. This is the primary element in defining what domestic property is. The word used in the definition is property and not hereditament. 'Property' can mean part of a hereditament rather than the whole. A room or part of a room within a hereditament can therefore be 'property' and be domestic. To be domestic that part has to be used 'wholly' for the purposes of living accommodation. A part which has a mixed use will not be wholly domestic and consequently will be non-domestic.

Note the word 'used'. The requirement for being domestic property is not one of character and design but the use to which property is put. Is the property used as living accommodation or not? If property is used as living accommodation it will be domestic. If a whole hereditament is used as living accommodation even for a short period it will cease to be rateable and should be removed from the rating list.

There is clearly no difficulty with houses or flats. They are normally used wholly for the purposes of living accommodation and are consequently treated as wholly domestic. Even if the whole or part of a hereditament is not actually being used at the present time, the Act provides a way to determine whether the whole or parts are domestic or non-domestic by stating that 'property not in use is domestic if it appears that when next in use it will be domestic' (s.66(5)).

Houses usually have yards, gardens, outhouses or other appurtenances. These appurtenances are treated as domestic under paragraph (b) if they are situated within the curtilage and belong to or are enjoyed with property that is wholly used for living accommodation. In *Martin* v *Hewitt (VO)* [2004] RA 275, the Lands Tribunal did not regard two boathouses situated some 500 m from the occupier's homes as appurtenant because the boathouses were a substantial distance from the houses, were separated by land not in the ratepayers' occupation and were not within the curtilages of the dwelling-houses.

Flats too can have yards, gardens, outhouses or other appurtenances, though more usually these will be appurtenant to the whole block or group of flats rather than each flat having its own appurtenance. The use of the word 'property' in section 66(1) is apt to include not just a single house or flat but can include several.

With paragraph (b) having added to the primary element of living accommodation those appurtenances that are enjoyed with the living accommodation, paragraphs (c) and (d) then go on to cover two other classes of property which do not need to be enjoyed with the living accommodation. These are private garages and private storage premises, which, though they may not be appurtenances because they are outside the curtilage of the living accommodation, and so not within paragraph (b), are nevertheless brought within the definition as constituting domestic property.

Private garages are domestic whatever their size providing they are used mainly to house a private motor vehicle. They do not need to belong to or be enjoyed with living accommodation. Where they house a business vehicle such as a van or a taxi then private garages are treated as domestic only if they have a floor area of 25 m² or less.

Private storage premises are domestic providing they are used to store articles of domestic use.

4.2.1 Cases on defining domestic property

Various cases have considered the section 66(1) definition, as discussed in the following.

The case of *Walker (VO)* v *Ideal Homes Central Ltd* [1995] RA 347 concerned whether show or view houses should be treated as domestic and be banded for Council Tax or whether, as the valuation officer contended, they were non-domestic and should be assessed for rating. The Lands Tribunal was satisfied the show houses were not 'used wholly for the purposes of living accommodation' as they were used for showing houses for sale and therefore did not satisfy the requirement in section 66(1)(a). The hereditaments were therefore relevant non-domestic hereditaments and required to be entered in the appropriate rating list.

Turner v *Coleman (VO)* [1992] RA 228 concerned a property comprising a mooring fronting the Thames (15.5 m), land to a depth of 30 m with a timber chalet, a car hardstanding and a garden. It was assessed as mooring, land and premises at a rateable value of £800. No one lived on the property and the appellant's boat was not used as a dwelling.

The appellant contended that the whole property should be deleted from the rating list as it was used for domestic and not business purposes. The Tribunal did not regard this as the proper test and examined the legislation. Garden land would be domestic if it belonged to or was enjoyed with living accommodation following the definition: 'garden, outhouse or other appurtenance belonging to or enjoyed with property ... used wholly for the purposes of living accommodation' (section 66(1)(b)).

However, it was clear the riverside garden land was not 'enjoyed with' the appellant's own abode because that was 1.25 miles away.

The appellant also argued that the definition of domestic property in section 66(1) was not exhaustive, but the Tribunal considered that the use of the conjunction 'if' at the beginning of the section indicated that for property to be domestic, one of the four conditions in section 66(1)(a)(d) had to be satisfied. Accordingly, the property was properly entered in the rating list as a composite hereditament.

Similarly in *Aylett* v *O'Hara (VO)* [2011] UKUT 418 (LC), the Tribunal considered whether a piece of riverside land situated many miles from the ratepayer's home and consisting of open land with a slipway and small summerhouse used by him and his co-owners for recreation was domestic property. The summerhouse was used to store a table, chairs, a kettle, cutlery, a fridge and a microwave oven. The Tribunal considered the principal use of the summerhouse was for storage, not living, though it might also be used for shelter. It was ancillary to the land and was used for the purposes of the recreational use of the land rather than for the purposes of living accommodation. It did not consider that the summerhouses or the like fell within section 66(1) (d) as private storage premises used wholly or mainly for the storage of articles of domestic use because it did not consider that the storage of articles of a domestic nature fell within section 66(1)(d) where the premises were part of a hereditament that was otherwise non-domestic and where the storage was ancillary to the non-domestic use.

In *Andrews (VO)* v *Lumb* [1993] RA 124, the ratepayer used a 319 m² warehouse building for the storage and restoration of his collection of vintage buses and associated memorabilia, including one fully restored and roadworthy bus with a private road fund licence. The ratepayer's house was some distance away. He argued the warehouse should be treated as domestic and not assessed for rates.

The Lands Tribunal said that the definition in section 66(1) paragraphs (c) and (d) for private garages and private storage premises did not require them to 'belong to' or 'be enjoyed with' a dwelling-house in order to qualify as domestic property. However, the main use of the accommodation was to house and work on a collection of transport artefacts and not 'wholly or mainly for the accommodation of a private motor vehicle'. The use of some 25% of the space to house the roadworthy bus was no more than incidental to the main use. The Tribunal thought the premises were probably 'private storage premises', but the artefacts could in no sense be described as 'articles of domestic use'.

In *Alford* v *Thompson (VO)* Unreported, 2001, the Lands Tribunal did not accept that a beach hut was 'private storage premises used wholly or mainly for the storage of articles of domestic use' in section 66(1)(d). Instead, the Tribunal regarded the storage of deck chairs, etc., in the beach hut as an adjunct to the recreational use of the beach hut in connection with the beach and not for storage for its own sake of articles of domestic use.

In *Martin* v *Hewitt (VO)* [2004] RA 275, the ratepayers sought deletion of three boathouses on the shores of Lake Windermere on the grounds that they comprised domestic property. Two boathouses were some 500 m from the occupier's homes, the third was some 50 miles from the owner's home, although the ratepayer also had a caravan a mile away.

The occupiers contended that the boathouses either were an 'outhouse or other appurtenance belonging to or enjoyed with property' used wholly with the houses for the purposes of living accommodation (section 66(1)(b)) or were 'private storage premises used wholly or mainly for the storage of articles of domestic use' (section 66(1)(d)).

The Tribunal, following an extensive examination of case law on the meaning of 'appurtenance', noted that in all cases the meaning was confined to the curtilage of the building in question. It could see no reason for treating the wording in the 1988 Act any differently. As the boathouses were a substantial distance from the houses or caravan and separated by land not in the ratepayer's occupation, they were not within the dwelling-houses' curtilages and therefore were not appurtenances.

Equally, the Tribunal found that the boats in the boathouses were not articles of domestic use but rather were articles stored for use on the lake. It could see that the storage in the house of articles used for recreation away from the house would be the storage of articles of domestic use. So also would be the storage away from the house of articles used for recreation at the house. However, it could not see that the storage in premises quite separate from the house of boats to be used for recreation away from the house could constitute the storage of articles of domestic use.

In *Reeves (VO)* v *Tobias and others* [2011] RA 149, the Lands Tribunal considered whether a number of open car parking spaces in a local authority car park let on annual licences to local residents constituted separate rateable hereditaments. The Lands Tribunal did not think the definitions in paragraphs (c) and (d) of section 66(1) overlapped. Paragraph (c) referred to 'garages' and not open car parking spaces. The Tribunal did not consider that a car parking space would normally be described as 'storage premises', that the parking of a car be termed 'storage' or that cars would normally be referred to as 'articles of domestic use'. In any case, as paragraph (c) dealt expressly with premises used for the accommodation of private motor vehicles, there was no justification for treating paragraph (d) as doing so as well. The spaces were rateable.

Various cases, as explained, established that to be an appurtenance the property, be it land, boathouse or other item, had to be within the curtilage of the 'property used wholly for the purposes of living accommodation'. In *Head (VO)* v *Tower Hamlets London Borough Council* [2005] RA 177, the Lands Tribunal examined the further question of whether it was possible for something to be appurtenant to more than one dwelling and consequently be regarded as domestic property.

The case concerned district heating systems serving various numbers of dwellings, ranging from 54 flats in a seven-storey block to one serving 1,298 dwellings in a number of blocks together with three shops and a tenant's meeting hall. The systems comprised a boiler room, supports, chimneys, flues and distribution pipe work. The Lands Tribunal found that the properties were not rateable, noting that section 66(1) was not worded so that to be an appurtenance, property must be appurtenant to a particular hereditament; rather the word used was 'property' and there was no constraint to look at whether a system was appurtenant to any single unit. The Tribunal commented that it would be contrary to the scheme of the legislation to hold that the systems, which were there to serve residential accommodation, were rateable.

It would seem for a property, such as a district heating system, to be appurtenant to more than one dwelling, the property would need to pass on a sale if all the dwellings were sold as a single lot without the need for express mention of the appurtenance. The later case of *Allen (VO)* v *Mansfield and Bassetlaw DC* [2008] RA 338 shows that it is necessary for there to be a building or buildings with a defined curtilage and for the appurtenance to be within that. A district heating system on its own and not within the curtilage of dwellings for which it provides heating will not be appurtenant and, therefore, is not domestic.

In *Winchester City Council* v *Handcock (VO)* [2006] RA 265, the Lands Tribunal found a small sewage works serving ten semi-detached houses was not appurtenant to them as each house had its own garden and clear boundaries each forming its own curtilage.

In *The Collection (Management) Ltd* v *Jackson (VO)* [2013] UKUT 0166 (LC), the Upper Tribunal considered whether a concierge room serving a high class estate of 15 houses was domestic as an appurtenance and found it was, together with other land used in common by the houses such as communal gardens, a plant room and an access ramp, appurtenant to the houses. At first sight this seems a little out of line with the Winchester decision above. However, the contrast seems to be between estates of houses where each house has its own curtilage and where there is no shared communal property and the situation with blocks of flats and, indeed, developments like The Collection, which have an intimate connection with various communal facilities and form a coherent whole.

In deciding what is or is not appurtenant in *Corkish (VO)* v *Bigwood [2019] UKUT 191 (LC)* the Upper Tribunal adopted the Court of Appeal's approach in *Clymo* v *Shell-Mex* that 'the question to be answered is whether the land is properly to be described as an appurtenance in all the circumstances of the case'. And the Tribunal said in considering that question 'we take into account the nature and function of the buildings and other facilities themselves, their proximity to each other, and the general layout of the site'. This was very much followed by the Upper Tribunal in another stables case, *The Propane Company Limited* v *Bunyan (VO)* [2022] UKUT 0237 (LC). In that this case the Tribunal also considered whether separate ownership from property wholly used for living accommodation prevented the stables in the case from being appurtenances, and therefore domestic property. It regarded this as very much a matter of degree, depending on the identity of the different owners. It noted the wording talked about 'enjoyed with' as well as 'belonging to' and held it was possible.

The legislation appreciates that people also live in caravans and house-boats and includes these as domestic property providing they are used as the sole or main residence of individuals (section 66(3) and (4)).

4.3 Hotels

Bedrooms in a hotel can reasonably be described as living accommodation, but a hotel as such is clearly a business property. The legislation makes provision to ensure that hotels are treated as non-domestic hereditaments.

The Local Government Finance Act 1988 provides, in section 66(2), that where property:

is wholly or mainly used in the course of a business to provide individuals whose sole or main residence is elsewhere with accommodation for short periods it is not domestic property.

This removes hotels and similar property from being classed as wholly domestic. However, care has to be taken with the definition, e.g. a hostel for homeless persons despite providing accommodation for short periods would not become non-domestic under this provision as the individuals do not have their sole or main residence elsewhere. They do not, unfortunately, have any residence elsewhere.

Property which is self-contained, self-catering accommodation provided commercially, is specifically excluded from the above provision. However, such property may still be treated as non-domestic if it satisfies the requirement of section 66(2B) for commercial self-catering accommodation (see the following section).

4.4 Holiday cottages and other self-catering accommodation

The statute brings together various types of self-catering accommodation let commercially, e.g. holiday cottages, flats and chalets, whether single properties or in blocks or complexes, and treats them as non-domestic and, therefore, rateable providing a number of conditions are met. These are set out in section 66(2B) and require the accommodation to be:

1. self-catering;
2. self-contained;
3. available for letting for short periods;
4. available to individuals whose sole or main residence is elsewhere;
5. intended to be available for letting for at least 140 days both within the coming year and the immediately preceding year;
6. let commercially (i.e. on a commercial basis with a view to the realisation of profits);
7. in the preceding year, short periods of actual letting totalled at least 70 days.

While the requirements are quite lengthy when set down, they mean that much self-catering holiday accommodation will be assessed for rating. A typical holiday cottage let for a few weeks at a time to holidaymakers will be in rating for that time, providing it is intended by the owner to be available for holiday lets for 140 days or more (i.e. about four and a half months), was so available in the preceding year and was actually let for more than 70 days in the preceding year.

The requirements relating to the previous year – to have been available for letting for 140 days and actually let for 70 days – were added with effect from 1 April 2023. Before then the requirement was simply for it to be available in the coming year for 140 days.

The object of this was to close what was seen as a potential loophole allowing owners of what were really second homes to, effectively, reduce the tax liability on their properties by becoming liable to pay non-domestic rates instead of Council Tax. They could do this by claiming their property was available to let for short periods totalling at least 140 days in a year, but making little or no realistic effort to actually do so, by, for example, not actively marketing the property, asking for unrealistic rents or restricting the dates that the property is actually available to let. The 2023 changes in England had already been made in Wales for the 2010 revaluation from 1 April 2010 and have been in effect in Wales since then.

The legislation therefore has a mixture of intention and actuality.

Intention – the basic 'making available for 140 days' is a matter of intention into the future: What has to be considered is the owner's intention to make the property available for letting. The 140-day period is viewed from the day the assessment is being considered, e.g. the date of a proposal or valuation officer's alteration of the rating list. The year looked at is, therefore, not necessarily the year from 1 April.

Actuality – the additional requirements necessitate examining the actual situation in the preceding year. The property must both have been made available to let for at least 140 days looking from the day the assessment is being considered AND actually have been let for at least 70 days in the preceding year.

Because the 70-day requirement has been in operation in Wales since 2010, the Upper Tribunal has already considered a case on the requirement. In *Gaer Ltd* v *Williams (VO) [2018]* UKUT 0327 (LC) the question whether it was necessary, in a complex of holiday cottages, that each holiday cottage had to be looked at individually to see whether the 70-day requirement had been met for it, or whether if one cottage within the hereditament satisfied the requirement then this covered the whole hereditament. The Upper Tribunal could not accept the wording applied to the whole hereditament if merely part satisfied the test and agreed with the VO that the 70-day test had to be applied to each self-contained unit within the hereditament individually. It was not sufficient that just one unit satisfied the test. Those self-contained units satisfying the test would be non-domestic property and those not satisfying would be domestic.

A holiday cottage made available for letting for less than 140 days a year, perhaps because the owner makes extensive use of it or because it will be let to someone as a sole or main residence for most of the year, will not be in rating. Instead, it will fall within Council Tax and appear in the local council tax valuation list.

In *Godfrey and Godfrey* v *Simm (VO)* [2000] 3 EGLR 85, a self-contained flat in the Lake District was made available to let by the owner as self-catering accommodation and to the valuation office it appeared to be available for much of the year. The owner, though, explained that he did not intend to let the flat for more than 139 days in a year. If bookings for more than 139 days were actually received then he would not accept them. He produced letting details for the last ten years showing lettings had never exceeded 139 days in any one year. The Lands Tribunal accepted that the flat was not intended to be made available for letting for 140 or more days within the year and therefore should not have a rating assessment but should be banded for Council Tax.

It should also be noted that a holiday cottage will not cease to be rateable merely because it is not let commercially and, therefore, does not satisfy the section 66(2B) tests if, in fact, it satisfies the section 66(2) 'hotel' test of being wholly or mainly used in the course of a business for the provision of short stay accommodation. The legislation provides that a 'business' includes:

* any activity carried on by a body of persons, whether corporate or unincorporated;
* any activity carried on by a charity.

Holiday cottages run, say, by a charity, and not on commercial lines, will therefore be non-domestic property if the other tests are satisfied.

4.5 Composite hereditaments

The type of property mentioned so far has generally been either wholly domestic or wholly non-domestic. The only problem has been to decide which. A further complication arises when a property is not wholly domestic or non-domestic but is a mixture of both. These are called composite hereditaments and are defined in section 64(9) of the Local Government Finance Act 1988 as follows:

a hereditament is composite if part only of it consists of domestic property.

Section 66(1)(a) defines property as being domestic:

if it is used wholly for the purposes of living accommodation ...

Composites are hereditaments where part of the property comprising the hereditament is used wholly for the purposes of living accommodation. Note again the word 'used'. To be composite there must be domestic use. The mere fact of part of a hereditament being set out as a flat will not make it composite unless it is actually used as living accommodation, or, if it is unused, unless it appears that its next use will be domestic (section 66(5)). Similarly, even though a hereditament such as a shop with living accommodation may appear at first sight to be partly non-domestic, it will not be composite if, in fact, the shop part is no longer used as a shop but as living accommodation. The 'use' test will judge this hereditament to be wholly domestic and, therefore, out of rating.

A very wide variety of properties actually contain domestic accommodation and will be composite. Examples include shops with living accommodation, public houses, hotels with permanent residents and live-in staff, boarding schools, hospitals with staff accommodation, possibly offices with overnight or caretaker accommodation, universities and colleges.

4.6 Valuation of composites

A sizeable proportion of composite hereditaments comprise shops which have living accommodation, used as living accommodation, on upper floors. Looking at a typical turn of the century parade of shops, it is easy to imagine quite a variation in actual domestic use between one shop and another (see Figure 4.1):

- Shopkeeper A may use several rooms on the first floor for stock and a room on the second as an office while living in the rest of the first and second floors.
- Shopkeeper B may have a large family and no requirement for large areas of storage and uses the whole of his (identical) first and second floors as living accommodation.
- Shopkeeper C may only use one room on the second floor as a bedsit for her son (who acts as night watchman as well as helping in the shop) and the rest of the first and second floors are largely unused.

Such a variation would cause valuation officers considerable difficulty in maintaining and establishing a correct rating list if the value of the actual parts used for living accommodation had to be excluded from the rateable value. This could be different for each shop and might well change regularly during a year as the need for stock space increased or decreased. Instead, once a property has satisfied the use test and is composite, the actual distribution of domestic and non-domestic use is ignored and a notional usage based on the prevailing pattern for similar accommodation in the locality is used. This follows the well-established rating principle of vacant and to let. The principle of notional assessment is, therefore, one of imagining a tenant coming fresh on the scene and deciding what his pattern of occupation will be, having regard to the existing pattern in the locality.

Figure 4.1 Composite hereditament (shaded areas are non-domestic use; unshaded areas are domestic use).

In the case of shopkeepers A, B and C, as already mentioned, the notional usage or normal pattern of occupation would probably be to treat the whole of the second floor of each shop and a room on the first floor as domestic. This would, of course, also depend on an examination of other shops with living accommodation in the locality. The normal pattern of occupation of a particular type of composite property, e.g. shops, can and will vary from locality to locality. The pattern for town centre shops is likely to be quite different from those in a suburban shopping parade.

Having established the parts which are notionally to be treated as non-domestic and those to be treated as domestic, the rateable value to be attributed to the non-domestic part can be found. Schedule 6(1A) of the Local Government Finance Act 1988 (as amended) provides that:

> **The rateable value of a composite hereditament none of which is exempt from local non-domestic rating shall be taken to be an amount equal to the rent which, assuming such a letting of the hereditament as is required to be assumed for the purposes of sub-paragraph (1) above, would reasonably be attributable to the non-domestic use of property.**

This definition does not mean that the non-domestic part is valued ignoring the existence of the domestic part of a composite hereditament. The definition is worded so that regard can be had to facilities which exist within a composite hereditament but are treated as domestic.

In *Williams* v *Gregory (VO)* [1998] RVR 139, the Lands Tribunal rejected the ratepayer's contention that there should be an allowance of up to 50% because his composite shop did not have lavatory and kitchen facilities. The ratepayer said these facilities were present but on the first floor as part of his living accommodation and therefore, the ratepayer reasoned, domestic and not part of the shop. The Tribunal said:

In the present appeal I am satisfied that the respondent valuation officer has correctly applied the rule in making his assessment. His valuation … is confined to the non domestic part of the hereditament i.e. the ground floor shop and showroom. As he explained there is no specific addition in respect of the parts of the residential accommodation used for non domestic purposes, the upstairs bathroom and kitchen, but the availability of these facilities to the occupier is reflected in the zone A valuation of £65 psm.

In *Atkinson* v *Lord (LO)* [1997] RA 413; [1997] 11 CL 448, the Court of Appeal considered whether it was beholden on the listing officer for Council Tax to value the whole of the hereditament on a capital value basis and then apportion this relevant amount to the domestic use. In reality the listing officer had valued the domestic use (the farmhouse) by reference to other similar sized houses not with farms, less 10%. The Court considered there was no obligation to determine the value of the whole farm, and then apportion it, as no particular method of valuation was prescribed. Rather the requirement was to arrive at a result – the value of the whole property that could reasonably be attributed to domestic use.

While it is necessary to have regard to the relevant amount (the value of the whole) it is not always necessary to determine it.

In *Ludgate House Ltd* v *Ricketts (VO) [2023]* UKUT 36 (LC) UTLC, the Upper Tribunal considered a situation where 'property guardians' lived in parts of a vacant office building in order to protect it from squatters and vandals. The Court of Appeal, in an earlier case on the same premises, had already decided the property comprised a single hereditament with the guardians living there but not being in rateable occupation: rather occupation was with the owners. The Tribunal decided it was not correct to assume distribution of the property guardians around the building, as was the case in reality but, instead, they could be 'corralled' into one or two floors. The correct assumption to make was that the hypothetical tenant would be taking the building as an office building and would have the same rights as existed in reality to terminate the scheme or to consolidate the occupation of the property guardians into a particular part or parts of the building, in order to make the most efficient use of space within the building.

4.7 De minimis uses

It is easy to imagine cases where there is a small amount of non-domestic use of an otherwise wholly domestic hereditament, e.g. someone working at home one day a week or even bringing work home on occasion, undertaking a shirt ironing service from home or doing a little hairdressing for payment in the front room. Equally, minor domestic uses of otherwise non-domestic property also occur, e.g. sleeping at the office during train strikes or when there is a major job to be completed.

At first sight such minor uses would appear to classify the hereditament as composite. The practical problems of administration and possible unfairness of assessing for rates minor non-domestic uses was recognised and during the Report Stage of the Local Government Finance Bill 1988 Lord Caithness said: 'where the use of domestic property for a non-domestic purpose does not materially detract from the domestic use, that should not result in that property being rated.'

It was originally proposed that regulations would be made to provide for de minimis non-domestic uses to be disregarded. In the event, regulations were not made but the Government's intention was made clear in a Consultation Paper:

The question of whether the use of a domestic property for non-domestic purposes is material is to be decided having regard in each case to the extent and frequency of the non-domestic use and to any modifications made to the property to accommodate that use.

This approach has been followed by valuation officers and composite hereditaments are not assessed for minor or de minimis domestic or non-domestic uses. A helpful definition of what is minor or not material has been provided and, while there will be cases where it is not obvious whether the use is sufficiently material to result in the hereditament being composite, generally it will be clear whether a hereditament is composite or not.

Additionally, section 67(4) and (5) of the Local Government Finance Act 1988 provides that whether a hereditament consists of non-domestic or domestic property is to be determined on the basis of the state of affairs existing immediately before the day ends. This may assist in determining whether a hereditament is composite if the non-domestic use of rooms only occurs during the daytime and some domestic use is made of them during the evening/night.

In *Fotheringham* v *Wood (VO)* [1995] RA 315, the Lands Tribunal considered whether an accountant's house was wholly domestic property or whether it was a composite hereditament. The ratepayer operated an accountancy business from his house which was advertised in the *Yellow Pages*. A part-time secretary and an accountancy student were employed at the premises. One room was wholly or mainly used as an office; two other rooms were used partly for office purposes and files were stored in a bedroom. Three to four clients visited the property each week and one to two consultations took place in the dining room.

The Lands Tribunal said that the test of whether a hereditament is domestic is whether it is used wholly for the purposes of living accommodation (section 66 of the Local Government Finance Act 1988). A hereditament which is wholly or partly used for non-domestic purposes is a non-domestic hereditament and required to be shown in the rating list. The Tribunal imported a de minimis rule into the words 'used wholly' commenting that where the non-domestic use is insignificant then the property remains domestic. It was happy to accept the guidance outlined above, including noting the suggested test of whether the use did not prevent the accommodation from being used for domestic purposes at any time.

It decided, looking at the hereditament as a whole, that it was not used wholly for the purposes of living accommodation on the material day and confirmed the entry in the rating list.

It seems, following the case of *Tully* v *Jorgenson (VO)* [2003] RA 233, that for an otherwise domestic property to be regarded as composite it will not be sufficient simply to find that part is used for business purposes. Working from home is becoming an increasingly normal use of houses and flats and use of a room as a study whether for private correspondence or business can be seen as a normal use of living accommodation. There needs to be something more than this to result in part of the dwelling being regarded as having a non-domestic use: an adaptation of the premises so that part of it actually loses its domestic character; the advertising of the business, e.g. in the *Yellow Pages*; the use of equipment of a non-domestic sort to a significant extent; or visits by employees or clients to the premises. As in many areas of rating valuation the question will always be one of fact and degree.

Tully concerned whether a bedroom fitted out and used as an office by a disabled ratepayer constituted domestic property within the meaning of section 66(1) of the Local Government Finance Act 1988. The ratepayer had agreed with her employer, the Inland Revenue, that she should work from her home because she was disabled. She used the third bedroom as an office fitted out by her employer with a desk, filing cabinets, a computer and fax. She did not receive work visitors at home. The room was also used as a private study, ironing room and occasional bedroom.

The Tribunal noted that increasing numbers of people worked at home. The Tribunal considered that the purposes of living accommodation was not confined to the satisfaction of basic bodily needs but could, for example, also include recreation and leisure requirements such as a billiard or TV room. Similarly, when a person working at home uses accommodation, furniture and equipment of the kinds commonly found in domestic property this, in its view, might also constitute use for the purposes of living accommodation.

It noted that accommodation could cease to be wholly used for the purposes of living accommodation where the accommodation is adapted so as to lose its domestic character, equipment of a non-domestic sort is used or employees or clients come to the premises.

The Tribunal decided the bedroom was used for the purposes of living accommodation (section 66(1)). It regarded it as part of the ordinary accommodation of the house. There had been no structural alterations, the furniture and equipment were the sort that might be found in a private study and no one visited for business purposes. The Tribunal saw no significance in Mrs Tully being employed as opposed to self-employed. It did not regard the use of the room for other domestic functions, such as sleeping and hobbies, as decisive though this served to emphasise that the room was part of the ordinary domestic accommodation.

4.8 Short-stay accommodation and the 'six person rule'

The rateability of hotels has already been mentioned, in particular the exclusion from being domestic property of properties, such as hotels, that are wholly or mainly used in the course of a business to provide individuals whose sole or main residence is elsewhere with accommodation for short periods.

This definition may also include guest houses, boarding houses and 'bed and breakfast' accommodation. It is likely that many such premises will be composite either because there are live-in staff or long-term guests who use the hotel or boarding house as their sole or main residence, or because the proprietor lives on the premises.

The Government did not wish to bring into rating properties where the occupier did a little bed and breakfast letting. To prevent this, a de minimis test was introduced to prevent limited bed and breakfast uses being rateable (section 66(2A) of the Local Government Finance Act 1988). Properties will not be rateable if:

- the proprietor intends to make short-stay accommodation available in the coming year for no more than six persons at any one time; and
- the proprietor intends to have his or her sole or main residence in the hereditament; and
- the use for short stay accommodation is subsidiary or secondary to its use for domestic purposes.

The test is one of availability not of actual letting. A rather unsuccessful bed and breakfast provider who let a room only 25 times in a year would still find the premises rateable as a composite hereditament if accommodation were available to let for at least six persons at a time in the coming year.

The tests do not mean that any bed and breakfast use where not more than six bed spaces are provided will be in Council Tax rather than rating. The subsidiary test means that if in reality a house or flat is being mainly used for a bed and breakfast business there will be a rating assessment. A three-bedroomed semi-detached house where the proprietor sleeps in one bedroom, lets the other two bedrooms and uses the dining room for guests' breakfast and allows the guests to use the lounge as a sitting room would have a bed and breakfast use which was more than just subsidiary or secondary to the proprietor's domestic use. This subsidiarity test was examined by the Lands Tribunal in *Skott* v *Pepperell (VO)* [1995] RA 243.

This case concerned a detached three-storey, six-bedroomed house used as a guest house by the resident proprietors. The parties agreed that not more than six bed spaces were made available for letting at any one time but differed on whether the use was subsidiary to the domestic use. The Lands Tribunal agreed with the valuation officer that the provision of en suite facilities, smoke alarms and self-closing fire doors went beyond that which would be provided for a private residence. The occupancy levels, the Tribunal's acceptance of the valuation officer's view that 50% rather than 33% of the living area was used for business purposes, and the existence of two advertising signs and car parking led the Tribunal to decide the use was not subsidiary to the hereditament's use as the ratepayer's residence and determined an assessment of a rateable value of £3,000.

If a hereditament is already composite and the proprietor decides to do some room letting, then the letting accommodation may be rateable even though rooms for less than six persons are offered. This is because the use for short stay accommodation is unlikely to be merely subsidiary to the use for domestic purposes, e.g. with a public house the letting will be secondary to the public house business, or indeed part of it, rather than subsidiary to the licensee's living accommodation.

4.9 Rating list entries

Wholly domestic properties do not appear in a rating list; hereditaments which are wholly non-domestic or composite should be included unless exempt. Composite hereditaments (or hereditaments which are part exempt) can be identified in the rating lists by the letter 'C' in the 'part domestic' column.

When a hereditament ceases to be non-domestic or composite because it becomes wholly domestic, it should be removed from the rating list. This applies as much to a farmhouse which has ceased, for the time being, to be available for bed and breakfast letting for more than six persons in the coming year or a seaside boarding house that has actually become wholly occupied as domestic for a few months by the proprietor and his family because the season is over and it is winter, as for a hereditament which has ceased to be non-domestic or composite forever.

Merely because rooms are left unused in a boarding house does not mean they are domestic. Unused property is only domestic if when next in use it appears it will be domestic.

It is, therefore, quite possible for a hereditament under the 'six-person rule' for bed and breakfast accommodation, or the 140-day rule for self-catering accommodation, to be in rating for only part of a year. The rules do not necessarily mean that once the tests are satisfied rating applies for the whole year. The primary principle of the Local Government Finance Act 1988 is if at any time the hereditament can be classed as 'wholly domestic' then it should not appear in a rating list.

4.10 Wales

As already mentioned, the Non-Domestic Rating (Definition of Domestic Property) (Wales) Order 2010 (2010/682) amended section 66 for Wales and added to the requirements for holiday cottages in Wales to be treated as non-domestic under the '140 day rule'. The additional requirement was that they had to have been available for letting for 140 days in the preceding year and to have actually been let for 70 days. While this approach has now been adopted in England, the number of days have been increased in Wales by the Non-Domestic Rating (Amendment of Definition of Domestic Property) (Wales) Order 2022 as from 1 April 2023.

The basic requirement is now that they are intended to be available for letting for at least 252 (replacing 140) days in the coming year, plus the additional requirements that they should also have been made available for 252 days in the preceding year and that they should actually have been let in that year for a total of at least 182 days.

5   Exemptions and reliefs

5.1 Introduction

In previous chapters an examination was made of those factors which go to make a property a rateable hereditament and the identification of the rateable occupier. Some properties or parts of properties are specifically exempt from non-domestic rates or subject to some relief from rates.

Exemptions are contained in Schedule 5 to the Local Government Finance Act 1988, as amended. In the majority of instances the exemptions have been carried over from the pre-1990 rating act, the General Rate Act 1967. Consequently, much of the legal precedent for exemptions remains valid.

In order to decide whether a hereditament or part of a hereditament is exempt from rating, section 67(5) of the 1988 Act provides that this is to be decided on the basis of the state of affairs existing immediately before the day's end, which shall be treated as having existed throughout the day. A non-exempt use during a day will not remove a hereditament from exemption if an exempt use has recommenced before the day's end.

Property not in use or unoccupied will be treated as exempt under paragraph 21 of Schedule 5 if it appears that when next in use it will be used for exempt purposes.

The Secretary of State has power under paragraph 20 to retain earlier exemptions or privileges which existed prior to the passing of the 1988 Act, although to date this power has not been used. Otherwise earlier exemptions were abolished by section 67(12) of the 1988 Act.

5.2 Exemptions

5.2.1 Agricultural land

The exemption of agricultural land dates from 1896 and has been successively modified since then to take account of developments in agricultural practice. Agricultural buildings, as well as agricultural land, are exempt. These can include the typical agricultural barns and sheds as well as modern intensive livestock breeding buildings. The wording of the exemption, while wide, does

DOI: 10.1201/9781003297369-5

not mean that any use of land or building on a farm is automatically exempt. The land, building or use has to fall within the wording of the Schedule to be exempt.

The definition of agricultural land is in paragraph 2:

2(1) Agricultural land is

    (a) land used as arable, meadow or pasture ground only,

    (b) land used for a plantation or a wood or for the growth of saleable underwood,

    (c) land exceeding 0.10 hectare and used for the purposes of poultry farming,

    (d) anything which consists of a market garden, nursery ground, orchard or allotment (which here includes an allotment garden within the meaning of the Allotments Act 1922), or

    (e) land occupied with, and used solely in connection with the use of, a building which (or buildings each of which) is an agricultural building by virtue of paragraph 4, 5, 6 or 7 below.

  (2) But agricultural land does not include

    (a) land occupied together with a house as a park,

    (b) gardens (other than market gardens),

    (c) pleasure grounds,

    (d) land used mainly or exclusively for purposes of sport or recreation, or

    (e) land used as a racecourse.

The definition provides that land used as arable, meadow or pasture land only is exempt. Fields used for the growing of crops or hay or for pasture will be exempt provided that is all they are used for. A field used to pasture a pony or horse used for leisure riding will be exempt because it is used as pasture land. There is no requirement that the pasture has to be used for a working horse on the farm. If arable, meadow or pasture land has another purpose then exemption will be lost. An example of this would be if a farmer grows a field of maize but in the summer by creating pathways forms a maze for members of the public to solve and charges for entrance. While the farmer is still using the field for the arable crop of maize, this is not the only use and exemption will be lost. Plantations, coppice and woodland are exempt; market gardens, nursery grounds, orchards and allotments are exempt. Land used for poultry farming is exempt if it exceeds 0.1 hectares.

Agricultural land does not include land occupied with a house as a park, gardens, land used mainly or exclusively for sport or recreation, or land used as a racecourse. For exemption to fail on the grounds of sport or recreation, the use has to be mainly or exclusively for sport or recreation. However, for exemption to fail on the ground that the land is used as a racecourse the land merely has to be used as a racecourse. This was illustrated in *Hayes (VO) v Lloyd*

[1985] 1 WLR 714 where the Court of Appeal held that land used for a substantial point to point meeting one day a year, but with permanent jumps, was not exempt as agricultural land despite it being used as pasture and meadow land for the rest of the year.

If land fails to be treated as agricultural land, e.g. because it is used as a racecourse, in assessing its rental value the value of both the non-exempt use and any agricultural value should be included. This is because it is the land that is being valued not merely the non-exempt use. For example, in valuing land used as a racecourse, its value as a racecourse should have added to it any value attributable to grazing or arable use for the rest of the year.

5.2.2 *Agricultural buildings*

Agricultural buildings are treated as a separate exemption from agricultural land and indeed the two definitions are mutually exclusive. The exemption of agricultural buildings has been extended over the years. The original exemption now in paragraph 3 provides that a building or part of a building is an agricultural building if it is not a dwelling and is occupied together with agricultural land and used solely in connection with agricultural operations on that or other agricultural land. While the building has to be occupied with some agricultural land, the exemption permits its use to be with other agricultural land not in the building occupier's occupation. So a building used for contract farming can be exempt providing it is occupied with agricultural land.

A building can also be an agricultural building if it is, or forms part of, a market garden and is used solely in connection with agricultural operations at the market garden. In *Tunnel Tech* v *Reeves (VO)* [2014] UKUT 0159 (LC), it was decided by the Court of Appeal that the definition of agricultural buildings did not include plant nurseries. The particular wording in Schedule 5, paragraph 3, only mentioned market gardens and not nurseries. This has been changed by the Non-Domestic Rating (Nursery Grounds) Act 2018 so the definition of 'agricultural buildings' now includes plant nurseries. This provision came into force on 1 April 2015 in England and 1 April 2017 in Wales.

To take account of farmers sometimes having barns and other buildings away from the land they farm which would not, therefore, form part of the same hereditament, the exemption does not require the buildings to be part of the same hereditament as the land. Instead it takes a wider view requiring only that the buildings should be occupied by the farmer with the agricultural land and farmed as a single enterprise. The land and buildings can be many miles apart providing the buildings are occupied with the land. The exemption requires the buildings to be used solely in connection with agricultural operations on the land. This means the use of the buildings has to be subsidiary or ancillary to the agricultural operations rather than the other way around. If the land is subsidiary to the buildings then exemption will not be available to the buildings under this aspect of the exemption.

The sort of buildings which can qualify for exemption as used solely in connection with agricultural operations on the land include more than just traditional barns, silos and tractor sheds. Buildings used for drying, cleaning, grading or packing prior to marketing will qualify as being ancillary to the agricultural operations on the land. However, processing into foodstuffs, such as crisps from potatoes or the retail sale of produce, are treated as more than ancillary operations and will not be exempt.

5.2.3 *Livestock buildings*

Exemption is also available to buildings used for the keeping or breeding of livestock. Such buildings do not have to be occupied with agricultural land but have to be surrounded by, or contiguous to, an area of agricultural land that amounts to not less than 2 hectares. This agricultural land does not have to be occupied by the farmer. The object of the provision is to allow livestock buildings to be exempt only if they are in rural areas. An intensive chicken rearing building on an industrial estate would not be exempt unless there happened to be two or more hectares of agricultural land adjoining.

The exemption requires the buildings to be solely used for the keeping and breeding of livestock, unless they are occupied with agricultural land, in which case they can also be used in connection with agricultural operations on the agricultural land. A farmer can therefore store a tractor used for the farm's agricultural land in a building used for pig rearing without exemption being lost.

Buildings used in connection with livestock buildings, e.g. for storing feed, can also be exempt providing they are solely used in connection with the operations carried out in the livestock building or, if occupied with agricultural land, are also used in connection with agricultural operations on the agricultural land.

Livestock is defined as 'any mammal or bird kept for the production of food or wool or for the purpose of its use in the farming of land'. Buildings used for the intensive rearing of pigs or poultry are clearly exempt. Stables for horses would only be exempt if the horses were actually used for farming the land. The leisure use of stables is not an agricultural use. Birds such as partridge or pheasant bred on game farms to be released into the wild for sport are not classed as livestock as the primary purpose of rearing is sport not food even though, after they are shot, they will be prepared for the table (*Cook* v *Ross Poultry* [1982] RA 187).

5.2.4 *Syndicates of farmers and bodies corporate*

Sometimes farmers join together to use a building jointly because it would be too expensive or inefficient for each of them to run such a building on their own. These can be exempt providing the building is occupied by a syndicate of less than 25 persons and used solely in connection with agricultural operations carried out on agricultural land occupied by the syndicate members.

Instead of a syndicate, farmers may form a corporate body such as a co-operative or a limited company. A building occupied by the corporate body can be exempt if, first, the building is solely used in connection with agricultural operations carried out on agricultural land or used in connection with livestock buildings. The building must be occupied by a body corporate any of whose members are, or are together with the corporate body, the occupiers of the land or the livestock buildings. The members of the corporate body who are occupiers of the land must together have control of the body. The exemption does not require the building to be 'occupied with' the land or livestock building, nor is there a requirement that it be surrounded by 2 hectares of agricultural land.

5.2.5 Fish farms

Paragraph 9 of Schedule 5 of the Local Government Finance Act 1988 provides a similar exemption to agricultural land and buildings for fish farming. The exemption applies to land or buildings used solely for or in connection with fish farming. The exemption is for fish or shellfish for human consumption and therefore if the fish are purely ornamental or intended for pet food exemption will not apply. 'Shellfish' includes crustaceans and molluscs of any description.

Buildings used in connection with fish farming would include those used for killing, gutting, cleaning, packing or freezing fish.

5.2.6 Places of public religious worship

Exemption is available under paragraph 11 of Schedule 5 of the Local Government Finance Act 1988 for places of public religious worship, church halls and administrative offices for religious denominations. The wording of the exemption is:

11(1)   A hereditament is exempt to the extent that it consists of any of the following

  (a) a place of public religious worship that belongs to the Church of England or the Church in Wales (within the meaning of the Welsh Church Act 1914) or is for the time being certified as required by law as a place of religious worship;

  (b) a church hall, chapel hall or similar building used in connection with a place falling within paragraph (a) above for the purposes of the organisation responsible for the conduct of public religious worship in that place.

(2) A hereditament is exempt to the extent that it is occupied by an organisation responsible for the conduct of public religious worship in a place
falling within sub-paragraph (1)(a) above and

   (a) is used for carrying out administrative or other activities relating to
   the organisation of the conduct of public religious worship in such a
   place; or

   (b) is used as an office or for office purposes, or for purposes ancillary to
   its use as an office or for office purposes.

(3) In this paragraph 'office purposes' include administration, clerical work and
handling money; and 'clerical work' includes writing, book-keeping, sorting
papers or information, filing, typing, duplicating, calculating (by whatever
means), drawing and the editorial preparation of matter for publication.

Exemption applies where the whole or part of a hereditament is a place of
public religious worship, either certified as a place of religious worship by
the Registrar General or belonging to the Church of England or Church in
Wales. Exemption will not apply to buildings for non-religious meetings such
as humanist meetings.

The worship has to be public which means there has to be an invitation to the
public to attend rather than the worship being for members of the church only. In
Church of Jesus Christ of Latter Day Saints v *Henning (VO)* [1963] 3 WLR 88, the
House of Lords held that a Mormon Temple which had been certified as a place
of religious worship was not a place of public religious worship because neither the
general public nor members of the Mormon Church who were not recommended
by their bishop could enter the Temple. This decision was reaffirmed by the House
of Lords when considering the Preston Temple in *Gallagher (VO)* v *Church of
Jesus Christ of Latter-day Saints* [2006] EWCA Civ 1598.

Church halls, chapel halls or buildings serving a similar function to
synagogues, temples and mosques are exempt when used in connection with a
place of public religious worship for the purposes of the organisation responsible for the conduct of the worship. The wording of this exemption is quite
widely drawn and a wide variety of premises used in connection with a place
of public religious worship have been regarded as exempt. The typical church
hall of one large room, several small rooms and a kitchen is clearly exempt but
a seven-storey building comprising a large lecture hall, dining room, library,
meeting rooms, nursery, youth centre and offices was also held exempt in
West London Mission of the Methodist Church (Trustees) v *Holborn Borough
Council* (1958) 3 RRC 86.

The test concerns the purpose for which the building is used rather than its
architectural characteristics: whether the uses made are activities in connection
with the place of public religious worship. In the *West London Mission* case it
was noted that Methodism places particular emphasis on both spiritual and
social welfare, and therefore the mix of activities both purely religious but also
social welfare were for the purposes of the church organisation. In some cases

church social clubs have been held to be exempt; Christian Science reading rooms have been held to be exempt; however, a Jewish ritual bath house was held in *Gillet (VO)* v *North West London Communal Mikvah* [1982] 2 EGLR 229 not to be exempt because the Lands Tribunal was unable to detect any point of similarity between the bath house and a church hall, seeing it as 'a bath house, albeit used for a most sacred purpose ... If it has any affinity with any building it is with the synagogue itself'.

In *Gallagher (VO)* v *Church of Jesus Christ of Latter-day Saints* [2006] EWCA Civ 1598, the Lands Tribunal considered, in essence, a church or chapel hall was:

> a hall, often with other rooms and ancillary accommodation, which is used for functions and meetings by the congregation, and at times also by others, for the conduct of church business and sometimes for wider community purposes that reflect the nature and purposes of the ecclesiastical body that is in occupation. It is not itself a place of worship.

Offices and other premises used for carrying out administrative or other activities relating to the organisation of the conduct of public religious worship in a place of public religious worship are also exempt. An example of this would be diocesan administrative offices. In *The Church Of Scientology Religious Education College Inc* v *Ricketts (VO)* [2023] UKUT 00001 (LC) the Upper Tribunal considered the extent of the office exemption and said:

> Use as an office or for office purposes, provided it is use by an organisation responsible for the conduct of public religious worship in a place certified by law, is sufficient to gain exemption for so much of the hereditament as is so used. There is no further requirement of a connection or relationship between the office use and the place of worship.

5.2.7 Trinity House properties

Lighthouses including property within the same curtilage and occupied for the purposes of a lighthouse, buoys or beacons owned or belonging to Trinity House are exempt under paragraph 12.

5.2.8 Sewers

Sewers and their accessories such as a manhole, ventilating shaft, pumping station or pump are exempt under paragraph 13. Sewage works are, however, rateable. The function of a sewer is to drain and therefore ceases when some treatment is given to the sewage. This is because the effluent is no longer simply being drained but is being moved as part of a treatment process. In *Webster (VO)* v *Yorkshire Water Services Ltd* [2009] UKUT 199 (LC), the Lands Tribunal found that a sewage pre-treatment works and a syphon carrying the sewage under a stream to a sewage treatment works were not exempt. Had the sole purpose of the screening and de-gritting been to protect the syphon from silting up then the pre-treatment works and syphon would have continued to be treated as exempt as a sewer, but the dual function of it being essential pre-treatment as well as protecting the syphon made the decision problematic. On the facts, including the relative importance of the pre-treatment to the overall treatment process, the Tribunal found the pre-treatment works and syphon not to be a sewer.

5.2.9 *Property of drainage authorities*

Land occupied by a drainage authority and which forms part of a main river or watercourse maintained by the authority is exempt. Structures and appliances maintained by a drainage authority for the purpose of controlling or regulating the flow of water in, into or out of a watercourse which forms part of a main river or is maintained by the authority are also exempt under paragraph 14.

5.2.10 *Parks*

The history of the exemption of public parks goes back to 1897 and the case of *Lambeth Overseers* v *London County Council* [1897] AC 625 (the 'Brockwell Park case'). This case concerned a park acquired under a local Act which required the county council to lay out the park and maintain it for the perpetual use of the public for exercise and recreation. The House of Lords held that there was no rateable occupation of the park as the 'public' itself could not be a rateable occupier and the county council was not in occupation but merely acted as custodian or trustee holding and managing the park for the use of the public.

From this case it was accepted that if a park was irrevocably dedicated for free and unrestricted use by the public then there was no rateable occupation and therefore no liability for rates. The exemption has always been regarded as slightly anomalous because other facilities provided free of charge by a council, e.g. art galleries and public conveniences, are not exempt. The exemption, however, was made a statutory exemption in 1961 and is now contained in paragraph 15 of Schedule 5 to the Local Government Finance Act 1988. This states:

1   **A hereditament is exempt to the extent that it consists of a park which**

    (a) Has been provided by, or is under the management of, a relevant authority or two or more relevant authorities acting in combination, and

    (b) is available for free and unrestrictive use by members of the public.

2   **The reference to a park includes the reference to a recreation or pleasure ground, a public walk, an open space within the meaning of the Open Spaces Act 1906, and a playing field provided under the Physical Training and Recreation Act 1937.**

3   Each of the following is a relevant authority:-

(a) a county council.

(b) a district council …

The scope of the exemption is wider than the Brockwell Park principle. There is no longer a need for any dedication to the public in perpetuity. The exemption covers not just the traditional municipal park but includes recreation grounds and playing fields, public walks and open spaces such as beaches within the meaning of the Open Spaces Act 1906. A park does, however, have to be managed by a relevant local authority and to be available for the free and unrestricted use by members of the public. Relevant authorities include London Boroughs and district, county and parish councils.

5.2.11 Ancillary parts of a park

The exemption extends to any part of a park that is ancillary to the use by the public of the rest of the park. Brockwell Park itself included a refreshment room. The local act applying to Brockwell Park did not permit any letting off of any part of the park and the Court said in relation to the refreshment room and other parts that: 'they are and must be part of the park. They cannot be separated from it, and are within the principle … already explained.'

A later case, *Sheffield City Council* v *Tranter (VO)* [1956] R&IT 216, concerned a refreshment pavilion in a public park where the council let the refreshment pavilion under a tenancy agreement and at a rent. The Court of Appeal decided that, despite the pavilion being in the separate rateable occupation of the tenant, it was nonetheless part of the park and exempt. Lord Evershed said:

> *the principle to be applied is prima facie a park – i.e., a place dedicated to, and used as such by, the public – is to be treated as in the occupation of the public. Therefore, since the public are not rateable, the park is itself free from rating liability, and that exemption is not removed by the circumstances that the public may be excluded in certain respects and on certain parts of the park in the course of the ordinary necessities of proper management.*

Lord Evershed noted that exemption would not be removed if the refreshments were provided directly by the council itself and therefore exemption would not be removed by the council making arrangements with a tenant whereby he carried on his own business in providing the refreshments.

Lord Evershed said:

> *The provision of refreshments such as are provided in this pavilion can properly be described as an essential amenity of a public park; and it was also amply justified on the facts in finding that the tenant's conduct of this refreshment*

pavilion was nothing more than an ancillary activity of the conduct of the park itself … that is not to say that it follows that every letting off by a local authority of part of a public park, assuming power so to let it off, to a man to carry on there some business of his own, will produce the result that the part so let-off will be free of rating liability. I think it is in every case a matter of degree and of the facts to be found in the case … it must be a question of fact and degree in cases of this kind whether it can be said that the refresh-ment pavilion – or, to take an example cited in argument, a hut or a shed for storing other amenities of the park such as the park chairs – and the part of the park so occupied and used, is in reality still an inherent and essential part of the park as an entity, providing a necessary amenity for the park; or whether the hereditament has been so carved out as to acquire a distinct status from the park and to render itself liable for rating assessment.

From She*ffield City Council* v *Tranter (VO)* [1956] R&IT 216 it seems that there are various requirements to be met for a property within a park to qualify under the 'parks' exemption:

- It has to form part of a park.
- The property has to be an ancillary activity to the conduct of the park.
- The property has to provide an essential or necessary amenity to the park.
- The occupation has to be, in effect, occupation by the public.
- The property has to be used to supply the needs of visitors to the park.
- The property must not have been sufficiently carved out of the park to acquire a distinct status.

It is a matter of fact and degree in every case. Cases where private occupiers have found their hereditaments exempt under the parks exemption have, not surprisingly, been quite rare. Usually cases on parks exemption have concerned either whether a property is exempt as a park or whether some part of a prop-erty in the occupation of the local authority also qualifies for exemption along with the main park, rather than whether the occupier of a sublet property within the boundaries of a park is exempt.

In *Blake* v *Hendon Corporation (No 2)* [1965] RVR 160, the Court of Appeal decided a bowling green in a public park was a separate rateable hereditament because it was maintained for the benefit of the club rather than the public at large and its use was not ancillary to the rest of the park.

In *North Riding of Yorkshire Valuation Committee* v *Redcar Corporation* [1943] KB 114, the local authority had seven hereditaments on the foreshore at Redcar. The council argued that these hereditaments were exempt as part of the 'park'. They included a swimming pool, boating lake, concert hall, two lock-up shops and the foreshore itself which had an area used for the purposes of stalls and amusements. The hereditaments were all held to be rateable because of the extent to which charges were made and they could not be said to be available for the free and unrestricted use by members of the public.

A miniature railway in a country park was regarded by the Lands Tribunal in *Whitby* v *Cole (VO)* [1987] RA 161 as a separate occupation. Although it was within the park it was not part or an ancillary part of it, but constituted a separate recreational facility which the Tribunal considered should be regarded in the same light as the *Redcar* case. An earlier case, *Southern Miniature Railways* v *Hake (VO)* (1959) 52 R&IT 591, concerned a miniature railway which was held exempt. The Tribunal in *Whitby* noted that in that case the railway, which was clearly part of the park, was operated under a licence and the takings were paid over to the borough treasurer who then passed some of them back by way of remuneration to the operators.

The Lands Tribunal decided in *Oxford City Council* v *Broadway (VO)* [1999] RA 169 that open air swimming pools surrounded by a public park were not so carved out of the park as to acquire a distinct and separate status, but were used for a purpose which was solely to enhance the attractiveness of the park as a park.

5.2.12 Property used for disabled persons

Paragraph 16 of Schedule 5 of the Local Government Finance Act 1988 provides for exemption for certain property used for people who have disabilities or are or have been suffering from illness. The exemption has been expanded by case law over the years. It is not the most straightforward of exemptions and it is not always clear at first sight what is or is not exempt. The wording of the exemption is:

16.-(1)   A hereditament is exempt to the extent that it consists of property used wholly for any of the following purposes-

(a)   the provision of facilities for training, or keeping suitably occupied, persons who are disabled or who are or have been suffering from illness;

(b)   the provision of welfare services for disabled persons;

(c)   the provision of facilities under section 15 of the Disabled Persons (Employment) Act 1944;

(d)   the provision of a workshop or of other facilities under section 3(1) of the Disabled Persons (Employment) Act 1958.

Paragraph 16(1)(a) exempts any part of a hereditament which is wholly used to provide facilities for training or keeping suitably occupied people who are disabled or have been or are ill. The phrase 'keeping suitably occupied' means an occupation appropriate to the person's disability or illness rather than one that would be generally suitable. In *Chilcott (VO)* v *Day* [1995] RA 285, the Lands Tribunal decided four holiday chalets let as a matter of policy only to disabled persons and their carers were not exempt. The Tribunal did not consider that providing holiday accommodation was something that could come within the phrase 'keeping suitably occupied'. The Tribunal considered the phrase had to be read with the word training 'so as to impart the sense of providing "training or occupation"'.

Premises used to allow a person to carry on a business will not be exempt if they are the sort of premises the person might use for the business if not disabled or ill. In *O'Kelly* v *Davey (VO)* [1996] RA 238, the Lands Tribunal found that a disabled musical instrument restorer's workshop was not exempt. The Tribunal said:

> *Occupation, however suitable, which is for the purpose of reward is not in my judgment contemplated by the subparagraph although if the occupation, pursued for its own sake, incidentally attracted remuneration … that would not necessarily prevent the facilities as a matter of fact, being provided for the purpose of keeping persons who are disabled 'suitably occupied'.*

Facilities for training will include schools and colleges or any part of them used wholly for disabled or ill persons. In *Halliday (VO)* v *Priory Hospital Group of the Nottingham Clinic* [2001] RA 355, the Lands Tribunal considered that 'training' did not include training of a general sort designed to make an individual fit for a normal life but rather 'the training of a person who is disabled or who is or has been ill so that he can occupy himself in a way that is suitable to his condition'.

Exemption is also available where any part of a hereditament is wholly used for the provision of welfare services to disabled people. Unlike the exemption for training or keeping suitably occupied, the welfare services exemption does not extend to people who are or have been ill. The exemption is also limited to welfare services a local authority could in the past provide under section 29 of the National Assistance Act 1948 1948 before it ceased to apply to local authorities, although local authorities did have quite wide powers under this section.[1] In *Chilcott (VO)* v *Day* [1995] RA 285, the Lands Tribunal accepted that the holiday chalets did provide a welfare service. However, as in most cases the disabled people were accompanied by members of their family who were not disabled and some lettings (4.68%) were to people who were not disabled, it could not be said the chalets were wholly used for welfare services.

Sheltered workshops provided under the Disabled Persons (Employment) Act 1944 and 1958 are also exempt.

5.2.13 *Air-raid protection works*

Property is exempt under paragraph 17 of the Local Government Finance Act 1988 if it is intended to be occupied or used solely for the purpose of affording

---

1 For Wales 'of a kind which a local authority in Wales has power to provide, or arrange to provide, to an adult in the exercise of its functions under sections 35 or 36 of the Social Services and Well-being (Wales) Act 2014'.

protection in the event of hostile attack from the air and is not occupied or used for any other purpose.

5.2.14 Swinging moorings

Exemption is given under paragraph 18 to a type of mooring for boats or ships known as a swinging mooring. Exemption does not apply to other types of mooring such as quayside moorings. Swinging moorings are moorings to a buoy that has an anchor or weight resting on the bed of the sea, lake or river designed to be raised from that bed from time to time for maintenance or some other reason.

This type of mooring is commonly found at marinas, as well as fixed moorings, and will be an exempt part of the marina.

5.2.15 Road crossings over watercourses, etc.

Fixed road crossings such as bridges, viaducts or tunnels for road vehicles or pedestrians across rivers, estuaries and other watercourses are exempt. This exemption in paragraph 18A removes from rating toll bridges and tunnels such as the Dartford Crossing that would otherwise be rateable. Included in the exemption are buildings, other than office buildings, used in connection with the crossing, and any machinery, apparatus or works used in connection with the crossing. Ferries do not come within the exemption.

5.2.16 Property used for road user charging schemes

Exemption applies to any roads or property used solely in connection with the operation of a scheme in respect of which charges are imposed under Schedule 23 to the Greater London Authority Act 1999 or Part III of the Transport Act 2000. This is under paragraph 18B of Schedule 5.

5.2.17 Visiting forces

Hereditaments occupied for the purposes of visiting armed forces as defined in the Visiting Forces Act 1952 are exempt under paragraph 19A. This exemption includes the various US Air Force air bases in the country.

5.3 Reliefs

Whilst some reliefs are set out by statute, other reliefs, often of a temporary nature, are not. The Government has increasingly encouraged billing authorities to use their discretionary powers under section 47(3) of the Local Government Finance Act 1988 to give relief in certain circumstances. In return for their use of such discretionary powers the Government will agree to reimburse the authority for the costs incurred. This has been especially true during the recent COVID19 epidemic when a whole raft of provisions were made.

Appendix 2 contains details of any late breaking changes proposed for reliefs for various occupiers and hereditaments that are not included in this section.

5.3.1 NNDR multiplier

The provision is made with the legislation to regulate the amount by which the NNDR multiplier can increase each year. For 2023/24 the government has proposed that the multiplier will not be increased and will remain at the 2022/23 level. It is expected that the multiplier will continue to be increased for 2024/25 onwards.

5.3.2 Empty property

Rates are a levy on the occupation of land. The logic being: no occupier: no rates. However, since 1966 there has been an additional rate on certain specified unoccupied property payable by the owner, as there is no occupier. Properly then, there is no 'relief' from Empty Property Rates, only an imposition on those specified classes! The empty property rate does not apply when a property first becomes empty and no liability will arise for the first three months of vacancy. Other owners of property that are not liable include:

- industrial premises do not have a liability for the first six months of vacancy;
- listed buildings;
- buildings with a rateable value under £2,900;
- properties owned by charities – but only if the property's next use will be mostly for charitable purposes;
- community amateur sports clubs buildings – but only if the next use will be mostly as a sports club

Chapter 17 deals with the rating of unoccupied property in detail.

5.3.3 Hardship

Billing authorities have powers to grant rate relief in cases of hardship though can be reluctant to use such powers.

5.3.4 Rural shops

Billing authorities are obliged under section 43 of the Local Government Finance Act 1988 to give relief to qualifying general stores, food stores and post offices situated in rural settlements in designated rural areas. The mandatory relief is available at 50% for hereditaments with a rateable value of less than £8,500 (£12,500 for sole pubs or petrol filling stations). Rural areas are defined by the Secretary of State and rural settlements are settlements which appear to the local authority to have a population of no more than 3,000.

Local authorities can also give discretionary relief of up to 100% for qualifying general stores, food stores, post offices and other hereditaments in rural settlements providing the hereditament has a rateable value of less than £16,500 and is used for purposes which are of benefit to the local community, and it would be reasonable for the authority to make such a decision having regard to the interests of persons liable to pay its Council Tax.

Although mandatory relief is 50% the government is expecting authorities to grant 100% relief and it will fund the additional costs incurred by authorities.

5.3.5 Stud farms

Where a hereditament is used for the breeding and rearing of horses or ponies, the rateable value of the parts used for these purposes is reduced by up to £4,700 (£2,500 in Wales) (Non-Domestic Rating (Reliefs, Thresholds and Amendment) (England) Order 2017 (2017/102), article 5). This stud farm relief only applies where the stud buildings are occupied together with over 2 hectares of land that is predominantly used as exempt agricultural land other than for the pasturage of horses or ponies.

5.3.6 Charitable relief

Charities enjoy 80% relief for hereditaments used wholly or mainly for charitable purposes under section 43(6) of the Local Government Finance Act 1988. If a property is unoccupied, relief will also be available if it appears that when next in use it will be wholly or mainly used for charitable purposes. Local authorities have the discretion to increase this relief up to 100%. They can also reduce the rates payable for hereditaments used wholly or mainly for recreation by clubs and organisations not established for profit and for hereditaments occupied in whole or part by non-profit making organisations whose main objects are charitable, philanthropic, religious or concerned with education, social welfare, science, literature or the fine arts.

The relief also applies to registered community amateur sports clubs where the hereditament is wholly or mainly used for the purposes of the club or the club and other registered clubs.

It is not sufficient for a hereditament simply to be owned or occupied by a charity; the hereditament must be used for carrying out the charitable purpose. In *Oxfam* v *City of Birmingham District Council* [1976] AC 126, Oxfam claimed relief for its gift shops used for the sale of clothing and other articles donated to Oxfam. The profit from the sales being applied to the objects of Oxfam, the first question which arises, as Lord Cross of Chelsea said, is 'what are the "charitable purposes" of a charity as distinct from its other purposes?' The answer must be, I think, those purposes or objects, the pursuit of which make it a charity, that is to say in this case the relief of poverty, suffering and distress. The Court decided relief was available to uses which were wholly ancillary to or directly facilitated the actual charitable use but considered raising or earning money for a charity was not directly related to the achievement of the objects of the charity, and gift shops were therefore not entitled to relief.

Following this case, the law was changed to make charity shops entitled to mandatory relief if wholly or mainly used for the sale of goods given to the charity and the proceeds are used for the purposes of the charity (section 64(10) of the Local Government Finance Act 1988).

5.3.7 Small Business Rate Relief (SBRR)

Article 3 of the Non-Domestic Rating (Reliefs, Thresholds and Amendment) (England) Order 2017 (2017/102) makes provisions for relief as follows:

- To qualify for relief the property must have a rateable value of not more than £50,999. These properties will have their rate liability calculated using the small business rate multiplier.
- Where the ratepayer occupies only one hereditament and the rateable value is not more than £12,000, E shall be 5,000,000.
- Where the ratepayer occupies only one hereditament and the rateable value is more than £12,000 and not more than £15,000, E shall be the amount derived from dividing 3,000 by the figure reached by subtracting 12,000 from the rateable value.
- In any other case E shall be 1.

E in the above refers to E in section 43(4B) Local Government Finance Act 1988.

Essentially ratepayers who occupy a property with a rateable value which does not exceed £50,999 (and who are not entitled to other mandatory relief or are liable for unoccupied property rates) will have their bills calculated using the lower small business non-domestic rating multiplier, rather than the national non-domestic rating multiplier.

In addition, generally, if the sole or main property has a rateable value not exceeding £15,000, the ratepayer will receive a percentage reduction in the rates bill for this property of up to a maximum of 100%. For a property with a rateable value of not more than £12,000, the ratepayer will receive a 100% reduction in the rates bill.

Generally, this percentage reduction (relief) is only available to ratepayers who occupy either:

(a) one property; or
(b) one main property and other additional properties providing those additional properties each have a rateable value which does not exceed £2,899.

The rateable value of the property mentioned in (a), or the aggregate rateable value of all the properties mentioned in (b), must not exceed £19,999 outside London or £27,999 in London.

5.3.8 Embassies and diplomats

Common law and the Diplomatic Privileges Act 1964 exempt diplomatic missions from municipal dues and taxes such as rates. A small proportion of what would be the normal rates payment is actually paid as constituting a payment for services, such as refuse collection, rather than being a tax.

5.3.9 Public lavatories

The Non-domestic Rating (Public Lavatories) Act 2021 provides that a hereditament in England or Wales and 'consisting wholly or mainly of public lavatories' effectively gains 100% relief from non-domestic rates.

5.3.10 Improvement relief

The 2021 Fundamental Review of Rates noted that responses to the review stated that businesses could be disincentivised from improving their properties because doing so would lead to an immediate increase in their rateable value and thus their rates bills. The same could be said of occupiers moving to larger premises rather than improving their properties. Again, their new rateable value would reflect the benefit of the property they occupied. Nonetheless to address this, the Government announced that it would introduce 100% 'improvement relief', which would delay increases in bills for 12 months where properties had been subject to eligible improvements.

A similar relief has operated in Scotland since 2018/19, as part of the Scottish Government's 'business growth accelerator relief' scheme. In Northern Ireland, a scheme called 'Back in Business' offers 50% relief for 24 months on retail properties that have been newly occupied after having been empty for 12 months or more.

Introducing this relief in England requires primary legislation. The relief was originally to have been introduced in 2023, but the 2022 Autumn Statement announced that it would be introduced from 1 April 2024. At the time this edition was prepared provision for the relief was included in a non-domestic rating bill being considered by Parliament (clause 1 Local rating: liability and mandatory reliefs for occupied hereditaments).

5.3.11 Retail, hospitality and leisure business rates relief scheme

This scheme applies to England only.

A temporary relief scheme (announced in the UK Budget on 29 October 2018) provided for non-domestic rates relief for retail properties with a rateable value of less than £51,000 in each of the years 2019/20 (one third discount) and 2020/21 (50% discount). In response to COVID-19, in the Budget speech on 11 March 2020, the government announced the discount would increase to 100% from 1 April 2020 to 31 March 2021 and extend to include the leisure and hospitality sectors. The £51,000 RV limit was also removed. This was a very substantial relief resulting in occupiers of eligible properties simply not paying Non-domestic Rates.

The relief was available to certain occupied retail properties that were wholly or mainly being used as shops, restaurants, cafes and drinking establishments. Also to Leisure and hospitality, including cinemas, museums, theatres, gyms, hotels, music venues, nightclubs and sports grounds and clubs. The relief was extended to childcare nurseries.

The requirement for occupation meant that owners of unoccupied properties were not eligible. 'Unoccupied', though, did not mean 'not used'. So, a shop currently closed due to lockdown restrictions was nonetheless treated as occupied, albeit not actually open for business. Stock, etc., would still have been in place – the shop was not actually unoccupied.

Local authorities, as collectors of Rates, were fully compensated for the loss of income due to these non-domestic Rates measures. The scheme continues for 2023/24 and provides eligible, occupied, retail, hospitality and leisure etc. properties with a 75% relief, up to a cash cap limit of £110,000 per business.

This is a temporary measure for 2023/24; the government has not changed the legislation relating to reliefs available. It will be for individual billing authorities to adopt a local scheme and determine in each individual case when, having regard to government guidance, to grant relief under section 47.

The relief should be applied after mandatory reliefs and other discretionary reliefs have been applied.

The ordering should be applied in the following sequence:

1. Transitional Relief
2. Mandatory Reliefs (as determined in legislation)
3. S.47 Discretionary Relief in the following order:

    a. 2023 Supporting Small Business (SSB)
    b. Former categories of discretionary relief available prior to the Localism Act 2011 (i.e. charitable, CASC and rural top up, not for profit) should be applied first in the sequence of discretionary reliefs, after SSB
    c. Other discretionary (centrally funded) including, Freeport relief
    d. 2023/24 Retail Hospitality and Leisure Relief Scheme
    e. Other locally funded schemes (such as hardship).

5.3.11 Supporting Small Business Relief (SSBR)

SSBR helps ratepayers who as a result of the change in their rateable value at the 2023 revaluation are losing some or all of their Small Business, Rural Rate Relief or 2017 Supporting Small Business Relief and, as a result, are facing large increases in their bills.

Charities and Community Amateur Sports Clubs, which are already entitled to mandatory 80% relief, are not eligible for 2023 SSBR.

The Government wish billing authorities to use their discretionary relief powers, under section 47 of the Local Government Finance Act 1988, to grant 2023 Supporting Small Business relief.

It will be for individual billing authorities, to adopt a local scheme for SSBR and determine in each individual case when, having regard to government guidance, to grant relief under section 47.

To support these ratepayers, 2023 SSBR will ensure that the increase in the bills of these ratepayers is limited to a cash value of £600 per year. This cash maximum increase ensures that ratepayers do not face large bill increases in 2023/24 after transitional relief and small business rate relief (if applicable) have been applied.

Those on 2023 SSBR whose 2023 rateable values are £51,000 or more will not be liable to pay the supplement (1.3p) to fund small business rate relief whilst they are eligible for 2023 SSBR.

The 2017 SSBR scheme was provided to support small and medium ratepayers who had seen large increases in their bills at the 2017 revaluation. Since that time they have had six years of support to allow them to adjust to their full 2017 bills. Therefore, for those ratepayers receiving 2017 SSB relief in 2022/23, any eligibility for 2023 SSBR will end on 31 March 2024.

5.4 Wales

5.4.1 Retail, leisure and hospitality rates relief

The Welsh Government has announced that for 2023/24 it will provide 75% relief for eligible ratepayers.

The relief is aimed at supporting businesses and other ratepayers in Wales in the retail, leisure and hospitality sectors, e.g. shops, pubs and restaurants, gyms, performance venues, hotels and community halls which provide services to visiting members of the public. All eligible businesses will need to apply for the relief. The maximum amount of relief will be a discount of 75% on non-domestic rates bills for eligible properties, a 50% discount was available for 2022/2023. The total amount of relief available is £110,000 across all properties occupied by the same business in Wales. To be eligible for the relief, the premises should be wholly or mainly used for qualifying purposes. This is a test on use rather than occupation so properties that are occupied, but not wholly or mainly used for the qualifying purpose, will not qualify for the relief.

Table 5.1 Table of relief for small businesses

Rateable value	SBRR	SBRR post offices	SSBR child care	Until 31/3/2025
1,000	100%	100%	100%	
2,000	100%	100%	100%	
3,000	100%	100%	100%	
4,000	100%	100%	100%	
5,000	100%	100%	100%	
6,000	100%	100%	100%	
7,000	83%	100%	100%	
8,000	67%	100%	100%	
9,000	50%	100%	100%	
10,000	33%	50%	100%	
11,000	17%	50%	100%	
12,000	0%	50%	100%	
13,000	0%	0%	100%	

5.4.2 Small business rates relief scheme

SBRR in Wales provides mandatory rates relief to properties according to their rateable value and broad category of use.

- Eligible business premises with a rateable value of up to £6,000 receive 100% relief; and
- Those with a rateable value between £6,001 and £12,000 receive relief on a tapered basis from 100% to zero – see Table 5.1.
- Certain business categories benefit from an additional level of relief. For example post offices see Table 5.1 and registered childcare premises.
- From April 2018 the number of properties eligible for SBRR has been limited to two properties per business in each local authority. Properties which meet the childcare or post office conditions will not be affected by this restriction.

The exceptions to the SBRR scheme include:

- Hereditaments occupied by a council, a police and crime commissioner, or the Crown.
- Beach huts.
- Hereditaments used exclusively for the display of advertisements, parking of motor vehicles, sewage works or electronic communications apparatus.
- Properties occupied by charities – charities are eligible for charitable rate relief.
- Unoccupied property – unoccupied properties are eligible for empty property relief.

- The number of properties eligible for SBRR is limited to two properties per business in each local authority area.

Post offices and registered childcare premises qualify for a higher rate of relief as shown in Table 5.1.

5.4.3 Transitional relief

Details of the Welsh Transitional Relief arrangements can be found in section 16.13.

5.4.4 Hardship

As with England, billing authorities have powers to grant rate relief in cases of hardship though can be reluctant to use such powers.

6   The basis of valuation

6.1 Rateable value

Rating is a tax on the annual value of land and buildings. The need to find an annual value presupposes the existence of some form of tenancy or a lease setting out the assumptions or terms on which the value is to be determined. For rating valuation this letting is known as the hypothetical tenancy. This imaginary, or hypothetical, tenancy sets out the terms upon which each property to be valued for rating purposes is assumed to be let. Each property to be valued for rating is assumed to be available to let on these terms. This provides a common basis of valuation for all property. The terms are now found in Schedule 6, paragraph 2(1), of the Local Government Finance Act 1988. They are set out in a very brief manner and, as with much rating law, it has been for the courts over the years to interpret and expand on the basic definition.

The basic definition of rateable value in Schedule 6 of the Local Government Finance Act 1988 (as amended by the Rating (Valuation) Act 1999) is:

2.- (1) The rateable value of a non-domestic hereditament none of which consists of domestic property and none of which is exempt from local non-domestic rating, shall be taken to be an amount equal to the rent at which it is estimated the hereditament might reasonably be expected to let from year to year on these three assumptions-

(a) the first assumption is that the tenancy begins on the day by reference to which the determination is to be made;

(b) the second assumption is that immediately before the tenancy begins the hereditament is in a state of reasonable repair, but excluding from this assumption any repairs which a reasonable landlord would consider uneconomic;

(c) the third assumption is that the tenant undertakes to pay all usual tenant's rates and taxes and to bear the cost of the repairs and insurance and the other expenses (if any) necessary to maintain the hereditament in a state to command the rent mentioned above.

DOI: 10.1201/9781003297369-6

(8A) For the purposes of this paragraph the state of repair of a hereditament at any time relevant for the purposes of a list shall be assumed to be the state of repair in which, under sub-paragraph (1) above, it is assumed to be immediately before the assumed tenancy begins.

6.2 Interpreting the definition of rateable value

The courts have closely examined each part of the definition of rateable value. The following sections examine each phrase and then look at various subsidiary rules that have evolved to ensure uniformity in assessing rateable values.

6.2.1 The rateable value of a non-domestic hereditament none of which consists of domestic property

The basic definition concerns hereditaments that are wholly non-domestic and do not have any domestic property within them such as a flat. Some hereditaments do have domestic accommodation and are termed composite hereditaments and paragraph 2(1A) makes specific provision for determining the rateable value of hereditaments that have a mixed domestic and non-domestic use: 'The rateable value of a composite hereditament … shall be taken to be an amount equal to the rent … which … would reasonably be attributable to the non-domestic use of property.'

6.2.2 And none of which is exempt from local non-domestic rating …

Similarly, paragraph 2(1B) makes further provision for determining the rateable values of hereditaments which are partially exempt from rating:

6.2.3 The rateable value of a non-domestic hereditament which is partially exempt … shall be taken to be an amount equal to the rent which … would be reasonably attributable to the non-domestic use of property

There is no requirement to determine the rateable values of hereditaments which are wholly exempt from rating as, by definition, they are outside the statutory provisions.

6.2.4 Shall be taken to be an amount equal to the rent at which it is estimated the hereditament might reasonably be expected to let …

The rateable value to be adopted for any hereditament is to be equivalent to the rent at which it is likely the average landlord and tenant would be able to agree terms. This is the rent at which the hereditament might reasonably be expected to let. It is not necessarily a reasonable rent because what one person might regard as a perfectly reasonable rent is not necessarily what another might consider reasonable. The idea of something being reasonable is subjective in a similar way to a judgement about something being attractive. The question to be decided is not what is reasonable but what rent would reasonably be agreed. As the decision in *Bruce* v *Howard (VO)* [1964] RA 139, where the ratepayer was arguing that the rent paid, due to a scarcity in the market, was unreasonable, puts it:

> *It is not what might be regarded as a reasonable rent – a question on which many different opinions might be held but what might reasonably be expected to be paid – in other words what would be the probable rent in the market.*

Nor is the rateable value automatically the rent actually being paid by a tenant for the hereditament. The actual rent may not be on the statutory assumptions or agreed near the valuation date. In any case it may be out of line with the rents of comparable properties. In a parade of ten identical shops all let around the same date, on similar lease terms, it might be that nine have a rent of £10,000 per annum (pa), whereas the one to be valued has a rent of £9,000 pa or £11,000 pa. The rent reasonably to be expected for this shop would be the £10,000 pa achieved for the other shops and not the unusual rent actually paid. That is not to say that the actual rent of the hereditament is not important evidence to be considered in deciding on rateable value, but other evidence also needs to be looked at to decide the likely rent a tenant would reasonably be expected to pay.

The definition requires an estimate of the rent the hereditament would achieve if it was let. Not surprisingly, the courts have interpreted this as requiring the property to be viewed not only to let but as vacant and to let with the tenant able to take immediate occupation.

London County Council v *Erith and West Ham (Church Wardens and Overseers)* [1893] AC 552:

> *Whether the premises are in the occupation of the owner or not, the question to be answered is: Supposing they were vacant and to let, what rent might reasonably be expected to be obtained for them? The question is what is their rental value in the letting market, not the remunerative rental value of the premises, to any particular occupier.*

6.2.5 ... *From year to year* ...

The phrase 'from year to year' indicates that the standard tenancy envisaged for all rating valuations, the 'hypothetical tenancy', is a yearly tenancy and not a lease for a term of years. It is well established that this does not mean that the tenancy will last just for a year and then finish with the tenant leaving the premises. Instead, the assumption is that the tenancy is likely to last for a longer period and has a reasonable prospect of continuance.

This prospect was described by Lord Esher MR in *R* v *South Staffordshire Waterworks Co.* (1885) 16 QB 359 as:

> *A tenant from year to year is not a tenant for 1, 2, 3 or 4 years, but he is to be considered as a tenant capable of enjoying the property for an indefinite time, having a tenancy which it is expected will continue for more than a year and which is liable to be put an end to by notice.*

The expectation that the tenancy will continue for an 'indefinite time' is now usually referred to as the tenant having a 'reasonable prospect of continuing in occupation': *Townley Mill (1919)* v *Oldham AC* [1937] 26 R&IT 20.

What is a reasonable prospect will depend on the circumstances of the case. What is reasonable for a newspaper vendor's kiosk may not be reasonable for a manufacturer of motor cars. This was considered in *Humber Ltd* v *Jones (VO)* (1960) 53 R&IT 293. The hereditament comprised a 100,000 m^2 car assembly works. It was established that it took five years to set up a new car production line and to reach full production. The premises were held on a lease renewed in August 1957 for 21 years with the right of renewal for a further 21 years. There were break clauses allowing the tenant to terminate the lease at 7 and 14 years.

The ratepayers argued that a reduction in rateable value was appropriate because a car manufacturer would regard an annual tenancy as a very great disadvantage compared with a term of years. The judgement said:

> *I agree that in the every-day world the tenant of a factory such as this would not consider for a moment a tenancy from year to year; but rating is not the every-day world; it is the domain of a hypothetical tenant who is faced with the fact that, try as he may, he cannot become the owner of the premises he occupies, neither can he get a lease for a term of years, short or long. The only security which he can obtain is set out in R v South Staffordshire Waterworks Co (1885).*

In this case, the actual occupier has just entered into a tenancy for 21 years with an option for a further 21 years – 42 years in all – and I can find no reason why the hypothetical tenant should not expect to remain in occupation for a similar period, though he has the possibility – but not the probability – of getting six months' notice at any future time.

The Court Appeal confirmed that the prospect of occupation for the motor vehicle works was not simply a year but likely to be similar in length to the expected period in the real world. This case shows that it would not be correct to reduce a rateable value simply because of the apparent shortness of the hypothetical tenancy. The length of the hypothetical tenancy is not to be treated as for only a year but will have a similar length to that normally expected for the type of property being valued.

The annual tenancy is an unreal assumption but is nonetheless the assumption in the legislation. In *Npower Renewables* v *Grace (VO)* [2011] UKUC 230 (LC), the parties agreed the likely duration of the hypothetical tenancy of an hydroelectric power station would be five years. This enabled them to take account of a loss being inevitable in the first year (from the antecedent valuation date (AVD)) but not thereafter and they were able to smooth the projected five year figures to obtain what they referred to as an 'equivalent constant rent' for the period.

In *Hardman (VO)* v *British Gas Trading* [2015] UKUT 0053 (LC), the length of the hypothetical tenancy was examined in depth. The case concerned a gas fired power station. The case was not so much concerned with the reasonable prospect of continuance, which was somewhat a given, but the effect upon value of the risk that the rent might be reviewed upwards in the future. At the AVD the price for electricity was low and the ratepayers argued the premises had no more than a nominal value given the likelihood of losses in the order of £7.5 million in the first and second years of a tenancy. They suggested a hypothetical tenant who foresaw such early years' losses but a significant improvement in profitability thereafter would fear that a commercially motivated landlord, with the right to terminate, would give notice when he saw profitability improve, albeit with the intention of renegotiating the rent. A tenant would not be prepared to risk this as soon as profits started to materialise and would only offer a minimal rent for a period of one or two years.

To ignore the possibility of an early rent review was not, in the ratepayers' view, permissible as it was tantamount to determining the rent that might be obtained for a term of years. In effect, the valuation officer was making the mistake of implicitly valuing a reasonable fixed term of years.

The Tribunal regarded the assumption of a reasonable prospect of continuance as established beyond doubt in law. From this it considered that the correct approach required the valuer to consider all the facts and circumstances that related to the hereditament in order to judge what sort of length in terms of a prospect of continuance would be appropriate in each particular case.

The Tribunal considered it wrong, as a matter of law, to suppose that the possibility of a rent review must prevent the hypothetical parties from contemplating continuance much beyond two years, as suggested by the ratepayers, or, importantly, looking at likely future changes in profitability. It was also wrong to say that the consequence of valuing the hypothetical tenancy on the basis of continuance for some, or possibly many, years was implicitly to convert a tenancy from year to year into a tenancy for a term of years.

It was not suggesting that the possibility of a rent review could never be a relevant matter in considering the appropriate prospect of continuance and there might arise circumstances in different cases when it might be appropriate to take into account the prospect of a rent review in deciding as a matter of fact how far ahead the hypothetical parties would be likely to look in fixing the hypothetical rent.

The ratepayers had posed a difficult problem, once again, for the concept of the annual tenancy of the rating hypothesis. Why should a tenant pay a substantial annual rent while awaiting likely better times if there was no guarantee that as soon as profitability recovered the landlord would serve notice and seek to agree a higher rent on the basis of the then prospects for the hereditament? The ratepayer's contention, though, clashed with the idea of a reasonable prospect of continuance. Recognising that the hypothesis assumed the land-lord and tenant were reasonable people the Tribunal concluded they would agree a rent based on a reasonable likely period. It said:

> *The valuation hypothesis is preserved by assuming that, as reasonable parties, the hypothetical landlord and the hypothetical tenant would view the prospect of continuance in the light of the rent that was agreed; if the hypothetical rent is assessed on the basis of a tenancy from year to year with a prospect of con-tinuance of several years it is rational to assume that the reasonable hypothet-ical tenant and the reasonable hypothetical landlord contemplate a prospect of continuance that bears a fair relationship to the rent to be paid and is the same likely period in the minds of both of them. There is no gamble. There is no implicit term of years.*

Sometimes, there will not in reality be a reasonable prospect of continuation or indeed the prospect of a year's occupation. For example, the real landlord may have served a notice to quit, coming into effect in six months' time. This would be disregarded, as rating law requires the assumption of a tenancy from year to year with a reasonable prospect of continuance to be assumed, and the notice conflicts with the statutory definition of rateable value and so must be disregarded. However, where the likely length of occupation is reduced by some form of statutory restriction or the exercise of statutory authority, then the effect on rental value of that reduced prospect may be taken into account.

In *Dawkins (VO)* v *Ash Brothers and Heaton Ltd* [1969] 2 AC 336, the House of Lords held that the rateable value of a hereditament should be reduced to take account of the probability that within a year part of the premises would be demolished for a road widening scheme. Lord Pearce differentiated between matters which were accidental to the hereditament, e.g. the current owner, and those which were essential to it, e.g. physical characteristics and any statutory orders upon it:

> *it would be accidental to the hereditament that its owner intended to pull it down in the near future. For the hereditament might have had a different owner who would not pull it down. So the actual owner's intentions are thus immaterial since it is the hypothetical owner who is being considered. But when a demolition order is made by a superior power on an hereditament within its jurisdiction different considerations apply. The order becomes an essential characteristic of the hereditament, regardless of who may be its owner or what its owner might intend. That particular hereditament has had branded on its walls the words 'doomed to demolition' whatever hypothetical landlord may own it.*
>
> *… in the normal case one tells (the hypothetical tenant) that he will have an indefinite prospect of continuance although the tenancy can be determined at the end of one year. In the suggested particular case one has to tell him that*

although he is being given a tenancy 'from year to year' his actual occupation will almost certainly end in eight months. How much he would pay depends on what his view of the circumstances is.

6.2.6 ... On these three assumptions: the first assumption is that the tenancy begins on the day by reference to which the determination is to be made

The first assumption states the day on which the hypothetical tenancy is to be treated as starting. This is found in paragraph 2(3) of Schedule 6 to the Local Government Finance Act 1988 and is the valuation date for rating. It is explained later in this chapter in section 6.5.

6.2.7 The second assumption is that immediately before the tenancy begins the hereditament is in a state of reasonable repair, but excluding from this assumption any repairs which a reasonable landlord would consider uneconomic

Since 1966 when the unoccupied property rate was introduced, there has been a succession of cases concerning the effect of disrepair on rateable value. The early cases up until 1990 were generally concerned with the position for valuations to 'gross value' which is an assessment assumption that no longer applies. Under gross value the hypothetical landlord was responsible for all repairs and was in usual cases assumed to have done them prior to the tenancy starting. From 1990 all valuations were to 'rateable value' with the hypothetical tenant responsible for all repairs.

Under gross value, hypothetical landlords were usually assumed to undertake repairs even if the property was in very poor repair because it would be 'economically reasonable' for them to do so because they had an interest in the property in perpetuity and could justify the cost over the long term.

However, when valuing properties to rateable value terms, it was difficult to envisage hypothetical tenants, coming new to a property, being prepared to spend large sums on repairs without some significant diminution in the rent they would be expected to pay. The extent to which the hypothetical tenant would be prepared to spend money on such repair works was the subject of cases to the Lands Tribunal. Following these decisions, particularly *Benjamin (VO)* v *Anston Properties Ltd* [1998] 2 EGLR 147, the Government introduced the Rating (Valuation) Act 1999 which changed the definition of rateable value to that outlined at the start of this chapter, with the aim of clarifying the valuation assumptions concerning repair and disrepair.

Under the current definition of rateable value there is an assumption that the property will be 'in a state of reasonable repair ...'.

It is only in exceptional circumstances, where a property is in such a poor state of repair that it would be uneconomic for a landlord to remedy the disrepair, that regard is had to the actual condition of the property.

The overall effect of this is, in most cases, that repair will not be a significant consideration in a rating valuation. In most situations the hypothetical landlord of the rating hypothesis will regard it as economically reasonable to undertake any repairs. The approach, when valuing a property for rating purposes, is to assume a standard level of repair rather than take the actual state of repair except in cases where it is in an unusually poor state of repair and uneconomic to repair. What this standard is and when an actual rather than an assumed level of repair is to be taken into account in valuing property is discussed in section 6.8. Where the hypothetical landlord would consider it economically reasonable to undertake the repair work, it is assumed the work has been carried out prior to the hereditament being offered for letting. The assessment of rateable value is therefore undertaken on the assumption the repair work has already been carried out.

In the lead judgement of the Court of Appeal in *Newbigin (VO)* v *Monk* [2015] EWCA Civ 78, Lewison LJ explained that the statutory requirement involved three questions:

1. Is the hereditament in a state of reasonable repair?
2. If not, can the works which are required to put the property into a state of reasonable repair properly be described as 'repairs'? ('the repair question') and
3. Would a reasonable landlord consider the repairs to be uneconomic? ('the economic question').

If the answer to question 1 is that the hereditament is already in a state of reasonable repair then there is nothing to be assumed, and the hereditament can be valued on its actual state. If it is *not* in such a state, however, then it must be assumed to be in such a state providing the works which would be required to put the property into reasonable repair can properly be described as 'repairs', and providing such works are not uneconomic.

In *Archer Ltd* v *Robinson (VO)* [2003] RA 1, the appellants contended that it was not possible to envisage certain works of repair to the premises being undertaken due to the control of asbestos legislation preventing works being undertaken while a building was occupied. This contention was rejected by the Lands Tribunal. The Tribunal explained that no employee would be exposed to risk because, under the rating hypothesis, the property is deemed to be vacant and to let before the assumed tenancy begins and the works could therefore be envisaged as being undertaken before the hypothetical tenancy commences.

The Valuation Office Agency (VOA) issued a Practice Note on the working and interpretation of the Rating (Valuation) Act 1999. While this Practice Note, since updated, has no statutory standing it provides very useful reading as to how paragraph 2(1)(b) of Schedule 6 to the Local Government Finance Act 1988 should be interpreted. Interpretation is not an easy matter and there has been significant litigation and it still is not completely clear how disrepair should be treated post the Supreme Court's decision in *Newbigin (VO)* v *Monk* [2015] EWCA Civ 78.

6.2.8 *The standard of repair normally envisaged*

No doubt partly because of the considerable difficulty in considering each property's actual state of repair, the courts have long interpreted the legislation as requiring a common standard of repair to be assumed for rateable property. This standard has been taken as the general state of repair of a particular class of property in a particular locality.

In *Brighton Marine Palace & Pier Co.* v *Rees (VO)* [1961] 9 RRC 77, the Lands Tribunal commented on the standard envisaged when considering the valuation of a seaside pier and quoted from a landlord and tenant case:

> *The standard was aptly defined by Banks LJ, in Anstruther-Gough-Calthorpe v McOscar [1924] 1 KB 724. It is the cost of putting the premises into such condition as I should have expected to find them in had they been managed by a reasonably minded owner, having full regard to the age of the buildings, the locality, the class of tenant likely to occupy them, and the maintenance of the property in such a way that only the average amount of annual repair would be necessary in the future.*

In *Archer*, the Lands Tribunal did not think what was repair varied depending on the state of other buildings in the locality as suggested by the valuation officer following the arbitrator's formulation in *Anstruther-Gough-Calthorpe v McOscar* [1924] 1 KB 724. It examined a number of cases dealing with repair in a landlord and tenant context and noted that in none of them was the locality or likely class of tenant considered. It concluded that in deciding whether works intended to remedy a defect in a building constitute repair, various matters are to be taken into account, the number of which and the weight to be given to each will vary from case to case.

A non-exhaustive list would include:

- the nature of the building;
- the terms of the lease;
- the state of the building at the date of the lease;
- the nature and extent of the defect sought to be remedied;
- the nature, extent and cost of the proposed remedial works;
- at whose expense the proposed remedial works are to be done;
- the value of the building and its expected lifespan;
- the effect of the works on such value and lifespan;
- current building practice;
- the likelihood of recurrence if one remedy rather than another is adopted;
- the comparative cost of alternative remedial works;
- the impact of the works on the use and enjoyment of the building by its occupants.

6.2.9   *What constitutes repair rather than an improvement?*

The definition envisages the hypothetical landlord undertaking repair works but does not extend to the carrying out of works which are improvements, however desirable they might be. Repair works may include those necessitated by vandalism or other deliberate damage to a property, as well as that resulting from fair wear and tear and the effluxion of time.

In *Marcrom* v *Campbell-Johnson* [1956] 1 QB 106, the Court of Appeal considered the difference between repairs and 'improvements'.

Lord Denning said:

> *It seems to me that the test, so far as one can give any test in these matters, is this: if the work which is done is the provision of something new for the benefit of the occupier, that is, properly speaking, an improvement; but if it is only the replacement of something already there, which has become dilapidated or worn out, then, albeit that is a replacement by its modern equivalent, it comes within the category of repairs and improvements.*

The replacement of old items by their modern equivalent is repair if the old items needed repair/replacement. A slavish adherence to obsolete standards or materials is not required.

The renewal of parts of a building is within repair (repair is restoration by renewal or replacement of subsidiary parts of a whole: *Lurcott* v *Wakely and Wheeler* [1911] 1 KB 905) but does not include the renewal or replacement of substantially the whole of a building.

In a landlord and tenant case, *McDougall* v *Easington District Council* (1989) 87 LGR 527, the Court of Appeal considered past landlord and tenant cases on what constituted repair. The Court identified three tests to be considered:

1. whether the alterations go to the whole or substantially the whole of the structure or only to a subsidiary part;
2. whether the effect of the alterations is to produce a building of a wholly different character from that which is being let;
3. what is the cost of the works in relation to the previous value of the building and what is their effect on the value and life span of the building?

In the usual case it is fairly straightforward to decide if work is repair. The replacement of a roof covering which has become defective due to the ravages of time is clearly a repair as it is an alteration to a subsidiary part and does not produce a building of a wholly different character.

The more difficult cases are where work needs to be undertaken because some part of a building has failed rather than become gradually worn out. Where this is due to an inherent defect, this might be thought not to be 'repair' because the result of the work is a building different from that which existed before, i.e. one which is not characterised by having an inherent defect.

However, assuming the work does not go to the whole, the result may not be to produce a building which is of a 'wholly different' character and, moreover, the cost may not be substantial in relation to the value of the building.

In *Ravenseft Properties Ltd* v *Davstone (Holdings) Ltd* [1980] QB 12, the stone cladding of a 15-year-old block of flats began to move due to the absence of expansion joints and an insufficient fixing between the frame and the cladding – an inherent defect. The Court held the works were repair (and therefore the tenant's responsibility) because the works would not result in giving back a wholly different building to the landlord at the end of the leases. It should also be noted that the cost of the repair was a small fraction of the value of the property.

Where work to remedy an inherent defect is found not to be repair, then the hereditament must be valued with the defect and a decision taken as to whether not remedying the defect would affect the value.

In *Archer*, the first case before the Lands Tribunal under the Rating (Valuation) Act 1999, the appellants had taken a five-year lease of a factory in 1992 at £41,000 full repairing and insuring (FRI). The agreement provided for the landlord to repair the roof lights in the factory's asbestos sheet roof before the tenant took occupation. This was not done. The landlord did seek quotes for the work. One company offering repair works said it could not guarantee success and recommended complete treatment. Eventually, rather than replace, the landlord agreed to erect a new roof over the existing roof, effectively creating a double skin roof with insulation between the old and new roof.

The Lands Tribunal was in no doubt that the approach of building a new roof over the existing roof did constitute repair, assuming the insulation works were not undertaken as these would be an improvement.

6.2.9.1 If a property is actually in disrepair what effect does this have on the rateable value?

As mentioned earlier, the aim of the 1999 Act was to restore the position to that which existed when hereditaments were valued to gross value. As yet there have not been many Lands Tribunal cases under the 1999 Act's definition, though particularly *Thomas & Davies (Merthyr Tydfil) Ltd* v *Denly (VO)* [2014] UKUT 0146 (LC), mentioned on pp. 00–00, provides assistance.

Where a property is in repair to the 'standard' envisaged then clearly the valuation can be undertaken on that basis. However, where the property is actually in disrepair the valuer needs to consider whether this can be ignored or whether some allowance should be made to reflect the actual condition and the cost of bringing the premises up to the normal 'standard' of repair.

Conversely, where a property is in a better state of repair than would be normal for the locality and type of property, it should be assumed to be in the normal state of repair only.

In *Saunders* v *Maltby (VO)* [1976] 2 EGLR 84, the Court of Appeal explained that not all disrepair capable of remedy has to be disregarded but only disrepair that the hypothetical landlord might reasonably be expected to put right rather than accept a reduced rent.

The case concerned a 100–120-year-old brick house. The brickwork and pointing were fair for its age, an inexpensive overhaul was required to the roof covering, there was no significant rising or penetrating damp, the structure of the house was in fair condition, the joinery was reasonable, but the internal decorative condition was 'deplorable'.

Lord Denning said:

> *It seems to me that the Tribunal has to consider whether (the want of repair) is such that it would be reasonable in all the circumstances to expect a hypothetical reasonable landlord to do the repairs. If the cost of doing the repairs would be out of all proportion to the value of the house, so much so that even a reasonable landlord would not do them all, then it must not be assumed that he would do them. He would let the premises at a low rent. In those circumstances the low rent would be the basis on which to arrive at the rateable value.*
>
> *[In the present case] it seems to me that the member of the Lands Tribunal fell into error. He only asked himself whether the defects were capable of remedy. He did not go on to inquire 'How much is it going to cost?' If the expenditure was such that it would be out of all sense to do the repairs, then a hypothetical landlord would let it at a low rent. That fits in exactly with the statute, which says it has got to be in such a state of repair as to command that rent.*

The question to be answered is therefore not just 'are the defects remediable?' but also 'would it be economically reasonable to expect the landlord to carry out the works?'

On remission from the Court of Appeal, the Lands Tribunal decided that an expenditure of some £225 on repairs (in 1973 terms) to put the house in reasonable repair would not have been 'out of all proportion' for a house with a further 35–50 years' life to obtain an annual rent of £86 and 'that such repair would have been regarded reasonably as (the hypothetical landlord's) obligation under the statutory definition'.

In most cases the hypothetical landlord will regard repairs as economically reasonable. Even in quite severe cases the landlord, when notionally faced with obtaining no rent at all or spending several years' worth of rent on repairs, will be deemed to decide to bear the expense in order to enjoy the income stream in perpetuity. The hypothetical landlord faces a choice either to spend money repairing the property before letting or to undertake no repairs and let it as it is. The 1999 Act envisages this decision taking place prior to the letting and for the landlord to undertake the repairs if it is economically reasonable. It is possible the landlord may decide that only some of the repair works are economically reasonable.

In *Saunders*, the Lands Tribunal decided on remission of the case that the expenditure of £225 to achieve a rent of £86 pa would be an economically reasonable proposition – a return of 38% or 2.6 years' purchase (YP) on the money expended. In *Childs* v *Wood (VO)* [1980] RVR 200, the Lands Tribunal considered it would be well worth the hypothetical landlord's while to carry out necessary works costing £1,242, £779 and £1,587 to secure £270, £280 and £155 pa rather than let at low rents. In *Foote* v *Gibson (VO)* [1982] RVR 138, the Lands Tribunal concluded it would be reasonable to expect a hypothetical landlord to carry out repairs of £1,155 in order to obtain £135 – a return of 11.7% or 8.5 YP:

> *I am satisfied that the appeal premises have in no sense become derelict or reached the end of their economic life. In their existing state the premises are uninhabitable. By expending £1,155 on repairs the hypothetical landlord could ensure receiving a rent of £135. I think that this expenditure would be justified. Therefore, for the purposes of ascertaining gross value the appeal premises are deemed to be in a proper state of repair ...*

These cases concerned domestic property and the high YP figures may be a reflection of the expected security of rental income from domestic properties. This security of income may not be present with some non-domestic properties. In a Valuation Tribunal case, *16/18 Princes Street (Ipswich) Ltd* v *Bond (VO)* [2002] RA 212, the tribunal took the view that any reasonable landlord would look at the local property market, consider the location of the premises, the likelihood of finding a tenant for the actual property, the likely length of any lease and whether further tenants were likely and from these answers determine over what period the landlord would be prepared to spread repair costs. For a prime property in a buoyant market it could foresee a long period of occupation and, as a consequence, amortisation could be expected to be over a similar period. For a poor property where similar properties were vacant, only a year could be expected.

Evidence was given that the likely letting was for a ten-year lease with a five-year break clause. The Tribunal considered, having regard to the state of the market at the AVD, a likely landlord would amortise the cost of repairs over a five-year period to the first break clause. Amortising the repair cost figure gave £55,000 pa, after making an allowance for contingencies. This was the same as the total rateable value. On this basis there was nil profit to the landlord in undertaking the repairs for the first five years. The Valuation Tribunal noted, however, that there was no evidence to suggest the property would definitely not let after five years. A landlord who by his very nature is in the business of taking risks to make a profit would take this risk. Therefore, the costs could not be said, in the mind of the hypothetical landlord, to be uneconomic. The rateable values were confirmed.

In *Archer Ltd* v *Robinson (VO)* [2003] RA 1, the Lands Tribunal considered that a prudent landlord would regard it economically reasonable to carry out the works, as it would considerably improve the marketability of the property and the cost would be recouped by additional rent after a period which was comparatively short compared with the remaining life of the building. The Tribunal considered the rental value without the works was £8,000 pa and in repair £32,000 pa. As it accepted the cost of the works at £33,010, the cost was approximately 1 YP on the rent or 1.38 YP on the difference between repaired and unrepaired rent.

The approach in *Princes Street* appears to have been followed and developed by the Upper Tribunal in *Thomas & Davies (Merthyr Tydfil) Ltd* v *Denly (VO)* [2014] UKUT 0146 (LC). This concerned a car showroom, but it was not simply disused and in poor repair: instead it was in use and the actual owners did undertake the repair work after the material day. The cost of repair was very substantial (agreed at £442,431 at AVD) when compared to an assessment of £78,500 rateable value. The repair costs were therefore 5.64 YP on the rateable value.

The Tribunal found:

- The works, notwithstanding that they improved the premises, were nonetheless repairs because they were replacing things that were dilapidated or worn out.
- The fact the appellant did carry out the repair works was not relevant. What an owner–occupier under a long lease might do was of no assistance in deciding what a hypothetical landlord might do if the property was vacant and to let.
- The state of repair was such that the premises would have been occupiable for perhaps five years at a likely rent of 50% of the repaired rent – £38,500 pa – but the hypothetical landlord would have known he would need to do the repairs in the future and in no more than five years' time.
- The present value of the landlord's interest assuming the works were carried out immediately, less the cost of carrying them out, would have significantly exceeded the present value in disrepair.
- The possible difficulty of the actual appellant not being able to raise a loan for the works at the AVD was not conclusive as another landlord might be of stronger standing or use its own resources.
- A reasonable landlord would not have considered the works required to put the appeal property into repair uneconomic.

The case appears to support the approach taken by the valuation officer in *16/18 Princes Street (Ipswich) Ltd* v *Bond (VO)* [2002] RA 212. In that case a comparison was of likely rental flow compared to repair cost to see if the return on capital expended was economic. In *Thomas & Davies (Merthyr Tydfil) Ltd* v *Denly (VO)* [2014] UKUT 0146 (LC), an analogous approach was adopted. This treated the economic test in the definition of rateable value as requiring an examination of the value of the landlord's interest before and after repairs to judge whether the cost of repair was economically reasonable.

This is an Upper Tribunal case which appears to give some guidance on what quantum of repair cost, compared to the likely rental income flowing from the repair costs, might be acceptable to the hypothetical landlord. The additional complication in the case is that the hypothetical landlord could have been expected to receive 50% of the repaired rent for five years before the deterioration of the premises would have required the expenditure on repairs. So the judgement was not the simple one of the old domestic cases where the landlord had a choice between spending nothing and receiving no rent due to the state of the premises or undertaking the repairs and receiving the rent in repair. The actual premises could be used in a state of disrepair and the potential income in disrepair whilst less than in good repair would figure in the landlord's assessment of what was economic.

The same approach was followed in *Mahmod* v *Annamalai (VO)* [2018] UKUT 20 (LC) for a 100-year-old vehicle repair workshop in poor repair where the VO and Upper Tribunal compared the present value of the property with and without the repair works and found the work would have been economically reasonable for a landlord to undertake.

6.2.9.2 *If a property is in the course of a programme of works*

An owner may decide to alter a hereditament perhaps by converting it into different units, refurbishing it or extending it. Sometimes owners have deliberately damaged properties with a view to making them incapable of being occupied with a view to avoiding unoccupied rates.

In the case of *Newbigin (VO)* v *Monk* [2015] EWCA Civ 78, the Supreme Court examined the question of whether the situation was different when the reason a hereditament was in poor order was the result of it being subject to a programme of redevelopment or reconstruction works. The Supreme Court reviewed a series of rating cases and found that case law 'distinguished between a mere lack of repair, which did not affect rateable value because of the hypothetical landlord's obligation to repair, and redevelopment works which made a building uninhabitable'.

In *Rank Audio Visual* v *Hounslow Borough Council (VO)* [1970] RA 535, a factory was in a bad state of repair having been empty for two years without maintenance. It was in the process of substantial alteration including the demolition of internal partitions, some external walls and some outbuildings together with extension and modernisation of sanitary fittings. The Lands Tribunal decided its condition was such that it would have commanded no or a nominal rent. In *Paynter (VO)* v *Buxton* [1986] RVR 132, the Lands Tribunal said, dismissing the valuation officer's appeal:

> *It is accepted here that the works done and to be done were part of a programme. In my view these were alteration and modernisation, not repair although no doubt some items would fall under the last heading. In my judgment therefore the position here is closer to that of Rank Audio Visual … The question is not an easy one to decide even disregarding the length of time which the works took to complete. There was no definite evidence as to the period concerned*

but certainly it was less than two and a half years. I accept of course the valu-ation officer's submissions with regard to Dawkins v Ash, that the landlord's intentions are to be disregarded but what cannot be avoided (and is accepted here) is that there was a definite programme of works and that the programme had been started, not only in respect of the third floor flat but also in respect of the subject flats. Since the respondent owner was working from the top floor downwards, on 9 August 1983, the relevant date, it is to be expected that works to the third floor flat were more advanced than those to the second floor flat and that those to the second floor flat more advanced than those to the first floor flat. I agree with the submission made by the valuation officer that the position must be looked at on 9 August 1983 but it must be looked at in my judgment in the context, not of the landlord's intention, but in the context that it was during a period of a programme of works which undoubtedly included structural alterations and modernisation.

In the leading case of *Newbigin (VO)* v *Monk* [2015] EWCA Civ 78, a floor of a modern office building was found to have had the cooling system removed along with the majority of the ceiling grid and suspended floor. The existing lavatory provision had been removed and was being reconstructed in a new position and partitioning was being erected in line with the building contract which envisaged the construction of three new letting areas within the floor.

The Supreme Court identified a 'logically prior question' that needed to be asked when a building was undergoing redevelopment. This was whether the premises were 'undergoing reconstruction rather than simply being in a state of disrepair'. If so, the premises would be incapable of beneficial occupation and of no value. Clearly, from this decision, it is important to identify whether or not a programme of works of redevelopment is underway. The test is not a subjective one, such as what the actual owner may or may not have been planning, but an objective one. The Supreme Court said the question had to be 'assessed objectively'. In the *Monk* case the evidence showed the premises were, as an 'objective fact', in the process of redevelopment.

Identifying a programme of redevelopment will be fairly easy when builders are on site perhaps changing an office block to flats, building an extension or significantly refurbishing the accommodation or where the boundaries of the existing hereditament(s) are being changed.

It is unlikely that merely re-organising an office floor, refitting a shop or undertaking cosmetic works constitute the sort of works envisaged by the judgement.

Obviously, a programme of reconstruction works will include stripping out what was there before as a necessary precursor to changing the premises by upgrading, conversion or extension. The work undertaken by the building contractors in both stripping out what was there before and the new work will together constitute the programme.

Merely having plans, perhaps seeking planning permission or undertaking repairs or minor works will not be sufficient.

In *Meadowbank International Ltd* v *Allis (VO)* (2020) RA/49/2018, the ratepayers suggested that pre-planning enquiries and a planning application marked the start of work of reconstruction justifying a nil assessment, but the Upper Tribunal regarded that as untenable, 'Even the obtaining of planning permission for reconstruction does not make a building incapable of beneficial occupation; still less can enquiries or an application do so.'

In *Colour Weddings* v *Roberts (VO)* (2019) RA/12/2018 a warehouse had been taken for conversion to a wedding venue. The question was at what point should the assessment be reduced to nil? The Upper Tribunal said that 'once the work had reached a stage which no longer involved repairs or minor internal alterations compatible with continued beneficial use in its original mode and category of occupation, it ceased to be a hereditament liable to rating'.

6.2.9.3 Deliberate damage

Deliberate damage or where works of stripping out a property are undertaken without definite plans of what alteration works will follow would not seem to be something that can be seen as part of a programme of redevelopment. Remedying the damage probably needs to be viewed as repair work and the premises in disrepair. The usual approach to properties in disrepair should be applied to decide if the carrying out of repair works would be economically reasonable. This was the situation in *16/18 Princes Street (Ipswich) Ltd* v *Bond (VO)* [2002] RA 212 mentioned earlier and also in *De Silva* v *Davis (VO)* [1983] 265 EG 785. In this case the owner wished to convert a maisonette into offices, but the scheme was not certain and, indeed, before the hearing planning permission was refused. The property was in poor order with kitchen and sanitary fittings removed and the services cut off and needing complete renewal. The Lands Tribunal regarded the remedial works as repair.

Deliberate damage or where works of stripping out a property are undertaken without definite plans of what alteration works will follow would not seem to be something that can be seen as part of a programme of redevelopment. Remedying the damage probably needs to be viewed as repair work and the premises in disrepair. The usual approach to properties in disrepair should be applied to decide if the carrying out of repair works would be economically reasonable. This was the situation in *16/18 Princes Street (Ipswich) Ltd* v *Bond (VO)* [2002] RA 212 mentioned earlier and also in *De Silva* v *Davis (VO)* [1983] 265 EG 785. In this case the owner wished to convert a maisonette into offices but the scheme was not certain and, indeed, before the hearing planning permission was refused. The property was in poor order with kitchen and sanitary fittings removed and the services cut off and needing complete renewal. The Lands Tribunal regarded the remedial works as repair.

In *Jackson (VO)* v *Canary Wharf Ltd* (2019) UTLC RA/33/2018 the owner had stripped out and left vacant two floors of a tower block. It was its normal practice when a tenant vacated to take back to a shell state with the intention of fitting the floors out at a later date to a new tenant's requirements. The parties agreed the premises were incapable of beneficial occupation, but the VO did not consider this meant an automatic nominal assessment because the premises were not actually in course of reconstruction as there were no concrete plans and it was not clear what would be the eventual fit out. In effect the premises were simply in disrepair having been damaged by the owner.

The Tribunal examined what constituted a programme of works. It agreed with the respondents that the Supreme Court's decision in Monk neither required a detailed programme of works to exist, nor physical evidence on the ground of the eventual form of the reconstructed premises, before a property in disrepair could be distinguished from a building undergoing reconstruction.

The Tribunal explained that while, obviously, the existence of a detailed programme could be taken into account, its absence was not at all fatal. Similarly, whilst in Monk an outline on the ground of the future development could be identified this was not determinative in the case, nor a necessary condition as a matter of principle.

The Tribunal examined the question of how a valuation officer might ascertain whether premises were undergoing reconstruction rather than simply in a state of disrepair – perhaps caused by some stripping out by vandals or theft. It said this would be by assessing the known facts.

While the subjective intentions of the owner are irrelevant, the objective facts to which those intentions have given rise are not. Regard may obviously be had to a pattern of behaviour in relation to other similar property, whether, as in this case, by the respondent itself on other floors within the same building, or by the owners of similar high-quality office buildings.

It considered that even if the actions of the actual landlord were irrelevant (which they were not) the VO would have known it was common practice in the better office locations in London for landlords to fully refurbish prime office buildings every 10 to 15 years, to maintain the position of the building in the marketplace and stripping out prior to finding a tenant was not unusual.

6.2.9.4 The 'Monk' decision – unresolved questions

The Monk decision of the Supreme Court left a number of questions unanswered.

- Quite what constituted a scheme of works and how was this objectively ascertainable. This Canary Wharf case appears to answer that.
- Was a single programme of works required in which stripping out was followed immediately by a reconstruction project or was it possible to treat stripping out as part of a programme of works even if the actual reconstruction project had neither crystalised nor been contracted? Again, the Canary Wharf case seemed to confirm a single project was not required
- If the test was one of capability of beneficial occupation, was it a test that applied generally or only where there was a 'scheme of reconstruction'? Did it also apply to any property in disrepair if it was demonstrably not capable

of occupation in its current state without repair? This question seems still unanswered. It would seem surprising given the definition of rateable value assumes the hypothetical landlord puts the property into repair before letting, if comparatively minor damage could result in a nil assessment even if repair was obviously economically reasonable.

In the Canary Wharf case the Tribunal appeared, though, to take the view that the question was simply one of beneficial occupation. The Tribunal did not seem to consider that the Supreme Court had differentiated between simple disrepair and the situation where there was 'a building under reconstruction'. It appeared to consider that capability of beneficial occupation applied to any circumstance where a property was incapable of beneficial occupation. It rejected the idea of 'a building under reconstruction exception' to the repair assumption, therefore suggesting the same test must apply to disrepair only.

In para. 36 it made the bald statement that 'If premises are not capable of beneficial occupation they are not a hereditament'; and in para. 37 said 'The VO's acceptance in this case that the appeal property is not capable of beneficial occupation is therefore the beginning and end of the appeal', again suggesting the test in the Tribunal's view is simply one of capability of beneficial occupation irrespective of whether this can be remedied by mere repair.

The Tribunal did, though, go on at length to consider whether the state of the premises and the facts surrounding constituted a programme of works and, as explained above, decided that they did. This alone would have been sufficient to have confirmed the Valuation Tribunal decision, given the VO's acceptance that the premises were incapable of beneficial occupation.

The Tribunal's careful examination of the works *may* indicate that, in fact, the Tribunal did regard it as necessary to establish the premises were undergoing works of reconstruction before finding the fact of incapability of beneficial occupation (as agreed between the parties) meant a nil value was appropriate.

6.2.10 *The third assumption is that the tenant undertakes to pay all usual tenant's rates and taxes above ...*

Nowadays, this usually means simply that payment of the non-domestic rate is assumed to be the responsibility of the tenant. Tenant's drainage rates will also fall within this phrase and be assumed to be paid by the tenant.

Rateable value is, therefore, a rent exclusive of rates.

6.2.11 *... And to bear the cost of the repairs and insurance*

The responsibility for external and internal, including decorative, repairs together with insurance is the hypothetical tenant's responsibility after the initial question of what state of repair is to be assumed at the start of the tenancy. The rating tenancy is, therefore, a FRI lease.

6.2.12 ... And the other expenses (if any) necessary to maintain the hereditament in a state to command the rent mentioned earlier

This phrase is not usually of importance in considering any case, but has exercised the minds of surveyors, lawyers and the courts over the years.

It appears that the phrase does not merely include expenses of a similar nature to repairs and insurance. This does not, however, mean that any expense associated with occupying property becomes the automatic responsibility of the hypothetical tenant.

The expense must be one 'necessary to maintain the hereditament' and usually must be expenditure on the hereditament itself. In exceptional cases it has been held that expenditure on land outside the hereditament comes within the phrase where it is necessary to preserve the physical existence of the hereditament. Expenditure on maintaining a sea wall to prevent the hereditament and other land being submerged has been allowed. Expenditure on dredging a channel to a wharf but outside the wharf hereditament has not been allowed because, while failure to dredge would certainly affect the rental value of the wharf and indeed its use as a wharf, 'no one could say that the hereditament would disappear if the river outside were not dredged', as in *White and Bros* v *South Stoneham AC* [1915] 1 KB 103.

For this type of expenditure outside the hereditament it appears that the responsibility falls on neither the hypothetical landlord nor the tenant automatically. It is a matter to consider in each case whether the possibility and amount of such expenses would affect the minds of the hypothetical landlord and tenant in agreeing a rent.

In *Vaissiere* v *Jackson (VO)* [1968] RA 28, the Lands Tribunal held that the burden of maintaining an unadopted road fell on neither the hypothetical landlord nor the tenant of a house and the condition of the road was regarded as a relevant consideration in determining values. In a later case, *Piner (VO)* v *House* [1980] 20 RVR 123, concerning an estate road, the Lands Tribunal commented:

> *In the present case the access road is clearly not part of the hereditament and I am of the opinion, following Vaissiere v Jackson (VO) (1968), that the cost of its maintenance is not included in the expression 'the other expenses, if any, necessary to maintain the hereditament in a state to command that rent'. There is thus no statutory assumption that the hypothetical landlord is to be deemed to bear this expense nor is any assumption to be made in law regarding the liability of the hypothetical tenant. In the absence of any statutory assumptions it is necessary to return to the world of reality and to ascertain what effect, if any, the existence of the road, its condition and the need to keep it in repair, would have on the rental value.*

6.3 Additional rules

6.3.1 *Statutory and lease restrictions – are they relevant?*

The definition of rateable value is a brief one and provides only a small number of terms, albeit important ones, when compared to those found in a normal lease. The tenancy is, therefore, a simple one devoid of unusual or restrictive covenants, for example, as to user.

The actual property may be let but the terms of the real world lease are disregarded and the simple ones of the hypothetical tenancy adopted in their stead. The actual lease terms or covenants are private arrangements between the real landlord and tenant and are not relevant to determining rateable value. The hereditament is treated as vacant and to let on the statutory terms.

However, any potential tenant must be affected by statutory powers or obligations that attach to the hereditament or (as per *Dawkins (VO)* v *Ash Brothers and Heaton Ltd* [1969] 2 AC 336) are essential to it in whoever's hands it may be in. The exception to this is any statutory restriction on rent.

In *Poplar Assessment Committee* v *Roberts* [1922] 2 AC 93, the issue was whether account was to be taken of the temporary legislative restriction on the amount of rent that could be charged by the landlord to the tenant of a public house under the Increase of Rent and Mortgage Interest (Restrictions) Act 1920. The effect of the Act was that the rent was less than the true value of the occupation. It was held that the statutory restriction was not material to the determination of the valuation for the purposes of rating because, so far as an occupier was concerned, the occupation was not made less beneficial by the operation of the statutory restriction. 'It may be observed the phrase is "might be expected to pay" it does not go on to say "and the landlord might be able to extract".'

6.3.2 *Vacant and to let*

To encourage uniformity in valuation, all hereditaments are assumed to be vacant and available to let on the statutory terms of the hypothetical tenancy. This means the actual occupier and landlord are disregarded so their particular attitudes, needs or concerns are not taken into account in the valuation. Characteristics that would affect any likely landlord or tenant for the property are taken into account because they would affect the rental value if it was offered on the market 'vacant and to let'.

The assumption as to vacancy is not, however, taken too far. It does not mean the hereditament has necessarily been on the market sitting empty for a long time. Indeed, if this was the case then for those properties relying on custom, e.g. pubs and hotels, this would have a depressive effect on rents as the incoming occupier would have to build up the trade again. Equally, as far back as 1936 in *Railway Assessment Authority* v *Southern Railway Co.* (1936) 24 R&IT 52, it was held that in valuing a railway the difficulties of selling rolling stock at the end of the tenancy must be disregarded. The existing staff was assumed to be there for the new tenant and no allowance was to be made for the risk of not being able to assemble the necessary equipment, staff, etc.

All hereditaments are assumed to be vacant and to let when being valued. But what if the actual hereditament is indeed vacant. This might indicate that there is no demand for the property or that it just 'happens' to be vacant at the time. This is a matter for evidence. In *Telereal Trillium* v *Hewitt (VO)* [2019] UKSC 23 the Supreme Court examined the situation where a three-storey 1970s block of offices near Blackpool had become vacant and the ratepayers argued a tenant could not be found. There were, however, other occupied comparable office blocks in the locality which were occupied. The Supreme Court looked favourably on a decision of the Lands Tribunal many years earlier, *Lambeth LB* v *English Property Corporation and Shepherd (VO)* [1980] RA 297, where Mr Emlyn Jones had said:

> *I accept the argument put forward by the appellant rating authority that the mere fact that premises are unoccupied does not of itself justify a lesser value than that applicable to similar premises which are occupied. As counsel for the rating authority expressed it, in a parade of shops where one shop remains unoccupied one would expect to find similar values applicable to all shops possessing similar characteristics. I think in principle that must be right, but that presupposes that the hereditaments are broadly identical …*

(p. 313)

The parade of shops in the case was indeed very similar to the Telereal Trillium office block situation, and the value could be found by looking at the values on the other similar occupied blocks in the locality. Lord Carnwath said:

> *Whether the hereditament is occupied or unoccupied, or an actual tenant has been identified, at the relevant date is not critical. Even in a 'saturated' market the rating hypothesis assumes a willing tenant, and by implication one who is sufficiently interested to enter into negotiations to agree a rent on the statutory basis. As to the level of that rent, there is no reason why, in the absence of other material evidence, it should not be assessed by reference to 'general demand' derived from occupation of other office properties with similar characteristics.*

6.3.3 *The nature of the hypothetical landlord and tenant*

The assumption in valuing any hereditament for rating is that it is only available to let on the terms of the hypothetical tenancy. There may be many possible tenants for this tenancy or only one. This will depend on the type and location of the hereditament. The person or type of person most likely to take the tenancy is known as the hypothetical tenant. This could include an existing owner–occupier. All possible occupiers need to be considered as possible hypothetical tenants. No one is excluded from being potentially the hypothetical tenant. If an existing owner is the most likely occupier then the owner's rental bid needs to be taken into account. Even if the actual owner is forbidden by statute from being a lessee, he or she must be regarded as a possible tenant. If this had not been confirmed by cases, particularly *London County Council* v *Erith and West Ham (Church Wardens and Overseers)* [1893] AC 552, it could in some cases have excluded the bids of the people most in need of occupying the hereditament. In *Erith* the ratepayers were barred by statute from becoming tenants of the sewage pumping stations they owned. The ratepayers said that the rateable value should be determined on the basis of the lower rent another occupier would pay for the premises for a different use and their bid should be disregarded. The House of Lords rejected this.

Lord Herschell LC said:

> *It has never been doubted that the rent which is actually being paid by the occupier does not necessarily indicate what is the rent which a tenant might reasonably be expected to pay, or that an owner who is in occupation, and who may not be willing to let on any terms, is none the less rateable. The tenant described by the statute has always been spoken of by the court as 'the hypothetical tenant'. Whether the premises are in the occupation of the owner or not, the question to be answered is: Supposing they were vacant and to let, what rent might reasonably be expected to be obtained for them? So far there can, I think, be no difference of opinion. But then arises the question: Is the owner to be regarded as one of the possible tenants in considering what rent might reasonably be expected? Bearing in mind what was the object of the Legislature in prescribing this test of annual value, I cannot myself entertain any doubt that the owner ought to be taken into account.*

For some hereditaments, there may in reality be only one possible tenant.

The hypothetical landlord may also be the real landlord or someone else. Again, the hypothetical landlord represents the typical landlord expected for the type and location of hereditament. Clearly, as the landlord is regarded as having the hereditament available vacant and to let, he or she must be the landlord of the whole hereditament and have it all available to let as a single unit. In reality the hereditament may comprise parts owned by different landlords and even partly owned by the occupier. Even so, the hereditament will be treated as if it was available to let under a single tenancy by a single landlord.

In *Marks and Spencer* v *Sanderson (VO)* [1992] RA 63, the Lands Tribunal commented in considering a large store in a shopping mall on the situation where the real landlord was not just landlord of the actual hereditament, but adjoining properties as well:

> *Where there is a new development of a shopping mall ... it is perhaps usual to find one developer/landlord, with one letting policy, who provides common services and areas of common usage. It cannot be assumed in the rating hypothesis that there is necessarily only one landlord for the whole development nor*

*can it be assumed that there is necessarily one landlord for each hereditament.
The rating hypothesis is neutral. However the further one strays from reality
the less certain is the whole foundation upon which assessments are made.*

In reality the actual landlord or tenant may be a difficult person with idiosyn-
cratic ideas or particular needs. To ensure a common approach to valuing all
hereditaments, both the hypothetical landlord and the tenant are regarded as
being reasonable people.

Willmer LJ in *Humber* said:

> *I accept that one has to postulate that the hypothetical tenant and the hypo-
> thetical landlord will be reasonable people, and will behave as reasonable men
> in the real world would behave in the circumstances which have to be postulated.*

Both the hypothetical landlord and the tenant are to be viewed as seeking to
reach a bargain on rent but neither is viewed as desperate to agree to a tenancy.

Lord Denning MR in *R* v *Paddington VO ex parte Peachey Property
Corporation Ltd* [1966] 1 QBD 360 commented:

> *The rent prescribed by statute is a hypothetical rent … It is the rent which an
> imaginary tenant might reasonably be expected to pay to an imaginary land-
> lord for a tenancy of this (property) in this locality, on the hypothesis that
> both are reasonable people, the landlord not being extortionate, the tenant not
> being under pressure …*

6.3.4 Fresh on the scene

The hypothetical tenant, in agreeing the rent, is regarded as coming new to the
property or fresh on the scene. This is a natural consequence of the hereditament being viewed as vacant and to let. The actual occupier may remember
how the 'locality used to be' and regard the situation as worse and consider
that rateable value should be reduced. The hypothetical tenant coming 'fresh
on the scene' looks at the hereditament and judges the property on its merits
at the date of valuation. In *Finnis* v *Priest (VO)* (1959) 52 R&IT 372 it
was noted:

> *But I have to consider, not the attitude of these old residents, which is under-
> standable, but that of the hypothetical tenant. He must be considered as a
> reasonably minded person, arriving fresh on the scene without any prejudice
> which might derive from experience of conditions which obtained previously.*

As shown in *Morton (VO)* v *Jones* [1986] 31 RVR 25, the idea that the tenant
comes fresh to the scene applies equally to where a nuisance is temporary in
nature. The hypothetical tenant comes fresh to the scene unaware of how bad
the nuisance has been in the past and will only be interested, when forming a

view of rental value, in how bad it currently is, and how bad it is likely to be in the coming months.

6.3.5 *The Principle of Reality*

As with valuations generally for other purposes, all factors which affect the rental value, other than those which rating law ignores, are to be taken into account.

Scott LJ in *Robinson Brothers (Brewers) Ltd* v *Houghton and Chester-le-Street Assessment Committee* [1937] 2 KB 445 said:

> It is the duty of the valuer to take into consideration every intrinsic quality and every intrinsic circumstance which tends to push the rental value up or down, just because it is relevant to the valuation and ought, therefore, to be cast into the scales of the balance ...

Lord Buckmaster in *Poplar Assessment Committee* v *Roberts* [1922] 2 AC 93, 103:

> *Although the tenant is imaginary, the conditions in which his rent is to be determined cannot be imaginary. They are the actual conditions affecting the hereditament at the time when the valuation is made.*

Rating can appear as a very theoretical exercise in valuation, with a large number of imaginary assumptions loosely described as the hypothetical tenancy. Wherever the statutory assumptions do not intrude, the realities of the real world are taken in preference. The hereditament is assumed to be the actual one subject only to various statutory and other assumptions as to repair, being available to let on the rating tenancy rather than as actually occupied etc. Beyond this, the world outside the hereditament is assumed to be as it actually was at the set valuation dates; the valuation is in the real market. In *Hoare (VO)* v *National Trust [1998]* RA 391; [1997] 2 EGLR 229 (CA), the Court of Appeal said:

> *The statutory hypothesis is only a mechanism for enabling one to arrive at a value for a particular hereditament for rating purposes. It does not entitle the valuer to depart from the real world further than the hypothesis compels.*
>
> *I would emphasise the necessity to adhere to reality subject only to giving full effect to the statutory hypothesis, so that the hypothetical lessor and lessee act as a prudent lessor and lessee. I would call this the principle of reality.*

One of the important assumptions is the principle of *rebus sic stantibus* – that you take matters as they are, not how they might be. This requirement has been restated in many recent cases as the 'Principle of Reality'. For the moment we shall use the traditional term before returning to see how well the new term, 'Principle of Reality', fits the principle.

6.3.6 Rebus sic stantibus – *taking the thing as it is*

When preparing a valuation for a rent review, a valuer will examine the lease to see if there are any restrictive user clauses or restrictions on the tenant modifying the property and note the assumptions made as to the state of repair. For rating, the lease is found in the definition of rateable value which, while it sets out the repairing assumptions, is silent on user clauses and the ability of the tenant to modify the hereditament. To remedy this the courts have, in successive cases over the centuries, evolved a clear requirement to value the hereditament as if the hypothetical tenancy restricted the use of the property to its current mode or category of use and prevented a prospective tenant envisaging making more than minor physical alterations to the existing property. This requirement is known as the principle of *rebus sic stantibus*. The phrase *rebus sic stantibus* can be translated as 'taking the thing as it is' or, alternatively, 'as things stand'.

It takes the thing as it is rather than how it might be. The reality of the current mode and category of use and its broad physical state is taken as it is, not how it might be. Change of mode or category of use or more than minor physical changes cannot be envisaged.

The principle of *rebus sic stantibus* is a fundamental one. While the rateable value of a hereditament is assessed on the basis of an imaginary lease with a hypothetical landlord and tenant, the actual property is taken as it actually is at the valuation date. Only limited changes can be envisaged and the hereditament is valued on the assumption that:

- only minor physical changes can be envisaged
- the mode or category of occupation cannot be altered.

The idea of *rebus sic stantibus* can be traced back to very early rating case law but it has been given statutory expression in the assumptions in Schedule 6, paragraph 2(7), of the Local Government Finance Act 1988:

(a) matters affecting the physical state or physical enjoyment of the hereditament,
(b) the mode or category of occupation of the hereditament.

Rateable value is, therefore, a current use value disregarding any possible additional value which might exist due to the potential to enlarge or significantly modify the property or change the mode or category of use. This does not mean that such changes are ignored when they do happen. If a physical change occurs or the mode or category of use is altered then the rateable value can be reassessed, *rebus sic stantibus*, and the hereditament valued as it now is in its new state or with its new mode or category of occupation. But until this happens, the rateable value is assessed on the assumption that such changes can never be made.

Rebus sic stantibus, in restricting possible changes to the hereditament, is seen as having two limbs: the restrictions on mode or category of use and restrictions on what physical changes can be envisaged.

6.3.6.1 *The mode or category of use*

Mode or category of use is not defined in legislation. It does not mean the precise use by the actual occupier but rather one within the same class. A change of use can be envisaged so long as it does not change the use beyond its existing mode or category. What this means is set out in *Fir Mill Ltd* v *Royton UDC and Jones (VO)* [1960] R&IT 389. In this case the Lands Tribunal held that a cotton mill must be valued as a factory, rejecting the ratepayer's contention that it should be valued merely as a cotton mill, and the valuation officer's contention that it should be valued without limitation on user. The Tribunal said:

> *The mode or category of occupation by the hypothetical tenant must be conceived as the same mode or category as that of the actual occupier. A dwelling-house must be assessed as a dwelling-house; a shop as a shop, but not as any particular kind of shop; a factory as a factory, but not as any particular kind of factory.*

In *Scottish and Newcastle (Retail) Ltd* v *Williams (VO)* [2000] RA 119, the Court of Appeal confirmed the correctness of this statement.

The phrase 'mode or category' means a use for the same general purpose as the existing use of the hereditament. In valuing the hereditament for rating, any value attributable to the prospect of using it for another purpose has to be disregarded. The formulation in *Fir Mill Ltd* v *Royton UDC and Jones (VO)* [1960] R&IT 389 of 'a shop as a shop', etc., does not mean that a baker's shop has to be valued solely at the rental figure a baker would give for it. The bids of other prospective shopkeepers can be taken into account. However, the use of shop-type premises as a public house or wine bar, where the primary trade is the sale of alcoholic drinks, would not fall in the same general category as a shop. The case of *Scottish and Newcastle (Retail) Ltd* v *Williams (VO)* [2000] RA 119 mentioned above considered this particular question and decided a public house use was not in the same mode or category as a general shop or a restaurant.

It is probable that banks, estate agents and other financial and professional uses in shop-type premises constitute their own mode or category of occupation that is different from general shops in similar premises. These are quasi office-type uses where the services provided are principally to visiting members of the public. This appears to be confirmed by *Humphreys-Jones (T/A Cathedral Frames)* v *Welsby (VO)* [2001] RA 67, which concerned the assessment of an A1 shop and the weight to be attached to the rents of other nearby shops which had an A3 planning use. The Lands Tribunal, following *Scottish and*

Newcastle, considered that the shop's A1 use meant it was in a separate mode or category of occupation from A3 users. However, following the earlier Lands Tribunal decision in *Scottish and Newcastle*, it said there was no rule that evidence relating to another hereditament in a different mode or category of use was to be treated as irrelevant. It considered that any evidence may be taken into account providing it is relevant to the valuation. The Tribunal found the A3 users were not subject to a premium and the rents were thus no different from those applying to A1 retail uses.

In *Re the Appeal of Reeves (VO)* [2007] RA 168, the 'Truro College' case, the Lands Tribunal found that an office building used as an open learning centre was in a separate mode or category of use from an office and therefore the potential to use it as an office must be ignored. However, following the *Scottish and Newcastle* case, this did not mean that rents for other mode or categories of use could not be used to value where it was clear this evidence points to a correct valuation of the hereditament. In the case it was established that the ratepayers were paying a rent for the hereditament that equated to an office rent, and that there were some other colleges using similar office buildings that were paying office level rents or were assessed at office levels of value. The Tribunal preferred the broad description of mode or category of use as 'college occupation' to the narrower 'open learning centre', again confirming the *Fir Mill Ltd* v *Royton UDC and Jones (VO)* [1960] R&IT 389 approach of seeing mode or category of occupation as broad categories.

In *Cheale Meats Limited* v *Ray (VO)* [2012] RA 357, the Upper Tribunal was very careful to explain that an abattoir was not in the same mode or category of use as a standard industrial unit and therefore could not be valued simply by comparison with industrial units as the valuation officer had clearly done – and, indeed, seemed to be industry practice. The Tribunal was careful to prefer the evidence of settlements reached on other abattoirs and use them to value the subject hereditament as they were within the same mode or category. While this would appear to have had the same result as valuing by reference to the local industrial tone, the comparison with other abattoirs (albeit valued by reference to local industrial levels) maintained the correct form in following the *rebus sic stantibus* rule.

In *BNPPDS Ltd & BNPPDS Ltd (Jersey) as trustees for Blackrock UK Property* [2022] UKUT 129 (LC) the appellants sought to differentiate between multi-storey car parks integrated with shopping centres and those that stood alone, seeking to see them as different mode or categories of occupation on the basis the motive for occupation was different, they served a different market and offered different facilities. The Upper Tribunal noted multi-storey car parks came in many forms and saw no reason to distinguish them in the way the ratepayers suggested. It preferred to avoid 'complicating the valuation process unnecessarily by creating ever narrower categories (of mode or category of occupation) with attendant issues around evidence', considering it should be well within the capabilities of the expert rating surveyor to make judgements on allowances for varying factors.

It seems, therefore, that by and large, mode or categories of use should be broad not narrow categories. The Fir Mill formulation of 'a shop as a shop, but not as any particular kind of shop; a factory as a factory, but not as any particular kind of factory' is very helpful in demonstrating this.

Some particular uses of properties, such as bus garages, form their own individual uses and are described as being *sui generis*.

Planning use classes and classes of permitted development should not be seen as indicating what is and what is not within a mode or category of occupation. In some cases a use class will be wider than a mode or category of occupation. It is likely that the B1 'business' planning use class encompasses a number of different rating modes or categories of use. Planning use classes form a background against which valuations are made. Rating is not a creature of planning. Indeed, rating has existed for many more centuries than planning control. However, where a potential use to which a property could physically be put would require planning permission, clearly the likelihood of obtaining that permission will affect the rental bid. It appears if permission is likely then there is no need to depress the rating assessment for any risk that permission might not be obtained. In many cases the simple need for permission will often indicate the proposed use is in a different mode or category:

> *As the Tribunal has said in various cases, it does not seem to be part of the role of the Tribunal to determine issues of planning. Planning legislation can be regarded as a modern innovation when compared with the ancient institution of the poor rate, and quite clearly it cannot be taken to have amended any of the principles upon which valuations for rating are to be made. Its significance, in my opinion, is that the existence of planning restrictions and the necessity of obtaining planning permission for certain changes of use are all factors which affect the minds of potential tenants in the real world and to that extent they must have an influence on value. However, the interpretation of these factors remains a matter of evidence and of expert valuation of evidence.*
>
> *(Midland Bank Plc v Lanham (VO) (1977) 246 EG 1117)*

The existence of an empty property rate since 1966 makes it sometimes necessary to decide on the mode or category of use for a vacant hereditament. In *Schofield (VO)* v *RBNB* [2007] RA 121, the owners of a vacant public house had asked for the description to be altered to 'Stores and Premises'. The Lands Tribunal noted that it had not been occupied for this purpose nor had steps been taken to make it suitable for this purpose, and considered this was not its mode or category of occupation. The Tribunal indicated that the correct approach in deciding the mode or category of a vacant hereditament was to look at the 'purpose for which it was designed or last occupied and for which it can be expected to be occupied in future'. This suggests that there are two possible modes that can be considered, *rebus sic stantibus*, for a vacant hereditament. The first is the purpose for which it was designed. The second is that for which it was last occupied.

6.3.6.2 *Physical changes*

The principle of *rebus sic stantibus* requires that any value attributable to the prospect of making substantial physical changes is to be ignored. The hereditament is valued as it is. The prospect of erecting an extension to a building or making new entrances are the sorts of change which cannot be envisaged when valuing for rating purposes.

This was considered in the *Fir Mill Ltd* v *Royton UDC and Jones (VO)* [1960] R&IT 389 case:

> *Some alteration to an hereditament may be, and often is, effected on a change of tenancy. Provided it is not so substantial as to change the mode or category of use, the possibility of making a minor alteration of a non-structural character, which the hypothetical tenant may be assumed to have in mind when making his rental bid, is a factor which may properly be taken into account without doing violence to the statute or to the inference we draw from the authorities.*

For many years valuers have taken the words 'non-structural character' to mean that the possibility of making any structural alteration, even if minor, had to be disregarded. In *Scottish and Newcastle (Retail) Ltd* v *Williams (VO)* [2000] RA 119, the Court of Appeal preferred simply to look at whether the proposed alterations were 'minor' rather than making a difficult distinction between minor structural and non-structural works.

The idea is that there are occasions when a significant value can be released by making a comparatively minor alteration. If the alteration can be regarded as minor then *rebus sic stantibus* does not prevent this being included as part of the hereditament's value. For example, a shop may have a partition wall separating the front from the rear part and positioned at quite a shallow depth from the shop front. If its removal can be regarded as a minor alteration this would allow the rear part to be regarded as valuable shop space and not as less valuable storage space.

It would also be unreasonable to envisage a prospective tenant being stuck with minor 'taste and fancy' features that could easily be replaced, such as shop fittings. The Court of Appeal in *Scottish and Newcastle (Retail) Ltd* v *Williams (VO)* [2000] RA 119 said:

> *the Lands Tribunal was clearly right, following Fir Mill, to allow for the possibility of minor alterations to the hereditament on the occasion of its hypothetical letting. The absurdity of any other view appears vividly from the circumstances of these appeals, with numerous very well-known retail chains seeking to establish their identities and brand loyalties by distinctive fascias and fittings installed in uniform, featureless units. (Rebus sic stantibus) cannot be applied so rigidly as to prevent (for instance) Burger King being considered as a possible bidder in competition with McDonald's.*

The cost of undertaking the works is a factor to take into account. However, the relative increase in rental value realised by the works is not determinative. The absolute cost is also not a deciding factor because the cost of even trivial alterations to large hereditaments is likely to be high.

An example appears in the recent case of *Shaw* v *Benton (VO)* [2018] UKUT 02 (LC) where a factory was without lavatory facilities. The VO regarded their provision as coming within 'minor works' under *rebus sic stantibus* and suggested the test was whether a reasonable observer would consider the alterations to be minor. The Upper Tribunal appeared to accept these were 'minor' in the context of the hereditament and considered the lack would have had a significant effect on rental value as the tenant would have had to bear in mind when bidding that it would have to provide facilities at its own expense, perhaps in the region of £8,450. The determined rateable value of £16,000 had a 10% allowance for lack of WCs.

6.3.6.3 Repair

The state of repair is not outside the *rebus sic stantibus* rule, but the statutory assumptions as to repair take precedence. It is necessary to consider, first, whether the hypothetical tenant would carry out some or all of any necessary repairs and only if he or she would not is the actual state, *rebus sic stantibus*, considered.

6.3.6.4 Rebus, reality and the Principle of Reality

Undoubtedly the courts are minded to rename *rebus sic stantibus* as the 'Principle of Reality'.

In *Newbigin (VO)* v *Monk* [2015] EWCA Civ 78, Lord Hodge said:

> For many years and long before Parliament enacted Schedule 6 to the 1988 Act, it had been an established principle of rating law that a hereditament is to be valued as it in fact existed at the material day. This principle, which in the past was described by the Latin phrase, rebus sic stantibus (ie as things stand), and is often referred to as 'the principle of reality' or 'the reality principle', was stated by Lord Buckmaster in Poplar Assessment Committee v Roberts [1922] 2 AC 93, 103, thus: Page 7 '[A]though the tenant is imaginary, the conditions in which his rent is to be determined cannot be imaginary. They are the actual conditions affecting the hereditament at the time when the valuation is made.'

And in *Merlin Entertainments* v *Cox (VO)* [2018] UKUT 0406 (LC) the President of the Upper Tribunal noted:

> The Supreme Court in the Monk case had chosen to refer to this principle in current day language as 'the principle of reality' or the 'reality principle' and

said 'We consider that this change of language was long overdue and that from now on we should adopt the term.'

Quite why this desire to rename should have come about – very probably the modern idea that Latin tags are no longer appropriate – given its widespread usage amongst valuation surveyors and lawyers is not quite clear. Certainly, the first mention of the Reality Principle in the National Trust case seemed to be merely a restating of the principle that rating valuation should be a valuation as close to the real world as possible and everything that can be considered should be considered – unless prohibited by statute. Though, of course, statute provides quite a lot of restrictions in Rating on what might be considered and when.

One of the problems with the new term is that, arguably, *rebus sic stantibus* is somewhat antithetical to reality – it is in many ways an unreal assumption that things stay the same when they do not! The *Principle of Unreality* indeed! You assume the hereditament will continue forever in its existing mode or category of occupation – which may well not be the case; and you assume only minor alterations can be made – when it may be obvious any sensible owner or tenant would carry out major alterations.

Further, it does seem the idea of taking reality wherever possible is a rather wider concept than the more limited set of factors understood by valuers as constituting *rebus sic stantibus* – valuing and envisaging a restriction to the existing mode or category of use and that significant physical change to the hereditament cannot be envisaged. In talking about these limited set of factors is how the term '*rebus sic stantibus*' is used in this book.

6.3.7 *Ability to pay – is it relevant?*

Usually the resources and ability of the actual occupier to pay rent are not relevant to a rating valuation. What the valuer has to determine is the most likely rent the tenant would reasonably be expected to pay. Where there is only one possible tenant or a limited number of possible tenants then ability to pay may be taken into account.

In *Tomlinson (VO)* v *Plymouth Corporation and Plymouth Argyle Football Co. Ltd and Plymouth City Council* (1958) 51 R&IT 815, the Court of Appeal referred to the bargaining power of the actual ratepayers 'by virtue of their being the only bidders for the hereditament'. In this case *Plymouth Argyle FC* was seen as the only possible tenant for the football club and if it ceased to exist any replacement club would in effect be Plymouth Argyle in a new guise and still the only possible tenant.

It may be that, despite being the only possible tenant, the actual ratepayer may be perfectly able to pay, though perhaps unwilling to do so. Nonetheless in negotiation with the hypothetical landlord a substantial rent may result reflecting the importance of its occupation to the ratepayers. Where inability is claimed it was held reasonable by the Lands Tribunal in *Marylebone Cricket*

Club v *Morley (VO)* [1960] 53 R&IT 150 to examine whether the ratepayer could organise his affairs in a different way to raise funds to pay. It was found that while the Marylebone Cricket Club could not increase ticket prices, it could increase its income from subscriptions and catering.

In *Eastbourne Borough Council and Wealden Borough Council* v *Allen (VO)* [2001] RA 273, the Lands Tribunal considered local authorities' ability to pay in relation to some council-owned indoor leisure centres comprising sports halls and swimming pools. It was argued that due to central government restraints the local authorities would have been unable to pay substantial rents arrived at using a contractor's basis of valuation. The Tribunal found the evidence showed that the local authorities could have afforded the rents, although this would have meant cutting down on other services or raising charges that they would have been reluctant to do. The Tribunal said:

> *We can see no evidence to suggest that either authority would have been so reluctant to take the necessary steps to find the money elsewhere that it would have closed the leisure centre rather than pay the rent demanded. On the contrary, it is clear that each authority attached considerable importance to the services provided by its leisure centre as a popular recreational amenity for its residents.*

6.3.8 If the remedy lies elsewhere

In a number of old domestic rating cases regarding external nuisances or disabilities the Lands Tribunal decided not to reduce rating assessments on the grounds that the remedy lay elsewhere. However, in *Gregory* v *Biddle (VO)* [1983] RVR 177, where a house was located opposite a double garage that was used, in breach of planning, for repairing vehicles and carpentry work, the Lands Tribunal declined to follow this approach:

> *Although this Tribunal has on a number of occasions … held that no reduction in the rating assessment can be given where the ratepayer has a remedy elsewhere, I am not prepared to so hold here. It seems to me that no tenant of the subject house from year to year would go to the trouble of starting an action for nuisance or of referring the matter to the Public Protection Department of the local authority: the latter might have no result. I think that any potential tenant would approach the hypothetical landlord for a reduction in the rent of the subject house and that any potential tenant, whose request for a reduction was met by the suggestion from the hypothetical landlord that he (the potential tenant) should take action in nuisance or should refer the matter to the Public Protection Department, would decide that he would not take a tenancy of the subject house until the landlord had remedied the matter.*

Similarly the Upper Tribunal when considering the effect on a warehouse's rental value of the use of adjoining premises by a refuse company in *Burvill* v *Jones (VO)* (2012) UKUT 0101 (LC) said that faced with such an un-neighbourly user, all that either the hypothetical landlord or the tenant could do would be to either approach the landlord of the adjoining unit and ask him to take appropriate action under the lease of that unit or, alternatively, contact the local authority to see whether a remedy was available under environmental health or planning legislation. It did not think the hypothetical tenant would be prepared to pay the full rental value on the assumption that either of these two approaches would be successful. The Tribunal was not at all happy with the idea that as there was a remedy elsewhere no reduction was appropriate and drew attention to *Gregory* v *Biddle (VO)* [1983] RVR 177.

It would seem the Tribunal, despite past reliance on the idea of ignoring nuisances where there are remedies available other than a reduction in rent, will be reluctant to accept arguments that 'the remedy lies elsewhere' and will follow the approach in *Gregory* v *Biddle (VO)* [1983] RVR 177.

6.4 Summary – the hypothetical tenancy

The terms of the hypothetical tenancy, which is the basis behind all rateable values, therefore are very simple. The idea is that each property to be valued for rating is vacant and available to let in the market on an annual tenancy which has only a few basic terms. The letting is on a FRI basis making the tenant responsible for repairs, although the hypothetical landlord is assumed to put the property into repair before the start of the tenancy if this is an eco-nomically reasonable thing for the landlord to do. The tenant is responsible for rates. The terms of the actual real world occupation are disregarded because the hereditament is viewed as vacant and on the market to let. That said, the assumption is made that the use will be restricted, *rebus sic stantibus*, to the existing actual mode or category of use and no alterations may be made to the property other than the minor alterations a tenant might make on taking over the tenancy.

6.5 Date of valuation

For most valuation purposes the requirement is for a valuation to be used at a fixed point in time. For inheritance tax the valuation is at the moment before death, for purchase it is the current open market value and for compulsory pur-chase it is the date of entry or, if earlier, the date of agreement or the vesting date. For rating, a rateable value set at a general revaluation may stay in use for the whole five years before the next general revaluation. Alternatively, it may change several times during the five years because of changes to the her-editament such as an extension or a change to its surroundings. Although it may change, a rateable value is in use and effective for a period of time rather than just at a single point in time. The way in which the rateable value has been made effective over a period of time has developed, particularly in response to property price inflation.

If property price inflation did not exist it would be possible to value all hereditaments at a revaluation and then make any new valuations to include extensions, allow for new disabilities or assess new hereditaments at the date the valuer made the new valuation. The reality of inflation has meant such an approach would produce unfairness. For example, if rental values had risen since a revaluation then a new, but identical, hereditament would have a higher current rental value and be given a higher rateable value than the identical earlier built property. If inflation increased values by a greater amount than a new disability reduced values then a new valuation for an appeal would, rather absurdly, arrive at a higher not lower rateable value.

This problem appeared in *Ladies' Hosiery and Underwear* v *West Middlesex Assessment Committee* [1932] 2 KB 679. On the face of it the ratepayer had a good case. The company's shop had been assessed at a higher level than seven nearby similar shops and *Ladies' Hosiery* wanted its assessment brought into line with the other shops on the grounds of fairness. However, their definite admission that the property would have let for a rent equivalent to the rateable value proved fatal to their case. The Court of Appeal held that correctness must not be sacrificed to uniformity. Eve LJ held:

> *The one concrete fact which emerges is the correctness of the assessment of the appellant's premises, and when that is once recognised, there is really nothing more to be said. It is quite impossible for the court in these proceedings to substitute for the correct figure another figure which, by their own evidence, they have proved to be incorrect. It may be that want of uniformity, and consequent unfairness, can be established, and that it is due to under-assessment of comparable premises, ... but, if so, this must be established by proper proceedings directed to the raising of the assessments of those other premises ...*

The ideal is, of course, for uniformity to be achieved through correctness. Clearly to achieve uniformity and at the same time overcome the problem in *Ladies' Hosiery and Underwear* v *West Middlesex Assessment Committee* [1932] 2 KB 679, it was necessary to change the rules of rating valuation. The modern rating system now achieves uniformity by setting a standard valuation date for all rateable hereditaments. At first sight this seems a simple approach, merely valuing all hereditaments at a fixed point in time. However, it is not possible simply to value all properties in the state they were in on that date. There will often be a need to alter the rateable values of properties for changes such as extensions or to value new hereditaments. At the standard valuation date these extensions or new hereditaments did not exist and would not be included in a valuation made at that date. To allow these changes to be included there has, in effect, to be two valuation dates, one when a general level of values is taken and another date, which will vary depending on the reason for the valuation, when the physical circumstances of a property are taken.

6.5.1 *General levels of value*

Another difficulty arising from inflation is ensuring valuations at a revaluation properly represent values at the standard valuation date when the general levels of value are taken and, in particular, to ensure all valuations across the country are at this standard date. This problem was considered by the House of Lords in *K Shoe Shops Ltd* v *Hardy (VO) and Westminster City Council* [1983] 1 WLR 1273. In this case it was demonstrated by the appellants that the general level of values in Oxford Street represented values in late 1970, whereas those in Regent Street, where the K Shoe shop was, had values close to the level when the list of rateable values had come into force in April 1973. They said this was unfair because Regent Street shops were being assessed by reference to a later valuation date than those on Oxford Street when, as a result of inflation, values had risen significantly. Instead, they suggested Regent Street shops should be assessed at the level of rental values existing at the earlier date to bring it into line with Oxford Street and other nearby roads. The House of Lords confirmed the correct valuation date was 1 April 1973 when the list came into force. Following *Ladies' Hosiery and Underwear* v *West Middlesex Assessment Committee* [1932] 2 KB 679, it was not prepared to reduce the 'correct' K Shoes rateable value at the 1 April 1973 level of rent rateable value to the 'incorrect' lower earlier level used in Oxford Street, as this would be placing uniformity before correctness. Indeed, the House of Lords suggested the better approach was to increase Oxford Street rateable values to the 'correct' 1 April 1973 level as a means of achieving uniformity.

To overcome the problem in *K Shoe Shops Ltd* v *Hardy (VO) and Westminster City Council* [1983] 1 WLR 1273, the general date of valuation is now set 'ante' or before a new rating list comes into force. This date is called the antecedent valuation date (AVD) and has been set two years before the coming into force of new rating lists. For the 1990 rating lists it was 1 April 1988, for the 1995 rating lists it was 1 April 1993, for the 2000 rating lists it was 1 April 1998, for the 2005 rating lists it was 1 April 2003, for the 2010 rating lists it was 1 April 2008 for the 2017 rating lists it was 1 April 2015 and for the 2023 rating lists it is 1 April 2021. The AVD remains constant throughout the life of a rating list. For valuations made for a 2023 rating list, whether for an original valuation when it was prepared, for settling a challenge or appeal against an original value, for including an extension built in 2024 or to reflect a new disability in 2025, the AVD is always the fixed valuation date of 1 April 2021.

6.5.2 *Physical circumstances*

The date on which the physical circumstances of the property are taken varies depending on the reason for the valuation. This date is called the material day. In the past, for many types of alteration to rating lists, this has been the date the appeal, known as making a proposal, was received by the valuation officer or the date the valuation officer altered the list on his or her own initiative.

From the 2005 Revaluation onwards, the Regulations have increased the situations where the material day is the date of the event that triggered the need to make the alteration. Adopting the date of the event as the date to take the physical circumstances is clearer and more obvious than using the date of the proposal or the valuation officer's alteration, as this could be many months after the event.

The date of the event is now adopted as the material day for all alterations instigated by valuation officers, and some alterations resulting from challenges made by way of proposals to alter the rating list:

- Where a valuation officer prepares a new rating list, the date the physical circumstances are taken is the date the new list comes into effect. For the 2023 rating lists this was 1 April 2023.
- Where the valuation officer later alters this original valuation or it is challenged because a ratepayer believes the value in the list at 1 April 2023 is wrong, the material day is also the day the list came into effect.
- Where a new hereditament comes into existence or ceases to exist, becomes or ceases to be wholly domestic or exempt, then the material day is the date this happens whether the alteration to the rating list is as a result of an alteration by the valuation officer or as a result of a proposal.
- Where the local billing authority serves a completion notice stating that a nearly complete building will be treated as complete and as a hereditament from a specified date then that date is the material day unless another date is agreed or determined by a tribunal (see Chapter 15: when correcting an earlier inaccurate alteration in the list, the material day of the earlier alteration is adopted).

In other circumstances the date the physical circumstances are taken varies depending on whether the change is instigated by the valuation officer or by a ratepayer or owner.

If it is the valuation officer then it is the date of the event triggering the need to make the alteration.

If the change is sought by the ratepayer or owner challenging the entry then the date is different. Until the 2017 rating lists this would have been the day the challenge, called a 'proposal', was received by the valuation officer, e.g. to reduce the assessment due to the demolition of part of the property or because of some roadworks outside the property. Following the new Check, Challenge and Appeal procedure introduced for the 2017 revaluation (see Chapter 0, section 0) the date is now the 'date on which the VO received a confirmation under regulation 4C of the Non-Domestic Rating (Alteration of Lists and Appeals) (England) Regulations 2009 (2009/2268) (as stated in an acknowledgement served by the VO under regulation 4D(1) of those Regulations)'. To make a formal challenge by way of a proposal it is necessary to first 'check' the details the valuation officer holds about the hereditament and advise the valuation officer of any matters that are wrong and incomplete, and it is only then that the formal challenge by way of a proposal can be made. Because, in some cases, it is important to 'fix' the physical circumstances before they change, it was not possible with the new Check, Challenge and Appeal system to retain the date of the proposal as the material day as this might be too late and the particular physical circumstances might have changed again. The Regulations, therefore, make the material day for these other ratepayer or owner challenges by way of proposal as the day the valuation officer receives confirmation that the valuation officer's records are accurate or need amending.

In simple terms, the material day is always the date when the change occurred requiring the alteration except for some changes resulting from proposals, where the material day is the date the confirmation under Regulation 4C is received by the valuation officer, or where there is a completion notice when the material day becomes the date in the notice.

The variations in material day are a little complex but are important and can significantly affect the rateable value. The material day determines what physically is to be included in the valuation. What is included in the valuation in turn affects the date the rateable value arrived at by the valuation first becomes effective for billing purposes. This is known as the effective date of an entry in a rating list.

The material day rules can prevent a proposal having effect for a happening in the past when there has been a subsequent, significant change in valuation. This can happen when the material day is the date of the confirmation under Regulation 4C. The valuation can only take account of the current material day circumstances and not those that existed in the past. For example, if a ratepayer wishes to seek a reduction in rateable value for a temporary disability, but only starts the Check process after the works have finished, or are nearing completion, then the material day rules will prevent a reduction. This is because, in these circumstances, the material day would be the date of the confirmation under Regulation 4C, as it is not a correction of the original 1 April 2023 entry, the insertion or deletion of an assessment, or treating the hereditament as wholly domestic or exempt. At the date of the confirmation under Regulation 4C, the works had either finished or were nearly complete and would not affect the current rental value on the rating basis of a rent for a tenancy from year to year.

Under the new rules the valuation officer can, however, make an alteration, if considered appropriate, because valuation officer alterations adopt the date of the event as the material day.

The sequence of action in altering a rateable value is, first, to establish what the reason is for the proposed alteration; second, to establish the material day at which the physical circumstances are taken; third, to prepare the valuation, taking account of material day physical circumstances but imagining them existing in the AVD world; and, finally, to establish the effective date from when the new rateable value will be used to calculate the rates bill.

6.5.3 Factors to be taken at the material day

In arriving at any valuation there are many factors a valuer has to consider. For rating, where there are, in effect, two valuation dates, the AVD and material day, the valuer needs to know what factors are to be taken at one date and what are to be taken as they were at the other date. The rules are set out in the Local Government Finance Act 1988, Schedule 6, paragraph 2(7). This provides a list of factors to be taken at the material day. All factors not listed are taken to be as they were at the AVD. As explained earlier, broadly speaking the factors to be taken at the material day are physical factors concerning the physical state of the hereditament and the physical state of its locality. The remaining factors, which are taken at the AVD, are non-physical factors such as state of the economy, interest rates, people's attitudes or fashion and levels of value.

The matters listed in Schedule 6, paragraph 2(7), and to be taken as they are at the material day are:

(a) matters affecting the physical state or physical enjoyment of the hereditament;
(b) the mode or category of occupation of the hereditament;
(c) the quantity of minerals or other substances in or extracted from the hereditament;
(cc) the quantity of refuse or waste material which is brought onto and permanently deposited on the hereditament;
(d) matters affecting the physical state of the locality in which the hereditament is situated or which, though not affecting the physical state of the locality, are nonetheless physically manifest there; and
(e) the use or occupation of other premises situated in the locality of the hereditament.

In making a valuation, the valuer takes the physical circumstances at the material day and asks the question: what rental value would the hereditament have possessed at the AVD had the material day physical circumstances existed at that time? This is fairly straightforward if the only change is the building of an extension to a property. All the valuer has to do is imagine the extension existing at the AVD and assess how much higher the rateable value would have been if the extension had existed at the AVD.

It is more difficult when the change or changes are to the locality around the hereditament. The change in the locality may be a fairly complex change that significantly alters relative value levels between hereditaments. In *Barlow (H) & Sons Ltd* v *Wellingborough Borough Council and James (VO)* (1980) 255 EG 461, a new covered shopping mall was constructed in a town. This had the effect of moving the focus of shopping in the town away from the street containing the ratepayer's shop, resulting in it being in a relatively less attractive pitch. The Lands Tribunal accepted that had the covered shopping mall existed when the list was prepared rental values in the street would have been 'appreciably lower' and the rateable values reassessed on this basis.

A good example of separating physical changes and their results from non-physical changes was the agreement made in 1992 to resolve a large number of rating appeals for the 1990 rating list in the City of London. At the AVD, 1 April 1988, the economy was enjoying a boom and office rental values in the City were at a peak. By 1992, values had fallen dramatically. The reasons for this were, first, the economic recession resulted in a low demand for offices and, second, there was a greater supply of new offices entering the market, due to extensive new construction (1 million m^2 under construction at the AVD), which depressed rental values. The assumptions in Schedule 6, paragraph 2(7), meant any effect on rental values resulting from the recession had to be ignored because this was an economic factor and not a physical factor. The change to the amount of modern office floor space available constituted a physical factor and was something that had to be taken into account. In making a valuation, a valuer had to assume this amount of extra floor space had actually existed at the AVD. It was agreed that had the extra floor space been in existence it would have resulted in lower rental levels because of the effects of supply and demand. The view was taken that the effect would have been minimal on modern office values because demand in 1988 would have absorbed the new buildings but the effect on older offices would have been substantial, up to 12.5% for 1960s buildings. In the real AVD world, 1960s offices had let readily because there had been limited modern space available. However, had the extra modern space actually existed in 1988 this would have been occupied in pref-erence to the 1960s space and, as a consequence, rental values for 1960s offices would have been appreciably lower.

A useful clarification came in *Merlin Entertainments v. Cox (VO)* [2018] UKUT 0406 (LC) with the Tribunal holding subparagraphs (a) and (b) and (e) and (f) were distinct; with the former concerned with the hereditament and the latter the locality. The Tribunal's view was the term 'locality' in (e) and (f) plainly excluded the actual 'hereditament' being valued.

The demand for a property is taken at the AVD except when demand itself would be affected by the physical changes. For example, the development of a housing estate near a parade of shops after the AVD will affect the level of demand for the shops and, being a physical change, can be taken into account. The question being, had the housing estate already existed at the AVD, what would the rental value of the shops have been? In *Leda Properties* v *Howells* [2009] RA 165, the ratepayers suggested the demand for an ageing computer centre should be considered at the material day. The Lands Tribunal said that demand fell to be judged at the AVD, not at the material day, and as the her-editament was occupied and fully functional at that date, there clearly was demand.

6.5.4 Physically manifest

The wording of Schedule 6, paragraph 2(7), is not always straightforward or clear as to its meaning. Particular difficulty may be found in the wording about something whilst not actually affecting the physical state of the locality is still 'physically manifest' in the locality. In *Re The Appeal of Kendrick (VO)* [2009] RA 145, the Lands Tribunal considered the meaning of the words 'though not affecting the physical state of the locality, are nonetheless physically manifest there' in examining whether the destruction of the World Trade Center in New York on 11 September 2001 constituted a factor that could be taken into account in reviewing the rateable value of some first class lounges at London Heathrow Airport. It did not consider that the destruction of the towers, as such, could be physically manifest as they were a past happening, not a matter existing on the material day. The Tribunal could see that the consequences of a past event could constitute 'matters', providing they were physically manifest in the locality. It was prepared to accept that the attitude of passengers to air travel could constitute a matter and that the changes in high value passenger numbers footfall and the change in long haul movements would have been visible. However, it found it difficult, as a matter of impression, to see how mere observation of movements of passengers or aircraft could 'reveal anything about the factors that had caused the numbers to be as they were. The level of movements must inevitably be the outcome of a vast range of economic and other factors'. It rejected a technical analysis of data regarding footfall or aeroplane movements to decide if there had been a physically manifest change. It appeared to regard something as being 'physically manifest' if it could be identified by simple observation rather than being measurable in a scientific sense.

The Valuation Tribunal had accepted that the impact of the destruction of the World Trade Center would have been masked by the wider recession affecting the global economy at the time. The President said it was 'a contradiction in terms to say that an effect is masked but that it is nonetheless manifest'. It therefore did not see the effects as being physically manifest but merely an undefinable contributor to the observed decline in movements and found there was no change that could be taken into account under Schedule 6, paragraph 2(7).

In *Merlin Entertainments* v *Cox (VO)* [2018] UKUT 0406 (LC) what was 'physically manifest' was examined further. The appellants sought a reduced assessment to reflect the considerable reduction in visitors to a theme park following a crash on a ride, which had seriously injured five passengers. The appellant sought to apply the approach taken in the Kendrick case. It suggested the attitude of the public to thrill rides resulting from the crash was a matter that might qualify as a 'matter' in paragraph 2(7)(d) by being 'physically manifest', in the same way as the attitude of passengers to air travel as a result of the terrorist attacks on 11 September 2001.

In the Tribunal's judgment, it was clear the crash and the public's reaction to it were matters concerned with the way in which the leisure park business was operated on the hereditament. The appellant had not sought to provide any evidence to show that the crash involved some attribute of the hereditament itself, or that the crash had nothing to do with the way in which the appellant operated the ride or conducted its business at the park. It had sought to disguise the true nature of its case, by presenting it as a reduction in traffic in the locality of the hereditament (and visitors to the park) resulting from an adverse reaction of potential customers to the crash.

It considered the circumstances upon which the appellant relied were irrelevant because they were simply concerned with the way in which it operated its business on the hereditament and the reaction of potential customers thereto. They had nothing to do with any intrinsic or essential characteristic of the hereditament itself or the locality in which it was set. Such success (or failure) was no more than an attribute of the actual occupier and not a characteristic of the property being valued or of the locality in which it is set.

Indeed, there was no change to the locality, rather the reduction in customer demand was simply a reaction to an event which occurred *within* the hereditament. The proposal was, therefore, in substance to do with the hereditament not the locality and there had been no change to the hereditament.

Examining the question of 'physically manifest', the ratepayers submitted that in paragraph 2(7)(d) 'matters … which, though not affecting the physical state of the locality, are nonetheless physically manifest there' included economic or intangible matters which resulted in changes in the locality which were physically manifest there.

The Upper Tribunal pointed out that the wording did not simply refer in isolation to just any matter that was physically manifest in the locality, but it applied to matters which, though not themselves affecting a locality's physical state, nonetheless were 'physically manifest there'. It did not allow the valuer or tribunal to take some economic or other intangible matter which has changed since the AVD simply because that matter has caused or resulted in some physical change in the locality since the AVD. It had to be physically manifest.

The valuer was not required to investigate the economic causes of matters; 2(7)(d) did not allow the assumption that those economic factors also subsisted at the AVD, if different from those which did in fact exist on the AVD.

The Tribunal went on to clarify certain aspects:

- Traffic on a road network or aeroplanes on flight paths might produce noise or other emissions to the environment which represent an integral part of the presence of such traffic in a locality. Such matters themselves may be said to be physically manifest in a locality.
- The Tribunal did not think the VO was correct to submit that something was only 'physically manifest' if it was apparent visually (or for that matter to any of the other senses), as distinct from a change which is measured or even calculated or estimated. For example, noise measurements or calculations to show the effect of changes in the level of traffic on a road may assist a tribunal when assessing the degree of change in that factor, rather than being restricted to subjective descriptions.
- The Tribunal rejected the requirement in the Kendrick case that for something to be physically manifest it must be observable to people and was happy for it simply to be measurable by instruments.

- Sometimes it might not be clear at first sight whether a matter which is physically present in a locality has anything to do with its characteristics. It may then be appropriate to examine what has caused or led to a physical change as an evidential tool to see whether that change is truly a characteristic of the locality, or instead 'accidental' to it. This was the case in the subject appeal where the appellant's case was not concerned with the level of traffic as such, but rather with the cause of that reduction in traffic.
- The matter upon which the appellant relied was broadly similar to footfall in an airline's lounge in the Kendrick case or to footfall in a town centre. Changes in footfall can reflect different types of change. Footfall may decrease or increase because, for example, a railway line serving the town is built or closed down or because the train services on that line are increased or reduced. Such changes may not involve a significant change in the physical structure or state of the locality but are nonetheless matters which themselves are physically manifest in a town centre as something which determines one of its intrinsic qualities, namely its accessibility. On the other hand, changes in footfall may simply be the product of changes in economic matters which are not themselves physically manifest, such as changes in unemployment or spending power, whether on a national, regional or local basis.

In *Wigan Football Club* v *Cox (VO)* [2019] RA 41, Wigan FC argued, perhaps not unreasonably, that its rates should reduce on account of relegation from the Premier down to League One – with a consequent enormous effect on club income. The Upper Tribunal agreed with the VO that for the rateable value to reduce a material change of circumstances had to have occurred and a drop in league status within or without a league was not a MCC. There was nothing intrinsic to the hereditament about the league status of the occupying club, and the physical changes in the locality were caused by the league status. The list could only be altered 'by reason of a MCC'.

In *The appeal of Moore (VO)* [2018] UKUT 0324 the Upper Tribunal had to consider whether the risk of future flooding to a property constituted a material change in circumstances permitting a proposal to be made. It found no evidence that the effect of allocation into flood zone 3 (prone to flooding) physically affected the state of the locality, nor was it physically manifest at the material day. It considered some physical effect observable on the ground was needed, and none was present. It noted the proposal mentioned a 25% rent reduction but did not consider this something physically manifest. It explained levels of rent are considered at revaluations, and do not give rise to an MCC: if they did the rating system would be awash with MCC-based proposals.

6.5.5 Conclusion

It would seem, therefore, that what the valuer needs to do is to imagine that the physical factors at the material day, including matters that are 'physically manifest,' are transported back to the AVD and imagine them replacing the actual physical situation at the AVD. Value effects that flow from these changes, including economic ones, are taken into account but the non -physical cause, e.g. an economic change, is not taken back – just the physical change. So, for example, the vacancy of shops or a change in transport services are taken back but set in the economic world of the AVD. At the AVD in the then economic situation it might well be that a vacant shop or shops would easily have been let and occupied and so no value effect would result.

In summary, in arriving at an assessment, a number of assumptions need to be made:

- Rental levels are as at the AVD.
- The matters contained in Schedule 2, paragraph 7, of the Local Government Finance Act 1988 are as they are at the material day.
- All other matters, not mentioned, are taken as at the AVD.

The following diagrams set out the dates for taking both physical and non-physical factors when compiling a new list and when making alterations to it.

Valuation Officer compiles a 2023 Rating List

1/4/2021	1/4/2023
Take level of values and other matters not mentions in para. 2(7)	Take mentioned matters as they are on this date (valuation officer compiles the rating list)

Rating proposal made or valuation officer alters a 2023 compiled list

1/4/2021	1/4/2023	
		WHEN MAKING AN ALTERATION TO CORRECT AN INACCURACY IN THE ORIGINAL LIST.
Take level of values and other matters not mentioned in para. 2(7)	Take mentioned matters as at the date (which was the material date for compiling the list)	Take mentioned matters as at the material day

a) where including or deleting a hereditament from the list, the material day is the day it came into existence or ceased to exist (this includes splits and mergers);

b) where the whole or part of a hereditament **becomes or ceases to become domestic property** or **is exempt** the material day is the date this happens.

c) where a **Completion Notice** is served the material day is the day specified in the notice (unless another day is agreed or determined by the tribunal);

d) in **other circumstances** the material day is:
   i) in the vase of a valuation officer initiated alteration, the date of the event triggering the need to make the alteration.
   ii) the date of a confirmation under Regulation 4C unless it is by way of an objection to a valuation officer alteration when it is the date the valuation officer altered the list (i.e. as i) above)

EXCEPT where an alteration is made to correct an **inaccuracy of in the list**;

EITHER in the originally compiled list or which arose in the course of making an earlier alteration;

OR due to a proposal being made disputing the accuracy of an earlier alteration.

In these cases the material day is the day on which the 'mentioned matters' fell to be assessed when the incorrect entry was inserted into the list. This is 1/4/21 for a 2023 list compiled inaccuracy.

6.5.6 *Coronavirus – COVID-19 measures*

An obvious question arose during the course of the COVID-19 pandemic as to whether the enforced lockdown and inability of shops and other premises to trade amounted to material changes of circumstance that might justify reductions in rateable value. The issues are quite complex and perhaps rather well illustrate the restriction to 'physical changes' within the Schedule 6, paragraph 2(7) matters.

Rather than test this in the tribunals, the government decided to confirm in legislation that, by and large, the effects of the pandemic were not something to be taken into account in the valuation. This culminated in the Rating (Coronavirus) and Directors Disqualification (Dissolved Companies) Act 2021. Its effect is, in effect, limited to 2017 lists and not 2023 lists because with 2023 lists the reality of the situation at the 2021 AVD including the pandemic is taken into account.

The act says:

'In making a relevant determination, no account is to be taken of any matter (whether arising before or after the passing of this Act) that is directly or indirectly attributable to coronavirus.' This is somewhat broad brush and does not focus in on what any matter might be though does include anything done by any person–

(a) **with a view to compliance with any legislation which concerns the incidence or spread of coronavirus,**
(b) **with a view to compliance with any other legislation for reasons relating to the incidence or spread of coronavirus, or**
(c) **in response to, or otherwise in consequence of, any advice or guidance given by a public authority relating to the incidence or spread of coronavirus.**

The government indicated in its November 2021 Technical Consultation on Rates that it intended to further clarify what is or is not a material change of circumstances, noting in the future there will be other events, not dissimilar to the pandemic, 'occurring between revaluations which by their nature concern the economic or market conditions or the general level of rents rather than, for example, property specific effects such as changes to the physical state of the property (such as internal demolitions) or the locality (such as roadworks)', and it planned to bring forward similar legislation to that for COVID-19. At the time this edition was prepared the provision was included in a non-domestic rating bill being considered by Parliament (clause 14 Alterations to lists: matters not to be taken into account in valuation).

6.5.7 *Tone of the list*

The expression 'tone of the list' or 'tone' is often used to mean, in general terms, the level of values established in a rating list. The term 'valuation officer's tone' may well be used to indicate the levels of value used by a valuation officer in compiling a new list. The term 'settled tone', on the other hand, may be used to refer to the level of values which later becomes settled either by agreement with ratepayers or by acquiescence.

Once a rating list comes into force, the valuation officer will seek to establish 'the tone of the list'. This means that the valuation officer will produce rental evidence to show that the level of assessment he or she has put on property is correct. For example, the valuation officer will seek to show that, for a particular parade of shops, a Zone A of £800 is the correct rental value as at the AVD. Once this general level has been established it will no longer be necessary to prove value by reference to rents, but rather by reference to agreed assessments or assessments determined by tribunals.

6.6 Temporary reductions in assessments

A temporary disability may be either external to the hereditament, e.g. adjacent noisy and dusty building works, or internal, e.g. structural alterations to the building. A different approach to considering these two types of disability for rating valuation has to be adopted because the *rebus sic stantibus* rule only applies to the hereditament: it does not apply to its surroundings. *Rebus sic stantibus* requires a hereditament to be valued in its existing physical state (other than repair) and its existing mode or category of use. This means the hereditament has to be viewed as if the physical state of the hereditament and its mode or category of use can never change – a frozen state. This does not apply to the locality of the hereditament, and future physical changes, insofar as they would affect the rental value, can be taken into account.

6.6.1 *Works or disabilities internal to the hereditament*

Where improvement works are being carried out to a hereditament, e.g. the building of a rear extension, it may be necessary to consider whether the rateable value should be temporarily reduced to reflect the disadvantages to the occupier of the dust, noise and loss of use of space. The question being: would these disadvantages affect the rental value of the hereditament?

If the works are only repair works then the actual nuisance would be disregarded on the basis that the repair works are deemed, under the statutory hypothesis, to be in effect already completed by the hypothetical landlord prior to the letting. Works, other than repair, may render the buildings constituting the hereditament completely unusable or, as in *Paul Rocky and Co. v Morley (VO)* [1981] RA 208, merely make the premises less useable for a time. In this case the premises comprised a shop with offices above. The landlord commenced works to remedy structural defects, involving the erection of scaffolding, the rebuilding of the front elevation and the removal of most of the roof. This rendered some rooms and the rear car park unusable. The landlord agreed a 50% reduction in rent during the works, which were expected to last for six months from the date of the proposals. The valuation officer offered a

25% reduction on the basis that completion of the works could be anticipated in six months and this was the annual equivalent of the reduction offered by the landlord to the tenant. The Tribunal rejected the valuation officer's approach:

> *In the instant case the anticipated reinstatement of the hereditament does not stem from the exercise of any statutory powers. In the language of Dawkins (VO) v Ash Brothers and Heaton Ltd (1969), reinstatement is accidental and not essential to the premises. So it is necessary to value the hereditament, as it exists, at the valuation date, and assume a tenancy from year to year. In the present cases the circumstances show clearly that at the dates of the relevant proposals the physical states of the two hereditaments resulted in 50 per cent reductions in value.*

6.6.2 *Works or disabilities external to hereditament*

As future events external to the hereditament can be taken into account, the likely duration of temporary disabilities outside the hereditament, such as road works, is a factor in considering the effect of the disability on the annual rental value of the hereditament. Unfortunately, this means that while the effect of a temporary disability on the rent from year to year determines the level of any temporary allowance, the reduction in rates payable only lasts for the duration of the works or disability. So, if it is agreed that the effect of a temporary disability lasting six months would reduce the rental value for that six-month period by 20%, it is likely this will only affect the annual rent by half that, i.e. 10%. The rateable value will therefore be reduced by 10% but this reduction will only last for the six-month period of works before it is returned to the full amount. The rates payable will therefore reduce by 10% per month during the period, giving a total rates saving of 5% over the whole rate year.

There are two factors to be considered in deciding whether an external temporary disability has had an effect on the annual rental value of the hereditament:

- the severity of the disability;
- the duration of the disability.

If a disability is not very severe, it is unlikely to affect rental value. Equally, a very severe disability but of short duration is unlikely to affect the rental value of the premises when looking at the annual rent.

6.6.3 *The severity of the disability*

In *Sheppard (VO), Re the Appeal of* [1977] 21 RRC 32, the valuation officer suggested that no reduction was appropriate because the disability lasted for less than a year. The case concerned two adjoining houses affected by noise and dirt from sewer laying work over a six-month period. The works involved the creation of a corrugated iron 75 sq ft, 6 ft high compound within 25 ft of the houses, and the sinking of a 40 ft deep shaft within the compound. Compressors and cranes were used and lorries came and went. Work continued from 7 am to 7 pm Mondays to Fridays for a period of six months.

The Tribunal commented that the degree of importance attaching to transience must vary with the circumstances of the particular case. It noted that the 12-month period was a convenient administrative practice when rental value was not greatly affected. However, for severe disabilities which reduce the rental value substantially, administrative convenience must take second place. The Tribunal suggested it had to put itself in the situation of a reasonable man considering his rental bid. If he would offer the same rent and 'shrug off the nuisance', the rateable value would be unaffected. However, if he would consider that when the nuisance passes, the property will be worth so much rent but meanwhile the nuisance 'is sufficiently severe and/or is likely to continue sufficiently long' to affect his rental bid for a year to year tenancy, then the assessment should be reduced accordingly:

> *In my judgment he would not reduce his bid by as much as £24 [in this case] because that would presume the nuisance to be continuing indefinitely, or at any rate for, say, a further 12 months or so from the proposal date. But neither would he, I think, shrug off the nuisance and offer the full rental value. In my opinion he would have depressed his bid by say, £10, and that reduced bid would represent the reasonably expected [annual] rent on the rating hypothesis.*

6.6.4 The duration of the disability

When considering a rating valuation, the physical state of the locality is taken to be as it was on the material day. The valuer has to consider what would be the rental value of the hereditament taking the physical state of the locality on the material day and imagining these physical circumstances existing at the AVD. Would there be a difference in the rental value had the AVD physical circumstances been not as they were on the day when the list was compiled but as they were on the material day? The likely duration of the works is therefore considered, looking from the material day and not the date of the start of the works. It also follows that it is not possible to consider any period of nuisance prior to the material day and if a disability has nearly finished at the date of the proposal a reduction will not be appropriate. This is illustrated by *Morton (VO)* v *Jones* [1991] 31 RVR 27 where four shops suffered disturbance of a varying intensity over a period of 12 months from carriageway, drainage and pavement repairs. The valuation officer submitted that although the works were carried out over a period of 11 months, the real interference arose from those done to the carriageway itself which were over five months, and only during three and a half months of that period was there real interference. He suggested that having regard to the small degree and short period of disturbance, no reduction was warranted, particularly as at the dates of the proposal the works were virtually at an end. The Lands Tribunal said:

It is I think unfortunate that the proposals in these appeals were lodged at the end of the total period concerned but on the authority of Barratt v Gravesend Assessment Committee [1941] 2 KB. 107; [1941] 2 All ER 308; 105 JP 331; 110 LJKB 552; 165 LT 286; 57 TLR. 590; 85 Sol. Jo. 273; 39 LGR 179; 38 Digest (Repl.) 691 the valuation dates must be taken as being in March 1984 (the date of the proposal) when the works were almost at an end and thus the disturbance had nearly ceased. A tenant coming fresh to the scene would take that fact into account and would not, in my judgment, on the valuation officer's evidence, reduce his bid for the subject premises. It may be cold comfort for me to say that even had the proposals been lodged at the very beginning of the works, on the evidence, in my judgment, the degree of disturbance suffered, which varied in intensity throughout the total period of about 12 months, was not sufficient to depreciate the value although it must have brought matters very close to the point at which a reduction would have been warranted.

In *Berrill (t/a Cobweb Antiques)* v *Hill (VO)* [2000] RA 194, the ratepayer sought a greater allowance for the temporary disabilities caused by the construction of a tramway outside a shop than had been decided by the Valuation Tribunal. There was no dispute that an allowance was appropriate. The Tribunal said it had had to have regard to the works affecting the property at the date of the proposal, which was the material day for this case, but regard had to be had to possible future changes in the scale of the works as they would be in the minds of the hypothetical landlord and tenant when negotiating the annual rent.

6.7 Wales

The Valuation for Rating (Prescribed Assumptions) (Wales) Regulations 2023 SI255 (W.34) applies the post COVID19 restrictions on what constitutes an MCC mentioned in section 6.5.5 in a different way to England. The Wales' approach would appear to be more restrictive and applies for the 2023 rating lists.

Where the RV for a hereditament would be affected by legislation, advice or guidance by a public authority of any country or territory this is deemed not to have occurred if the RV change is due to a matter listed in Schedule 6 para 2(7) of the 1988 Act, other than matters affecting the physical state of the hereditament or locality, or the two 'Minerals' matters in sub paragraphs (c) and (d).

7 Rental analysis and valuation

Note: Whilst some of the lease terms used in the examples in this chapter may seem 'dated' and out of line with current practice, they have been chosen to illustrate the approach to analysis in different market conditions.

7.1 Introduction

In the previous chapters the definition of rateable value and the principles that have to be adopted in arriving at a rateable value have been considered. Rating valuation is based on rental evidence and this chapter will consider:

* the selection of rental evidence on which to base a rating assessment;
* the adjustment of rents to conform with the definition of rateable value;
* the analysis of rents to form the basis of a rating list.

Before going any further, it is worth reminding ourselves of the definition of rateable value as provided by Schedule 6, paragraph 2(1), of the Local Government Finance Act 1988:

2.-(1) **The rateable value of a non-domestic hereditament none of which consists of domestic property and none of which is exempt from local non-domestic rating shall be taken to be an amount equal to the rent at which it is estimated the hereditament might reasonably be expected to let from year to year on these three assumptions**

  (a) **the first assumption is that the tenancy begins on the day by reference to which the determination is to be made;**

  (b) **the second assumption is that immediately before the tenancy begins the hereditament is in a state of reasonable repair, but excluding from this assumption any repairs which a reasonable landlord would consider uneconomic;**

  (c) **the third assumption is that the tenant undertakes to pay all usual tenant's rates and taxes and to bear the cost of the repairs and**

DOI: 10.1201/9781003297369-7

insurance and the other expenses (if any) necessary to maintain the hereditament in a state to command the rent mentioned above.

7.2 Rental selection

Rental selection is the process of deciding, out of all the rental evidence available to the valuer, which rents should be used as the basis for a rating assessment. Ideally, the valuer would like a good number of leases and rents agreed close to the antecedent valuation date (AVD) and on the same terms as the definition of rateable value available as evidence. The closer a lease conforms to the definition of rateable value, the more regard can be had to that lease and its rent.

For some types of property, the valuer may have difficulty in finding any rents at all and may have to resort to other methods of valuation. While it is true that all rental evidence should be considered, the real challenge for the valuer is to consider how much 'weight' should be attached to any piece of rental evidence. The valuer's job is to select the rents which carry the most weight and to base the valuation on those rents.

Some rents may have to be rejected without further consideration, as follows:

- Rents which require so much adjustment that the adjusted rent should be discarded as unreliable if there is other evidence.
- Rents which do not conform and cannot be made to conform to the rating hypothesis.
- Rents fixed between connected parties – in these cases the reliability of the evidence is immediately suspect as the two parties involved are connected in some way and they may not have agreed the best rent obtainable for that property. Examples of connected parties could be where a businessman owns a property and then lets it to his own company. Alternatively, it could be owned by one company and let to another company which is a subsidiary of the first company. In either case it is possible the rent agreed is the best rent obtainable but due to the closeness of the parties to each other the letting cannot be used to show this, though it might be of some use in supporting conclusions from other rental evidence.
- If there is insufficient information available concerning the lease. It is not uncommon for a valuer to know the rent for a property, but be unable to find out all the information regarding the size of the property, the lease terms or some other information which is needed to use the rent or to analyse the letting. In such cases the rent will have to be rejected on the grounds of the lack of reliable information.
- Although not common in the market today, rents fixed on long leases and without rent review will be of little assistance in arriving at the rental value at the valuation date.
- The rent was fixed some time ago and in different market conditions to those that existed at the valuation date. Where the rent is old and market

conditions have changed, it is difficult to reliably adjust the rent to the valuation date. In such cases these rents will have to be rejected.

- Sale and leaseback rents occur when a property owner sells the interest in the property, usually to a financial institution and then immediately takes a lease of the property. The rent agreed in such circumstances is generally related more to returns in the money market than to the true rental value of the property. Consequently, these rents are normally rejected (see *John Lewis & Co. Ltd* v *Goodwin (VO) and Westminster City Council* (1979) 252 EG 499).

In other cases it may be more difficult for the valuer to decide how much weight should be attached to a particular rent:

- The rent on a new letting of a new property – while such a rent may conform very closely to the definition of rateable value as regards date and lease terms, its reliability can still be questioned. When a tenant initially agrees a rent the tenant needs to make some assumptions as to the trading potential of the property and also that other adjoining new property would also be fully let. In some instances the adjoining properties may not let, or may have been let to a lower grade of tenant than was expected or indicated by the landlord. Furthermore, the trading potential of the new development may not reach the expected level. While there is no dispute as to the rent paid, how reliable that rent is in demonstrating the true value of the property is less certain.
- Often in large retail developments, landlords will seek to attract an anchor tenant, or tenants, for the development, who are well-known high street names. The rent agreed with such a tenant may well be lower than the true value of the property, as the landlord hopes that, with such a tenant, not only will the demand for other units in the development increase but also the rent that will be paid for those units will potentially increase. The analysis of the rent of anchor tenants is difficult. It may involve a far larger unit than other units in the development and the question of quantity allowances may have to be considered. It is often difficult to compare a unit in one centre with one in another centre some distance away. It is possible that after the first rent review, the rent agreed may reflect more the true rental value of the anchor property than it did for the initial letting, though this would depend on the construction of the rent review clause.
- Charities can, in some circumstances, qualify for an 80% reduction in rates payable. This could result in the rent they pay being higher than the true rental value of the property due to what is known as the 'equation theory'. The idea of this is a tenant is not particularly concerned with the amount of rent or the amount of rates paid but is concerned by the total figure. If the amount of rates is depressed by a relief then this leaves more money available to pay rent and a prospective tenant enjoying such relief may outbid another possible tenant which does not enjoy the relief. The rents of such

properties need to be carefully considered to see whether charitable relief has impacted on the rent agreed. In the rating hypothesis, we must assume the property 'vacant and to let' and such relief would not be available to any other tenant in the market other than a charity.

- Tender rents – these rents can fall into two categories. The first can apply to retail property in prime positions where there may be substantial competition to gain a foothold. Rather than let the property in a traditional manner, some landlords may decide to put the lease out to tender with a view to achieving the highest rent possible within a given timeframe. In such cases the prospective tenants who bid for the property will be advised by surveyors and will make their bid in full knowledge of market conditions. These rents are generally good evidence of rental value. At the other end of the scale, tender rents are sometimes used for parades of shops in residential areas, owned by the local authority. In these cases the property may also have a user covenant restricting its use to a given trade – greengrocer, chemist or newsagent, for example. The prospective tenants often do not seek professional advice as to the rent they should offer and, coupled with the user covenant, such rents are not regarded as good evidence of value.
- Rent review rents – rents fixed at rent review may not be reliable for rating purposes. Each rent review clause will require the valuer to make certain assumptions in arriving at the new rent. These assumptions may not correspond with the statutory definition of rateable value and the rents will need adjusting to bring them into line with the definition.

Typically, a rent review clause will require the valuer to make certain assumptions:

1 The length of the lease to be assumed. This may be of a similar length to the lease being reviewed, or be for the remainder of the term, or some other specified provision. How long the lease is for may affect the value of the rent which is agreed between the parties. For example, on a ten-year lease, the rent review clause may require the valuer to find the rental value on the assumption that there are still ten years outstanding on the lease when, in fact, the lease may only have five years remaining. Depending on market conditions, there may be a difference in rental value depending on whether the lease is for five or ten years.
2 How improvements are to be treated. Provision may be made to either have regard to or disregard improvements on review.
3 Upward only rent increases, i.e. the rent cannot fall below current rent even if rental values have fallen; therefore, the rent agreed may be above market value.
4 A willing tenant provision. This assumes that the tenant is willing to take the property. This may have a significant effect on the valuation of certain types of property. For example, where there is no demand for the property (even the tenant may find it surplus to requirements), it must be assumed

for the purposes of the review that the tenant is willing to take the property, thereby creating an artificial market.

The other feature of a rent fixed at review is that the rent, by definition, has not been tested in the market. It has been fixed by the landlord and tenant in an artificial market and as a result it needs to be treated with some caution.

5 Rents on lease renewal – where a rent is fixed following the renewal of the lease it may on first examination appear to be an open market rent. Closer examination may reveal that the rent cannot be accepted as an open market rent as the tenant may have carried out improvements to the property which cannot be reflected in the rent due to the Landlord and Tenant Act 1954. Furthermore, the rent will not have been 'tested' in the market but rather will have been fixed between the parties or even determined by the courts.

Recession, the COVID-19 pandemic and the upheaval in the retail rental market have all had an effect on rents. It can be difficult for the valuer to know what reliance or weight can be placed on revised rents when either tenants or landlords are under particularly stress. An example is a Company Voluntary Arrangement resulting in a reduction in future rent obligations. A CVA is a legally binding agreement with a company's creditors which allows a proportion of its debts to be paid back over time. Some 75% of the creditors who voted (by value) need to support the proposal. If approved all unsecured creditors, are bound by the arrangement. The company can carry on trading with the directors remaining in control. It is a good rescue tool for a company that can trade viably were it not for the burden of historic debt. Part of the arrangement may be that the arrangement involves paying a lower rent, but the landlord may not have been one of the creditors who agreed to the scheme perhaps rendering the reduced rent not good evidence of market rental value at the time

7.3 Weight to be attached to rental evidence

For many years, the prevailing view on weight of evidence was that expressed by Scott LJ in the Court of Appeal judgement in *Robinson Brothers (Brewers) Ltd* v *Houghton and Chester-le-Street Assessment Committee* [1937] 2 KB 445:

When the particular hereditament is let at what is plainly a rack rent or where similar hereditaments in similar economic sites are so let, so that they are truly comparable, that evidence is the best evidence and for that reason is alone admissible; indirect evidence is excluded not because it is not logically relevant to the economic enquiry but because it is not the best evidence.

However, with the passage of time this view came under increasing criticism.

In *Garton* v *Hunter (VO)* [1969] 2 QB 37, the Court of Appeal rejected Scott LJ's dictum in the *Robinson* case. Lord Denning said:

> *I would amend the dictum so as to say that when the particular hereditament is let at what is plainly a rack-rent or when similar hereditaments in similar economic sites are so let, so that they are truly comparable, that is admissible evidence of what the hypothetical tenant would pay: but it is not in itself decisive. All other relevant considerations are admissible. We admit all relevant evidence. The goodness or badness of it goes to weight, and not to admissibility.*

This and other decisions enabled the Lands Tribunal in *Lotus & Delta Ltd v Culverwell (VO) and Leicester City Council* (1976) 239 EG 287 to set out the following propositions as a properly established procedure for considering evidence:

> *In the light of the authorities, I think the following propositions are now established:*
>
> *(1)   Where the hereditament which is the subject of consideration is actually let that rent should be taken as a starting point.*
> *(2)   The more closely the circumstances under which the rent is agreed both as to time, subject matter and conditions relate to the statutory requirements contained in the definition of gross value in s 19 (6) of the General Rate Act 1967 the more weight should be attached to it.*
> *(3)   Where rents of similar properties are available they too are properly to be looked at through the eye of the valuer in order to confirm or otherwise the level of value indicated by the actual rent of the subject hereditament.*
> *(4)   Assessments of other comparable properties are also relevant. When a valuation list is prepared these assessments are to be taken as indicating comparative values as estimated by the valuation officer. In subsequent proceedings on that list, therefore, they can properly be referred to as giving some indication of that opinion.*
> *(5)   In the light of all the evidence an opinion can then be formed of the value of the appeal hereditament, the weight to be attributed to the different types of evidence depending, on the one hand, on the nature of the actual rent and, on the other hand, on the degree of comparability found in other properties.*
> *(6)   In those cases where there are no rents available of comparable properties a review of other assessments may be helpful but in such circumstances it would clearly be more difficult to reject the evidence of the actual rent.*

The Upper Tribunal almost invariably refers to the principles set down in *Lotus & Delta Ltd* v *Culverwell (VO) and Leicester City Council* (1976) 239 EG 287 when considering rental evidence. Examples are outlined below.

Hodges Ltd v *Howells (VO)* [1993] RA 236 concerned whether the actual rent agreed as from March 1988 (£12,000 per annum (pa)) should be adopted as the rateable value or, as the valuation officer contended, whether this rent was not the rent which might reasonably be expected in the open market and the rateable value should be higher and in line with rents on other nearby shops and agreed assessments. The valuation officer supported the assessment of £18,000 rateable value.

The Lands Tribunal adopted the propositions from *Lotus & Delta Ltd* v *Culverwell (VO) and Leicester City Council* (1976) 239 EG 287 which indicate how conflicting rental evidence should be considered. The Lands Tribunal accepted that the actual rent was less than the rent which might reasonably be expected to be paid in the open market and, following the various evidence, determined £16,375 rateable value.

In *Specialeyes Plc* v *Felgate (VO)* [1994] RA 338, the Lands Tribunal was asked to decide on the weight to be applied to various rental and tone evidence in order to determine the correct Zone A rate to be applied to a shop, and hence its rateable value. The ratepayer considered £400 per m² in terms of Zone A (ITZA) correct and the valuation officer considered £500 per m² ITZA.

Both parties referred to *Lotus & Delta Ltd* v *Culverwell (VO) and Leicester City Council* (1976) 239 EG 287, and the Lands Tribunal, in giving its decision, followed through the six propositions for dealing with different types of rental and tone evidence given in that case. The report of the present case includes useful examination of the various rating authorities on weight to be given to evidence.

The Lands Tribunal agreed with the valuation officer that an addition should be made in adjusting the rent of the appeal hereditament for tenants' improvements but adopted 7.5% rather than the valuation officer's 10% and the ratepayer's 0%, having regard to the standard of shop fitting. No evidence of cost was given.

The Lands Tribunal considered that as the rents of the appeal hereditament, together with the immediately adjoining shops, were agreed by one landlord with one firm of surveyors representing most of the tenants, the weight to be applied to this evidence was reduced (as it was effectively one transaction).

The Lands Tribunal examined other rents in the street away from the immediate shop and divided these into three categories:

1. **Rents agreed between 1986 and 1987**
   The Lands Tribunal found the evidence helpful in showing an upward trend in rental values over the period but otherwise attached little weight to them due to the need to adjust the rents to 1 April 1988 levels.

2. **Rents close to the AVD**
   The Lands Tribunal paid particular attention to these as indicating the correct level of value and showing the appeal hereditament's rent to be at the lower end of the band of values to be expected at the AVD.

3. **Rents after the AVD**
   Both parties accepted these rents as admissible but the Lands Tribunal only
   considered those close to the AVD due to the need, because of the rapid
   rise in rents in 1988/1989, to make large and uncertain adjustments to relate
   them to AVD levels.

The valuation officer attached significant weight to his agreements on value
in the street and considered a tone of the list was established. Both parties
accepted that the date of the hearing was the correct date for ascertaining
whether a tone of the list was established. At this date the Lands Tribunal
noted 39 assessments had been agreed, nine appeals were outstanding and no
appeals had been made on 27 shops. The Lands Tribunal regarded the tone as
established at £500 per m² ITZA.

Weighing up all the evidence the Lands Tribunal concluded at £475 per
m² ITZA.

The case of *Marks Trustees of T N Marks (Dec'd)* v *Grose (VO)* [1995] RA
49 concerned the disagreement on the interpretation of rental evidence. The
Lands Tribunal, following the principles in *Lotus*, found the rental evidence
just about supported the valuation officer's Zone A of £350 per m². It also
rejected the appellant's contention following *Marks* v *Easthaugh* [1993] RA 11
that a tone of the list had not yet been established. It found that there had been
sufficient appeal activity in the locality to be a challenge to the assessments
entered in the list, and the agreements reached with professional advisers
and occupiers, together with valuation tribunal decisions, demonstrated an
acceptance of the initial Zone A values adopted of £350 per m² ITZA.

In *ELS International* v *Prekopp (VO)* [2016] UKUT 0423 (LC), the Upper
Tribunal repeated an earlier comment that the principles from *Lotus & Delta
Ltd* v *Culverwell (VO) and Leicester City Council* (1976) 239 EG 287 should
not be treated as rules which impose a straitjacket on the analysis of the available evidence but provide a useful framework within which to consider the
valuation in the round.

7.4 Relative values in an old list

The case of *Lamb* v *Minards (VO)* [1974] RA 153, though dealing with a
dwelling-house, gave the Lands Tribunal view that there was a presumption
that the relative values of different properties were correctly expressed in the
previous rating list. Whilst values in the old list would be of no direct assistance,
the relativities might be of some help in arriving at the correct valuations in
the new list. However, valuation officers could, if they did not accept the old
relativities still applied, adduce evidence to show a change in relativities.

It would seem that new evidence as to up-to-date relativities can fairly easily
displace the relativities established for an old list.

More recently in *Lamb (VO)* v *Go Outdoors* [2015] RA 475, the valuation
officer sought to make a comparison with previous 2005 list relativities for a

2010 list appeal but the Upper Tribunal (Lands Chamber) (UTLC) was not at all happy with this:

To limit the amount by which the assessment of the appeal property should alter, by reference to the change between rating lists of other assessments which themselves are subject to appeal, and in doing so then possibly affect the ultimate result of those other appeals, would in my judgement be to let the tail wag the dog.

7.5 Rent in a changing market

In arriving at the rateable value, regard must be had to the rental values as at the AVD. Inevitably, the property market is continuously changing whether upwards or downwards. In order to ensure an accurate valuation, the valuer will need to understand the market for the property in that location. Rents will need to be adjusted to have regard to any changes that have taken place before or after the valuation date. It may well be that actual rents may have increased or fallen since the AVD, but what has to be determined is the rental value as at that date.

7.6 Consideration of the effect of disabilities on rents

In Chapter 6, section 6.6, the need, on occasion, to adjust assessments for temporary disabilities was discussed. Disabilities may also be permanent ones and need to be reflected in an assessment. The quantification of these disabilities is a job for the valuer. Ideally the valuer will be able to find rental evidence to give an idea of the effect. Rents may show what restricted access or an obscured shop frontage have by comparison with premises without the disability.

Trade accounts may also be useful when considering such material changes of circumstances. Where receipts are affected by a temporary or permanent material change this may indicate that a hypothetical tenant would reduce, or indeed increase, the rental bid to reflect the changed opportunity for profit arising from the material change. What has to be considered, however, is not whether trade is affected but whether the change would affect the annual rental value.

Care has to be taken with receipts when considering them for disabilities:

- Ideally they need to be from a number of traders who are representative of the locality. Just looking at the subject property's turnover is often not sufficient because the hereditament has to be viewed vacant and to let and not confined to the particular trade of the ratepayers, and not necessarily restricted to the volume of trade that they experienced.
- Trade may have been affected by other factors (particularly if only a few traders are examined and a full picture not gained) including changes in the

economy and the receipts figures may not isolate the Material Change of Circumstances (MCC) factor.
- Rents do not move in direct proportion to trade, as commented on by the Upper Tribunal in *Berrill (t/a Cobweb Antiques)* v *Hill (VO)* [2000] RA 194: 'There is no direct correlation between turnover and rent.'
- Is like being compared with like?

These factors were considered by the Upper Tribunal in *GPS (Great Britain) Ltd and others* v *Bird (VO)* [2013] UKUT 0527 (LC). This case considered the effect of the extension of a nearby town shopping complex on an existing out of town shopping centre. The Upper Tribunal explained that, since a hereditament is to be valued vacant and to let, it is not usually helpful for the valuer to base his valuation simply on the turnover of the occupier. That is because other potential tenants might be able to trade more profitably and to afford a higher rent.

However, in the particular case the appellants had not relied on a single set of turnover figures; rather they had produced details of turnover from half of the terraced retail units at the out of town centre, prepared in a variety of ways, but showing a consistent picture – namely that turnover performed less well after the opening of the nearby extended centre.

The Tribunal accepted that, since the figures compared turnover at the out of town park with that in other locations over the same period of time, the appellants' figures had not been artificially distorted by the then recession.

The Tribunal considered it would have made the evidence more robust if it had:

- covered a number of years before and after the extended centre opened;
- consisted of actual, rather than indexed, figures.

It did not base its conclusions solely on the turnover evidence but regarded it as tending to add support to other evidence provided.

Undoubtedly, turnover figures are a matter to be considered. They will not form primary evidence but cannot be merely dismissed as irrelevant. Care needs to be taken in deciding what they do in fact show.

The Tribunal would have liked to have seen evidence of trade for a number of years before and after the centre opened. Clearly this is not always going to be possible. There is considerable attraction, perhaps particularly from the ratepayer's point of view, in resolving a MCC challenge early on. One year's accounts may have 'to do'.

Even with valuations normally arrived at on the basis of the analysis of rents on an area or square foot basis, receipts can have a role to play. In *Lidl (UK) GmbH* v *Ryder (VO)* [2014] RA 23, the Upper Tribunal considered the compiled list assessment of a supermarket. It found turnovers useful as a check on valuations which had been reached by comparing the rents and relative merits of the location of the appeal property and the comparable supermarkets.

The ratepayer's expert said that the supermarket company would predict the revenue it could achieve from a store's location when planning a possible new store and would use that prediction as the basis for what it would be prepared to pay in rent. This suggested a strong relationship between revenue and rental value.

Valuers need to approach receipts with care. Changes in receipts are unlikely to see an exact correlation in rental movement and for the rating valuer there is the additional complication of having to take different factors at the AVD and material day which will often require the 'stripping out' of non-physical causes of receipt movements. Excluding these non-physical factors is vital if accurate consideration is to be given to rateable value.

What, if any, effect results from a drop, or indeed rise, in receipts is a matter for expert consideration. Rating valuers have sometimes used a 'rule of thumb' of taking 50% of the receipts change as the change in rental value when the effect is clearly the result of a particular physical change in the locality. In the Fosse Park case (*GPS (Great Britain) Ltd and others* v *Bird (VO)* [2013] UKUT 0527 (LC)), the Tribunal considered the opening of the competing Highcross Centre in Leicester would have reduced turnover at Fosse Park by between 10% and 12%. It stressed there was no direct relationship between turnover and rent, but was satisfied that the hypothetical tenant's rental bid would have significantly reduced as a result of the fall in trade and determined that the reduction of 10% in rateable value suggested by the appellants was reasonable in all the circumstances.

7.7 End allowances

Rental evidence needs to be considered, as above, to reflect any disabilities in the rented property that might not be present in those being valued. The aim being to arrive at a rental value free of such disabilities.

In actually valuing properties with disabilities, allowance will need to be made to reflect those disabilities as against the rental evidence of those properties without those value negative features or where they have been adjusted out of the rental evidence.

Allowance can be made either line by line where a particular disability only really affects that element, or, for other disabilities that affect the whole hereditament it may be best to make an adjustment at the end of the valuation as an 'end allowance.'

In *City of York* v *Sykes (VO)* [2017] UKUT 0230 (LC) the Upper Tribunal agreed with the valuation officer that the ratepayer's arguments for so many end allowances simply 'did not stack up'. If all were applied the resultant figure would be out of line with all the other hereditaments' assessments upon which the experts had relied. There was a need to 'stand back and look'. There is, indeed, as much a danger in over-stating the effect of disabilities as under-stating. While the 'Stand back and look' stage has traditionally been seen as the fifth stage of a Contractor's Basis valuation, it applies as much to a Rentals valuation, or indeed a Receipts and Expenditure valuation. It is always wise to look at the valuation and make a 'sense check.'

7.8 Adjustment of rents

As suggested, the best rental evidence is that which most closely matches the definition of rateable value as regards both date and terms. Ideally, it would be preferable if the rating list and rating assessments could be based on such evidence. In practice sufficient evidence may not be available to make this possible and the valuer may have to have regard to other evidence.

Such evidence will need to be adjusted in some way or other for a whole variety of reasons. These could include:

* repairs;
* insurance;
* rates;
* services;
* fixtures and fittings;
* user covenants;
* premiums;
* key money;
* tenant's improvements;
* rent-free periods;
* reverse premiums.

In this section we will examine some of the main reasons why rents have to be adjusted and consider some of the potential approaches to such adjustments.

7.8.1 Repairs

Under the definition of rateable value, repairs are the responsibility of the tenant. In the current market most properties are let on full repairing and insuring (FRI) terms. Where this is the case, no adjustment is required to conform to the statutory definition.

In some leases the landlord will undertake repairs but recoup the cost from the tenant under a service charge. This is often termed an effective FRI lease as, at the end of the day, the repairs are not included in the rent and the tenant pays the cost of repairs via a service charge.

Where repairs are actually the responsibility of the landlord and cannot be recouped from the tenant, then the rent will have to be adjusted to reflect this liability. As a rule of thumb 5% is usually allowed for internal repairs and a further 5% for external repairs – This seems to have a pedigree right back to the Central Valuation Committee's Resolution 10 (26 November 1926) which recommended '*as a general rule*' 5% for internal repairs and 10% where both internal and external repairs were undertaken. In some locations this may be 7.5% for external and 2.5% or 5% for internal repairs. While these figures are typical of those adopted, in practice the valuer must always look at the property that is being considered and decide what rate is appropriate in the circumstances. For the 2023 Revaluation, the Valuation Office Agency (VOA) has adopted 5% for external repairs and 5% for internal repairs.

7.8.2 Insurance

As with repairs, insurance is the responsibility of the hypothetical tenant. Today most properties are let on FRI terms and accord with the rateable value assumptions. Where insurance is paid by the landlord but recouped from the tenant by a service charge, again no adjustment is required. In those limited cases where the lease requires the landlord to pay for the insurance then it is usual to adjust the rent by 3.5%, which reflects changes in rates of insurance.

7.8.3 Example

An office suite is let for £12,000 pa inclusive of all repairs and insurance but not rates. Adjust the rent into terms of rateable value (ITRV).

Rent reserved under lease	£12,000
Less external repairs 5%	£600
Less internal repairs 5%	£600
Less insurance 3.5%	£420
Rent ITRV	**£10,380**

7.8.4 Rates

The hypothetical tenancy assumes that the tenant pays the rates on the hereditament. Where the rent is inclusive of rates and the lease provides for the rent to vary to reflect changes in rates payable, the amount to be deducted will be the amount currently payable. Where the rent does not alter to reflect changes in rates then the amount to be deducted is the amount payable for the year in which the rent was originally set.

7.8.5 Services

Where the rent includes payment for services provided by the landlord, a deduction will need to be made to reflect the cost of these services. Where the landlord provides services to the tenant and recovers the cost in the form of a service charge, no adjustment needs to be made. However, the level of the charge should be closely examined. The effect of an abnormally high or low charge may make the rent suspect as evidence of rateable value. It may be necessary to treat any exceptional profit made by the landlord on the service charge as in the nature of rent and add it to the rent in determining true open market value.

7.8.6 Fixtures and fittings

In some cases the rent of a property may include a payment for the use of fixtures and fittings. As these items are not rateable, the element of rent charged for these items will need to be deducted from the rent passing.

7.8.7 Plant and machinery

As with fixtures and fittings, the rent may include a payment for plant and machinery. In these cases the valuer must determine what plant and machinery is rateable and what is not (see Chapter 13). The value of any non-rateable plant should again be deducted from the rent passing.

7.8.8 User covenants

A user covenant in a lease will restrict the use of the property to a specified purpose. The user restriction may be very wide, such as use as an office or shop, or restrictive, such as use by consulting engineers or as a doctor's surgery.

The effect a covenant may have on the rent agreed will depend on the extent of the covenant and the market conditions for that type of property. It is possible for very restrictive user covenants to have little or no effect on rental value, as there may be a healthy demand for that use in an area. For example, in Harley Street a user clause restricting the use of the property to doctors' consulting rooms would have little impact as there is a healthy demand for such a use in that area. However, if such a covenant was imposed in another area where there is little or no market for that type of property, the effect may be very substantial.

Generally, a user covenant will reduce the rental value of the property as it limits the potential market. The more onerous the covenant, the more the rent will be below open market value. From a rating perspective, the valuer must assume that the property is 'vacant and to let' in the open market without such a restriction. Consequently, the amount by which the user covenant has depreciated the rental value must be 'added back' to arrive at the true open market value.

7.8.9 Premiums paid for the acquisition of a lease

Where a tenant pays a premium for the acquisition of a lease, this implies that the purchaser is obtaining something of value over and above the lease itself.

Payments which could be included in a premium are set out below:

- Goodwill – the value of the business as a trading entity. Here the incoming tenant would be undertaking the same trade as the outgoing tenant. As the business is already established, the incoming tenant will obtain benefits such as not having to develop a clientele as he or she will have the benefit of the existing customers. Goodwill as such is not rateable and the amount paid for goodwill will need to be deducted from the premium. Where an actual figure is not available, the valuer may have to make an estimate based on what limited information there is available. Care needs to be taken not to confuse goodwill built up as part of a business with inherent goodwill which forms part of the property and results from its special suitability or position.
- Fixtures and fittings – these do not form part of the property being valued and their value must be excluded from any premium before it can be analysed. As with goodwill, the amount paid will need to be deducted from the premium. Where an actual figure is not available, the valuer may have to make an estimate based on what limited information is available.
- Locational 'key money' – in the retail sector tenants will often pay what is termed key money to establish themselves in a particular trading location. In many retail areas there can be substantial competition to occupy a property in that location and retailers may have to pay substantial sums of money in order to secure a property. It can be argued that the amount paid in key money should be added to the rental value of the property, as really being in the nature of a premium representing capitalised rent in advance. Where such a payment is far in excess of market value in such a location it is difficult to fit such rents into the rental pattern in the locality. It is suggested that this should be excluded from the premium as the amount paid may not relate to the true rental value of the property but rather to the profits that particular business considers it can make if it can trade from that location.
- Capitalised profit rent – where a property was let some time ago the rent paid under the lease will be lower than the current full rental value of the property. The profit rent represents the annual difference between the rent paid under the lease and the current market rental value of the property. The capitalised profit rent will need to be analysed so that the true rental value of the property can be ascertained.

In the analysis of the profit rent element of a premium, the valuer will need to consider the rate which should be used for the analysis and the period over which the profit rent should be analysed.

The rate should be the market rate for the property at the time the premium was paid. The period of amortisation (how long the premium should be analysed over) will generally be limited to the next rent review, as presumably the rent will be reviewed to the full rental value at that time. If this is not the case then a longer period of amortisation will be required. This may be because the tenant has carried out improvements which cannot be reflected in the rent on review. This aspect will be covered in more detail later on in this chapter.

7.8.10 Example

A shop is let on a ten-year FRI lease with five-year rent reviews with a two-year unexpired term. The rent paid is £50,000 pa on FRI terms. Two years ago a premium of £40,000 was paid on assignment of the lease.

Analysis of £40,000 premium

Rent paid under lease	£50,000
Annual equivalent of £40,000 for 2 years (1) @ 8% (2) – 1.78 (3)	£22,471
Full rental value at date of assignment	**£72,471**

Notes:
1 2 years to the next rent review remaining under the lease, at which time the rent will be reviewed to the full rental value.
2 8% is the market rate for the property at the date of assignment.
3 1.78 is the years' purchase (YP) for 2 years at 8%.

7.8.11 Tenant's improvements

The approach to dealing with improvements undertaken by the tenant will depend on whether the improvements were undertaken as a condition of the granting of the lease by the landlord or whether they were to fulfil the tenant's own requirements for the property.

7.8.12 Improvements as a condition of the lease

An adjustment will need to be made to the rent where, as a condition of the lease, the tenant:

- pays a capital sum;
- spends money on repairs or decorations, the need for which arose prior to the start of the lease;
- spends money on alterations;
- otherwise spends money according to the wishes of the landlord.

Really, a requirement to spend money on repairs or alterations to the property is, in effect, a payment of a premium to the landlord. Whether the tenant pays a capital sum to the landlord as a premium or pays out money to improve the property, the tenant is still out of pocket by the sum involved as well as having to pay rent for the property. If instead the landlord had repaired or improved the property and the tenant merely had to pay a rent, then the tenant would be prepared to pay a higher rent for the property because the tenant does not have to pay the extra capital cost of the repairs or improvements.

7.8.13 Example

The tenant of a shop has just taken a new 15-year FRI lease with five-yearly rent reviews at £40,000 pa. As a condition of the lease the tenant agreed to carry out the improvements detailed below. The lease provides that the improvements will not be taken into account at the rent reviews.

New shop front	£30,000
Sprinkler system	£25,000
Internal alterations	£20,000

As the improvements were carried out as a condition of the lease, no decision needs to be made as to whether they add to the property's rental value. The rent agreed reflects the tenant's liability to pay the cost of the works.

The period of amortisation will usually be the shorter of the life of the improvement or either the next rent review (where improvements are not to be disregarded) or the end of the lease.

Rent agreed under the lease		£40,000
Add value of improvements		
Shop front – life considered to be 10 years	£30,000	
Sprinkler – life considered to be 25 years	£25,000	
Alterations – life considered to be 15 years	£20,000	
Total cost of sprinklers and internal alterations	£45,000	
Annual equivalent of £30,000 over 10 years (1) @ 8% (2)	6.71	£4,470
Annual equivalent of £45,000 over 15 years (3) @ 8% (2)	8.56	£5,257
Full rental value		**£49,727**
Say		**£49,250**

Notes:
1 The cost of the improvement is spread over the life of the improvement or the length of the lease, whi chever is the shorter.
2 8% is the market yield for this type of property in this area.
3 The life of sprinklers is greater than the length of the lease so the cost is spread over the life of the lease. For the internal alterations the life and the length of the lease are the same.

7.8.14 Tenant's voluntary improvements not as a condition of the lease

Where improvements are undertaken by the tenant not as a condition of the lease then a different approach must be taken and a number of matters need to be considered:

1. Is regard to be had to the improvements at the rent review? In many leases provision is made to disregard the effect of any tenant's improvements in similar terms to those that apply for lease renewal. In such cases the value of

any improvements cannot be reflected until the end of the lease. If the lease does not make such a provision then the value of the improvement may be taken into account at a rent review. In adjusting rents to take account of tenant's voluntary improvement expenditure, care has to be taken in the selection of the period for amortisation due not only to the effects of lease terms but also the Landlord and Tenant Act 1954. The length of time depends on:

a. the useful life of the improvements: this is obviously the maximum;
b. whether the lease specifies that such tenant's improvements are to be disregarded in determining the rent at a rent review. If they are to be disregarded then the period for amortisation may extend to at least the end of the lease;
c. on the renewal of a lease, the Landlord and Tenant Act 1954 provides (section 23(1)(c)) that any increases in letting value resulting from tenant's improvements are to be disregarded providing:

   • they were not carried out in pursuance of an obligation to the immediate landlord; and/or
   • they were carried out during the current tenancy or were completed not more than 21 years before the reference to court.

2. The period for amortisation may therefore be a considerable number of years. If the lease specifies that a tenant's voluntary improvements are to be disregarded then the amortisation period will be the length of the lease plus the likely renewal period the courts would grant for a new lease on the hereditament under the 1954 Act. This will vary depending on the terms of the original lease but can be up to 15 years. If the improvements were made at the start of a 21-year lease then this period could, therefore, be as long as 36 years. The rate adopted will usually be based on the market rate for the property concerned.

3. Do the improvements add to the value of the property and, if so, by how much? In most instances, a tenant's improvements will add to the rental value of the property. In some extreme cases it may well be that the improvements are so specific to the actual occupier that they may add no value in the market and could actually detract from value. The second element of the question, 'by how much?', can be more difficult to answer. The valuer must recognise that cost does not equal value. If you replace all your windows in your house with hardwood or uPVC replacement windows at a cost of, say, £15,000, it is unlikely that you would see a corresponding increase in the capital value of the property. The same applies to commercial property. The fact that a tenant has spent a given sum on improving a property does not always mean that the rental value will increase proportionately. Conversely, the rental value may even increase by more than the cost of the improvement. On the other hand, take, for example, shop fronts. On the high street, most tenants have their own style of shop front with their own corporate

identity. Such fronts can cost many thousands of pounds to install and yet have no value to another tenant and would be removed immediately once the property was occupied by another tenant. However, off the high street and where shop fronts are not specific to a particular trader or tenant, a good-quality shop front may well enhance the value of the property. The valuer must therefore consider each improvement individually and weigh up whether the improvement adds value and by how much.

7.8.15 Example

The tenant of a shop has just taken a new 20-year FRI lease with five-yearly rent reviews at £40,000 pa.

The new tenant of the shop has carried out the following 'improvements' which were not a condition of his lease:

New shop front	£30,000
Sprinkler system	£25,000
Shop fittings	£30,000
Internal alterations	£20,000

7.8.15.1 Stage 1

What has to be ascertained is the amount of rent, if any, that the above 'improvements' add to the rental value.

- Shop front – it is considered that tenants of 'prime shops' will all have their own 'house style' shop fronts and that no matter how good an existing front is they will replace it. In these cases it is suggested that the new shop front does not add to the rental value.

  In other shopping areas it is likely that new fronts will probably add something to the rental value as they will probably be better than what was there before.

  In this example it is considered that the front does not add to the rental value.
- Sprinkler system – this does add to the rental value and it is considered that the whole of the value should be analysed.
- Shop fittings – generally shop fittings are the tenant's chattels and not part of the hereditament and unless rateable (plant and machinery) do not add to the rental value. It is, however, necessary to ascertain exactly what is included under the generic term.
- Internal alterations – in this case the alterations include additional WC facilities, additional wiring, and removal and rebuilding of internal walls. It is considered that of the total cost of £20,000 only £12,500 adds to the rental value.

The period of amortisation will usually be the shorter of:

- the life of the improvement;
- the next rent review (where improvements are not to be disregarded);
- where a lease has not been renewed, the shorter of:
  - 21 years from the date of carrying out the improvement; or
  - the residue of the original lease plus 15 years, if less than 21 years, subject to a view being taken on whether the tenant will exercise the option for a lease extension. In this case it has been assumed that the tenant would not. If this was not the case then the amortisation period would be 36 years;
- the residue of the lease, either where it is known that the tenant will not renew the lease or where the lease has already been renewed.

7.8.15.2 Stage 2 – the analysis

Full rental value		£40,000
Add value of improvements		
Shop front	£ –	
Sprinkler – life considered to be 25 years	£25,000	
Annual equivalent 21 years @ 8%		£2,495
Shop fittings	£ –	
Alterations – life considered to be 15 years	£12,500	
Annual equivalent 15 years @ 8%		£1,460
Full rental value say (including improvement)		**£43,955**
	say	**£44,000**

7.8.16 Specific issues

7.8.16.1 Accrued repairs

Where a property is let in a poor state of repair and it is a condition of the lease that the tenant 'puts the property into repair', then the full cost of putting the property into repair should be taken into account and the rent paid adjusted to arrive at the true rental value of the property in a reasonable state of repair. Although the period of amortisation will vary from lease to lease, it is common for leases to provide, in these circumstances, that on review it is to be assumed that the required works have been carried out. Consequently, the cost of repairs should be amortised up to the first rent review.

7.8.16.2 Shell rents and fitting out

Many retail units in new developments are let on what are termed 'shell rents'. The tenant is provided with a shell, usually three walls and an open front, by the landlord and then the tenant will fit out the property to the tenant's own requirements. The valuer will, in these instances, ascertain what the cost of fitting out the unit was as a basis for finding out the rental value of the property in its fitted out state. While the actual cost may be a starting point, that cost may also include fixtures and fittings which are part of the tenant's chattels rather than part of the property. The cost of these must be deducted from the overall cost. The valuer again must not lose sight of the objective of the exercise – the rental value of the property. Fitting out may be specific to the actual tenant and may not be fully reflected in the rental value of the property 'vacant and to let'.

In *Bunyan (VO)* v *Acenden Ltd.* [2023] UKUT 17 (LC) the Upper Tribunal considered whether fitting out works to an office block increased the rental value. It referred back to two earlier cases:

- *Edma (Jewellers) Ltd* v *Moore (VO)* (1975) RA 343 (LT). It was not disputed by the parties in the case that the ratepayers' expenditure on fitting out should be taken into account. The dispute was over how that capital expenditure should be amortised to arrive at an annual equivalent which could be added to the rent reserved by the lease to give the RV. The classic exposition of the 'virtual rent'
- *Dorothy Perkins Retail Ltd* v *Casey (VO)* [1994] RA 391, which concerned an air conditioning system installed by the ratepayer in a retail shop. The Lands Tribunal was convinced the system conferred a benefit on the occupying retailer saying, 'It is difficult to conceive of a hypothetical tenant, faced with the choice of two shop units vacant and to let, one with an effective air conditioning system and the other without, who would not pay more for the acknowledged benefit of the air conditioning system.'

The Upper Tribunal in Acenden noted in both cases the tribunal had been prepared to assume the tenant's expenditure had increased the rental value. 'That proposition was one which in both cases the Tribunal regarded as obvious, so that the only question was by how much the rental value of the premises had been increased.'

The Tribunal preferred the VO's evidence on the willingness of the market to pay more if offered in the fitted out, 'Category B' condition. This was 'consistent with our own expectations of how a rational market would behave and reflects the approach taken in Dorothy Perkins and in Edma (Jewellers)'. It was satisfied the ratepayer's 'Category B' fit out was, in the main, sufficiently generic and unexceptional and would have had 'general market' appeal.

7.8.17 Annual tenancy

Under the interpretation of the definition of rateable value the valuer must envisage a rent for year to year with a reasonable prospect of continuance. However, in practice most properties are let on longer leases – 10 to 15 years being common – and with a five-year rent review pattern.

The valuer will need to consider whether an adjustment to the rent passing needs to be made to reflect the annual tenancy envisaged in the rating hypothesis compared to the more standard lease terms that exist in the market. The Lands Tribunal has considered the adjustment of rents to the terms of an annual tenancy. In *British Home Stores* v *Brighton Borough Council and Burton (VO)* [1958] RVR 665, the member stated:

> *Nothing I have heard convinced me that the rent for a term of 21 years fixed at or about the date of the relevant proposal is materially different from the rent which may be expected under the statutory terms. The hypothetical tenant is not an annual tenant for one year only, but has a reasonable expectancy of continuance ...*

It is generally accepted that there is no need to adjust rents for this fact, although the Lands Tribunal has stated that where evidence clearly shows a difference, then the rateable value should reflect this (*Black (L&A) Ltd* v *Burton (VO)* [1958] 51 R&IT 307).

7.8.18 Rent review periods

The statutory definition of rateable value requires the assumption of an annual tenancy, whereas most commercial rents are fixed under leases which provide for periodic review at intervals of usually three, five or seven years but sometimes longer.

In some cases, landlords have sought and achieved higher rents on review where the period specified in the lease is greater than normal. This is termed 'overage'. A longer period between reviews is, in times of rising rents, to the advantage of the tenant because the tenant will be able to enjoy a profit rent for a longer period than with a more frequent review pattern. Overage is the result of the landlord seeking to gain a share of this benefit.

Where it can be shown that a lease rent has been fixed at a higher level because of the longer than usual review pattern, a downward adjustment may be required. Generally, comparing rents fixed on the standard five-year rent review pattern and the other different pattern should show the difference that the market is making. In the absence of market evidence many valuers will adopt a 1% per year adjustment factor.

7.9 Living accommodation

Where a property comprises living accommodation then an apportionment of any rent will be required in order to determine the value of the non-domestic element and thus enable a full rental analysis to take place. The value of the residential part to be deducted will depend on the local market conditions.

7.10 Value Added Tax

While landlords may elect to impose Value Added Tax (VAT) on their properties, most tenants will be able to reclaim any VAT charged as part of their business. In such cases any VAT is unlikely to have any impact on the rental value of the property. Where the occupiers of the property are such that they are exempt for VAT, such as the financial services sector, then they will not be able to offset against their business any VAT charged by the landlord on rent. In such cases the imposition of VAT will effectively increase the rent charged.

From a rating and rental analysis viewpoint, the net of VAT rent should be the starting point and, where VAT is charged, it should be deducted to arrive at the net of VAT figure. Whether the charging of VAT on rent will affect the rental value will depend on the level of demand for the property and the nature of the potential tenants.

7.11 Incentives

7.11.1 Stepped rents

Most rent reviews of commercial and industrial property are to open market value. However, in some cases a series of annual or two-yearly reviews to rents fixed at the start of the term are provided for in the lease. When this happens the initial rent is likely to be below the open market rental value; however, to compensate for this the final rent is likely to be above it. The valuer's aim is to calculate the rent which would otherwise have remained constant during the period of the stepped rent.

A stepped rent is a form of incentive as it allows the tenant to pay a lower rent at the start of the lease when the tenant may have cash flow difficulties, although this is compensated for by the tenant paying an above market rent later in the lease.

The adjustment is made, first, by calculating the present value of the stepped rents and then by taking the annual equivalent of the sum over the full period. The YP is selected from the tenant's point of view as with premiums.

7.11.2 Example

A property has just been let on a ten-year FRI lease with a five-year rent review. A rent of £20,000 has been agreed for the first year, £30,000 pa for the next two years and £35,000 pa until the rent review.

This approach outlined below capitalises the value of the rent per annum and then amortises it over the period to the rent review. The discount rate applied is the market rate for the property in question.

£20,000	PV £1 1 year @ 10%	0.9090	£18,180
£30,000	PV £1 2 years @ 10%	0.8264	£24,792
£30,000	PV £1 3 years @ 10%	0.7513	£22,539
£35,000	YP 2 years 1.7355		
	def 3 years 0.7513 @ 10%	1.3039	£45,636
			£111,147
Annual equivalent of £111,147 5 years @ 10%		3.79	**£29,326**

7.11.3 Rent-free periods

A more extreme, but not unusual, agreement is for the tenant to have an initial rent-free period. This might be a matter of a few months or several years. Clearly this is attractive to a tenant but will mean the tenant will have agreed to a higher than FRV rent for the years after the rent-free period until the review. Indeed, if the rent review is upward only, the tenant may well be continuing to pay the high rent after the review.

The analysis of this situation is rather more complicated. Indeed, there are a number of quite complex adjustments that can be undertaken.

This is one of the most common incentives and a number of different approaches to the analysis of rent-free periods are possible.

Before undertaking any analysis of rent-free periods, an assumption will have to be made as to whether the rent to be paid at the end of the rent-free period and set at the start of the lease will equate to the full rental value at the time of the first review or will not, probably, have yet reached that level. This will have a considerable bearing on the resultant analysis as it will affect the period of time over which the rent-free period needs to be amortised. It is only in exceptional market conditions that the rent at first review will not be fixed at the then full rental value and often on an upward only review basis. If the rent paid has not yet reached FRV then, on an upwards only basis, the contractual rent will continue to be paid.

Rents where there are rent frees, and indeed other valuable incentives, are often referred to as 'headline rents'. They are called 'headline rents' because landlords may reveal, perhaps to the property press, the 'rent' they have achieved – the 'headline' – without disclosing the details of inducements such as rent-free periods. A proper analysis of the deal would show the headline rent not to be achievable without the inducements and therefore it is not representative of open market rental value. The headline rent is artificially high. It may, nonetheless, be used by landlords to boost the apparent value of their investment or as comparables in rent review negotiations.

7.11.3.1 Example

A property was let on a 15-year FRI lease with five-yearly rent reviews with the first three years being rent free. The rent will be £35,000 pa for the last two years before the rent review.

7.11.3.2 Approach 1

It is assumed that by the time the rent is reviewed in year five, the rent will be at least £35,000 pa. Thus, the incentive need only be spread over the period to the first rent review.

A number of different approaches to the analysis are possible. This approach capitalises the income flow for the two years for which it is receivable and discounts that value back to the present day. The capital value is then spread over, or 'annualised' or 'amortised' over, the period to the rent review (five years in this case).

The following layout is typical of that used in rating analysis:

£35,000 YP 2 years 1.7355	1.3039	£45,636
def 3 years 0.7513 @ 10%		
Annual equivalent of £45,636 for 5 years @ 10%	3.79	£12,041
Rental value		**£12,050**

Note: The return of 10% is the All Risks Yield (ARY) or market yield for that property.

7.11.3.3 Approach 2

It is possible to argue that in 'normal' market conditions it is usual to give any new tenant a rent-free period of between three and six months, usually to fit out the property, etc. During this time the tenant is not occupying the property for the purpose of the business but solely to fit out the property to a state where the tenant can conduct the business.

In this case, assuming a six-month rent-free period for fitting out, Approach 1 would be modified as follows:

£35,000 YP 2 years 1.7355	1.3039	£45,636
def 3 years 0.7513 @ 10%		
Annual equivalent of £45,636 for 4.5 years @ 10%	3.48	£13,113
PV 0.5 years @ 10%	0.9545	£12,516
Rental value		**£12,500**

7.11.3.4 Approach 3

In all the previous approaches it has been assumed that rents will have increased to either the rent agreed or a higher figure by the time the first review occurs. In Approach 2, the rent at review will be at least £35,000 pa.

It is possible to conceive of instances where the rent at the first review would not have reached such a level. In this situation, and assuming an upward only rent review clause, the tenant would have to pay a rent in excess of the market value after the rent review. In such a case the period of amortisation should be in excess of the rent review period, the actual period being the number of years that it takes for the true rental value of the property to equate to the rent actually being paid. The revised calculation might be:

Headline rent		£35,000
YP 7 years def 3 years @ 10.00%	3.66	£128,020
Annual equivalent of £128,020	6.14	£20,834
YP 10 years @ 10.00%		
Rental value		**£20,750**

7.11.3.5 Approach 4

This approach simply takes the arithmetical sum of all the rent received and divides it by the number of years for the rent review, in this case five:

4th year's rent	£35,000
5th year's rent	£35,000
	£70,000
Rental value (£70,000/5)	**£14,000**

Note: This is arithmetically incorrect but often used in practice.

7.11.4 Capital contributions

Sometimes a landlord will pay an amount of money to an incoming tenant as an inducement to take the property. Such a payment will usually be for a specific purpose such as to assist in the cost of fitting out the unit.

In looking at the capital contribution from an analysis viewpoint, the first issue to be determined will be the ascertainment of the amount of the actual contribution. While this may seem simple it is possible that it may also be tied up with other works the landlord is undertaking, particularly in a new building. Here, the cost may reflect some form of discount having regard to the total work involved. The apportionment of the cost may also cause further complications.

Where such a contribution is made the next issue is to consider whether the whole of the cost should be amortised or only the part that would increase the rental value. Traditionally, where a tenant carries out improvements as a condition of the lease then the whole of the cost should be amortised, whereas where the improvements are voluntary then only those that would affect the rental value should be taken into account.

7.11.4.1 Example

A landlord has offered the tenant the sum of £100,000 on the basis that the tenant signs a new 20-year FRI lease with five-year rent reviews at a rent of £35,000 pa.

In this example the capital contribution is simply a straightforward incentive for the prospective tenant to take the lease of the property. To that extent, the capital contribution can be amortised over a period of time. At least two alternative arguments are possible as to the actual time period over which the contribution should be amortised:

1 The £100,000 should be spread over the whole duration of the lease, that is the 20 years. This is supported by the incentive being to take the lease for that period of time. On this basis the analysis would be as shown follows:

Rent paid under lease		£35,000
Less annual equivalent of £100,000 20 years @ 10%	8.5136	£11,745
Rental value		**£23,255**

2 The £100,000 should be spread over the period to the first rent review. This is on the basis the rent determined at the five-year rent review will be uninfluenced by the £100,000 payment and will be determined on the basis of rental levels prevailing at the time of the review:

Rent paid under lease		£35,000
Less annual equivalent of £100,000 5 years @ 10%	3.7908	£26,380
Rental value		**£8,620**

Note: The return of 10% is the ARY for the property.

In the above example the capital contribution was to entice the tenant to take the property. In other cases the capital contribution may be towards the fitting out of the hereditament. The contributions can be considered as equivalent to a 'reverse premium', as discussed in the following section.

7.12 Reverse premiums

A reverse premium is a feature of a depressed market where outgoing tenants have to pay incoming tenants a premium to take a lease off their hands. This is usually when the true full rental value of the property has fallen below the current rent paid. In this case the annualised value of the reverse premium will be deducted from the rent paid to arrive at the true rental value of the property.

7.12.1 *Example*

A property is currently on a 20-year FRI lease with five-yearly rent reviews with 12 years unexpired term at a current rent of £50,000 pa. The tenant has assigned its lease and has paid the incoming tenant a reverse premium of £40,000.

In order to analyse this premium, it is necessary to know the provisions of the rent review clause, especially whether it is upward only.

7.12.1.1 Upward only rent review

On this basis on review the rent of £50,000 would continue to be paid after the next review, even if the current market rental was below that figure.

Rent paid under lease		£50,000
Less annual equivalent of reverse premium of £40,000 for 12 years @ 15%	5.42	£7,380
Current full rental value		**£42,620**

Notes:
1. As the lease contains an upward only rent review clause, the reverse premium is amortised over the unexpired period of the lease. This is on the basis that it is not expected that market conditions will improve over the next two rent reviews. If it was considered that the full rental value would have exceeded the rent paid by the time of the second rent review in year seven then the premium would be spread over the seven years rather than the 12 in the above example.
2. The rate at which the premium is to be amortised is the 'market rate' for that particular interest. Due to the unusual circumstances of the case this is likely to be high because of the nature of the risks involved.

7.12.1.2 Review to open market rent

In this case the rent of £50,000 would only continue to be paid until the next rent review, at which time it would then be reviewed to the current full market rental value.

Rent paid under lease		£50,000
Less annual equivalent of reverse premium of £40,000 for 2 years @ 15%	1.62	£24,691
Current full rental value		**£25,309**

A useful case giving a good example of the processes in adjusting a headline rent where there are complications of rent frees, break clauses and amortisation periods, is *Bunyan (VO)* v *Acenden Ltd* [2023] UKUT 17 (LC). In the case, the headline rent was amortised to the tenant's break clause (6 years) and not the end of the lease (11 years).

7.13 Surrender and renewal

The actual rent paid under a lease may be below market value as a result of the tenant surrendering an earlier lease.

   A tenant may surrender an existing lease so as to safeguard the future. For example, the lease may be coming to an end and the tenant may not be sure whether the landlord will be happy to offer a new lease or object to a lease renewal. Alternatively, the tenant may wish to dispose of the business, and the fact that the lease has only a limited time to run may affect the saleability or value of the business.

In such a situation the tenant will often agree with the landlord to surrender the existing lease and take out a new lease, thus securing the future for both the landlord and the tenant.

The rent agreed will reflect the value that the surrendered lease has to the tenant, and this should be reflected in the rent agreed for the new lease. The value of the profit rent surrendered is similar to a premium and is treated as such. The calculation is, however, slightly more complicated, as the full rental value (the unknown quantity) also appears in the calculation of the profit rent. The approach used is based on the reasonable assumption that neither the landlord nor the tenant would agree to terms making them any better or worse off after the surrender and renewal than before.

It should be noted that surrenders and renewals are uncommon in the marketplace today. This example is only included for the sake of completeness and to illustrate the approach that could be taken to the analysis.

7.13.1 Example – surrender and renewal

An existing lease on a shop was surrendered on 1 April 2012 in exchange for a new 20-year lease. The rent passing under the old lease was £40,000 pa. Both leases were on FRI terms with five-yearly reviews. The old lease was due to expire on 1 April 2015. The new lease rent is £50,000 pa.

Let the full rental value be £x.

The freehold capitalisation rate for this type of property and locality on a five-year rent review pattern is 6% and the leasehold yield is 7%. The analysis is given in Tables 7.1 and 7.2.

Table 7.1 Rental value from the tenant's point of view

Value of tenant's interest under old lease	
Full rental value	£x
Less rent under old lease	£40,000
Profit rent	£x − 40,000
YP 3 years @ 7%	2.62
Value of tenant's interest before renewal	£2.62x − £104,972
Value of tenant's interest under new lease	
Full rental value	£x pa
Less rent under new lease	£50,000 pa
Profit rent	£x − 50,000
YP 5 years @ 7%	4.10
Value of tenant's interest after renewal	£4.10x − £205,009
	£4.10x − £205,009 = £2.62x − £104,972
	£4.10x − £2.62x = £205,009 − £104,972
	£1.48x = £100,037
	£x = £67,592

Table 7.2 Rental value from the landlords' point of view

Value of landlord's interest under old lease		
Lease rent	£40,000	
YP 3 years @ 5.5%	2.70	£107,917
Reversion to FRV	£x	
YP in perpetuity deferred 3 years @ 6%	13.99	
Value of landlord's interest before renewal		£13.99x
Value of landlord's interest under new lease		£107,917 + £13.99x
Lease rent		£50,000
YP 5 years @ 5.5%	4.27	£213,514
Reversion to FRV		£x
YP in perpetuity deferred 5 years @ 6%	12.45	£12.45x
Value of landlord's interest after renewal		£213,514 + 12.45x
		£107,917 + £13.99x = £213,514 + 12.45x
		£13.99x − 12.45x = £213,514 − £107,917
		£1.54x = £105,597
		£x = £68,569
Full rental value from tenant's point of view		£67,592
Full rental value from landlord's point of view		£68,569
Say, full rental value		**£68,000**

7.14 Adjustment for date

The adjustment of rents to conform to the market situation at the AVD can only be done after careful analysis of the market. The valuer will have to use evidence of other rents to see by how much rents have changed from one date to another. General property indices are published but while they may be regional and specific to a broad class of property, they still remain general indices only and may well not be an accurate guide to movements in the local market.

7.15 Order of adjustment

In the adjustment of rents it is normal practice for any adjustment to reflect the responsibility for repairs and insurance to be made at the end after all other adjustments have been undertaken.

7.16 Property measurement

For rating, measurements follow The Royal Institution of Chartered Surveyors (RICS) *Code of Measuring Practice*, sixth edition notwithstanding the introduction of international measurement standards by the International Property

Measurement Standards Coalition (set up in 2013). The international standards are now contained in the 2023 IPMS All Buildings Standard.

Given all buildings had already been measured for rating and council tax to the *Code of Measuring Practice*, sixth edition standard it would be very costly and time consuming to remeasure to the international standard – at least in the short term.

The definitions of the main measurement bases in the Code of Measuring Practice, sixth edition are:

Gross External Area (GEA) is the area of a building measured externally at each floor level.

Gross Internal Area (GIA) is the area of a building measured to the internal face of the perimeter walls at each floor level.

Net Internal Area (NIA) is the usable area within a building measured to the internal face of the perimeter walls at each floor level.

The code goes into more precise detail about what these mean and what is included or excluded from the various areas.

7.17 Rental analysis

Having adjusted the rents to accord with the terms of rateable value, the valuer will then need to analyse the rent to some common unit of comparison.

The unit adopted will vary according to the type of hereditament and local practice, but the unit of devaluation must be the same as that used in valuation and the maxim 'as you devalue so you must value' is fundamental to the process.

For shops, the traditional approach since the 1920s for rating has been to use the zoning method of analysis. Clearly it would be wrong to value some shops on the basis of a 6M: 6M: Remainder zoning pattern if rents of comparable shops have been analysed on a 5M: 7M: Remainder or 9M: 9M: Remainder basis.

Most properties are usually analysed in terms of a price per m^2 either to net internal area (NIA) or gross internal area (GIA), e.g. offices, warehouses and industrial properties. However, analysis is not always related to floor area and for other property types other units of comparison may be adopted, such as a price per seat for cinemas and theatres or price per bed unit for hotels.

Many valuers will often summarise the evidence in the form of a table where the key features of the property and its rent can be clearly identified.

At this stage the valuer will seek to identify issues affecting rental value. For example, the valuer may be able to determine whether one side of a shopping street is more valuable than the other or whether the values of a street are higher in one part than another. The valuer may also be able to identify whether any quantity allowance has been given for a particularly large property.

It is from this detailed analysis of the rents that the valuer can start to establish the level of values to apply in the valuations.

7.17.1 Example 1

This example attempts to show the process that could be adopted when undertaking the revaluation of the main street of a small town. Odd number shops are on the opposite side of the road from the even properties. For the purposes of the example it is assumed there is no other relevant rental evidence available in the area other than that shown.

All leases are on the basis of a ten-year FRI lease with a five-year rent review pattern and all rents have been fixed as a result of rent review or lease renewal negotiations.

The following shows the rental evidence available.

Unit	Frontage (m)	Depth (m)	Rent £ pa	Date fixed
1	8.0	23	£72,500	2020
3	6.5	19	£53,200	2019
5	9.0	27	Owner occupied	
7	8.0	26	£78,200	2021
9	7.5	25	£73,800	2021
2	10.0	30	£91,000	2019
4	10.0	30	Owner occupied	
6	12.0	35	£99,500	2018
8	11.0	32	£109,000	2020
10	10.0	32	£105,300	2021

In undertaking the analysis for the 2023 Revaluation, the valuer should have regard to:

- the AVD of 1 April 2021; and
- the definition of 'rateable value' and its legal interpretation: paragraph 6(1) of Schedule 6 of the Local Government Finance Act 1988, as amended.

In order to undertake the revaluation exercise for this particular town, the following approach will be adopted:

- The analysis of the evidence – this analysis will analyse the 'facts' of each letting. It will not at this stage attempt to interpret the information. It will enable such interpretation at a later stage.
- The summarising of the evidence – this stage will summarise the evidence, highlighting the important matters that may have an effect on value.
- Drawing conclusions – this stage will involve the adjustment of the evidence to the terms of rateable value and weighing the evidence in order to arrive at Zone A rents for particular locations in the town centre.

Table 7.3 Zoning of units

Unit		Frontage (m)	Depth (m)	X factor	ITZA
1	Zone A	8	6.1	1	48.80
	Zone B	8	6.1	0.5	24.40
	Zone C	8	6.1	0.25	12.20
	Remainder	8	4.7	0.125	4.70
	ITZA				**90.10**
3	Zone A	6.5	6.1	1	39.65
	Zone B	6.5	6.1	0.5	19.83
	Zone C	6.5	6.1	0.25	9.91
	Remainder		0.7	0.125	0.57
	ITZA				**69.96**
5	Zone A	9	6.1	1	54.90
	Zone B	9	6.1	0.5	27.45
	Zone C	9	6.1	0.25	13.73
	Remainder	9	8.7	0.125	9.79
	ITZA				**105.86**
7	Zone A	8	6.1	1	48.80
	Zone B	8	6.1	0.5	24.40
	Zone C	8	6.1	0.25	12.20
	Remainder	8	7.7	0.125	7.70
	ITZA				**93.10**
9	Zone A	7.5	6.1	1	45.75
	Zone B	7.5	6.1	0.5	22.88
	Zone C	7.5	6.1	0.25	11.44
	Remainder	7.5	6.7	0.125	6.28
	ITZA				**86.34**
2	Zone A	10	6.1	1	61.00
	Zone B	10	6.1	0.5	30.50
	Zone C	10	6.1	0.25	15.25
	Remainder	10	11.7	0.125	14.63
	ITZA				**121.38**
4	Zone A	10	6.1	1	61.00
	Zone B	10	6.1	0.5	30.50
	Zone C	10	6.1	0.25	15.25
	Remainder	10	11.7	0.125	14.63
	ITZA				**121.38**
6	Zone A	12	6.1	1	73.20
	Zone B	12	6.1	0.5	36.60
	Zone C	12	6.1	0.25	18.30
	Remainder	12	16.7	0.125	25.05
	ITZA				**153.15**
8	Zone A	11	6.1	1	67.10
	Zone B	11	6.1	0.5	33.55
	Zone C	11	6.1	0.25	16.78
	Remainder	11	13.7	0.125	18.84
	ITZA				**136.26**
10	Zone A	10	6.1	1	61.00
	Zone B	10	6.1	0.5	30.50
	Zone C	10	6.1	0.25	15.25
	Remainder	10	13.7	0.125	17.13
	ITZA				**123.88**

- Preparation of the rating list – this is the valuation stage using the Zone A rents and the area ITZA to arrive at the rateable value of the individual hereditaments.

The first stage of the analysis is to zone the units so that their area ITZA can be calculated.

From the above the individual Zone A rents can be calculated (see Table 7.4).

Table 7.4 Zone A rental valuation calculations

Unit	ITZA	Date	Rent	Zone A
1	90.10	2020	£72,500	£804.66
3	69.96	2019	£53,200	£760.48
5	105.86	Owner occupied		
7	93.10	2021	£78,200	£839.96
9	86.34	2021	£73,800	£854.72
2	121.38	2019	£91,000	£749.74
4	121.38	Owner occupied		
6	153.15	2018	£99,500	£649.69
8	136.26	2020	£109,000	£799.93
10	123.88	2021	£105,300	£850.05

7.17.1.1 Conclusions

Starting with units 2–10, with the exception of unit 6, all the properties are very similar in size and consequently the issue of size and quantity allowances should not be an issue.

Rents have increased over the last four years:

- It is apparent that the rate of increase on the odd side of the road is less than on the even side, although the starting point was higher.
- Looking at the evidence there appears to be little difference between the two sides of the road at the present time. Unit 10 analyses at £850 Zone A, unit 7 at £839 and unit 9 at £854.
- Looking at the entirety of the rental evidence, it would seem that the Zone A rent at 2021 was around £850, which will be adopted as the basis of valuation for this shopping area. The evidence is rather heavily weighted to the higher numbers in the street as three shops in this area let in 2021 whilst the lower numbers were let earlier. There does not appear to be any evidence to suggest that the lower numbers in the street are worth less than the higher numbers (see Table 7.5).

Table 7.5 Rental increase

Unit	ITZA	Year	Rent	Zone	% change
6	153.15	2018	£99,500	£649.69	
2	121.38	2019	£91,000	£749.74	15.38
8	136.26	2020	£109,000	£799.93	6.67
10	123.88	2021	£105,300	£850.05	6.25
Looking at the odd numbered side of the road:					
3	69.96	2019	£53,200	£760.48	
1	90.10	2020	£72,500	£804.66	5.92
7	93.10	2021	£78,200	£839.96	4.35
9	86.34	2021	£73,800	£854.72	1.79

7.17.1.2 Valuation of main street

In the valuation of the town centre it should be remembered that each property is valued individually but that, vacant and to let, it is to be expected that each shop in each part of the shopping street will have the same Zone A rent (see Table 7.6).

This example was very simple; the evidence was very straightforward and was not clouded by many of the issues which have been discussed in this chapter. As with rental evidence in general it is open to different interpretation, and it should be noted that the example is to show the approach that should be adopted.

Table 7.6 Zone A rents

Unit	ITZA	Zone A	Rent	Rateable value (rounded)
1	90.10	£850	£76,585	£76,600
3	69.96	£850	£59,462	£59,500
5	105.86	£850	£89,983	£90,000
7	93.10	£850	£79,135	£79,100
9	86.34	£850	£73,392	£73,400
2	121.38	£850	£103,168	£103,200
4	121.38	£850	£103,168	£103,200
6	153.15	£850	£130,177	£130,200
8	136.26	£850	£115,823	£115,800
10	123.88	£850	£105,293	£105,300

Note: In the rateable values rounding has taken place and reflects market practice.

7.17.2 Example 2

This example attempts to show the process that could be adopted when undertaking the revaluation of the main street of a small town and involves the need to adjust the rental evidence for a range of different factors.

Table 7.7 List of rental evidence

Unit	Area ITZA	Rent pa	Date fixed	Remarks
1	135.00	£65,390	1/4/2018	In 2018 the tenant spent £20,000 fitting out the shop and repairing the property as a condition of the lease. These improvements are to be taken into account on review.
2	85.00	£45,750	1/4/2021	Property let with a two-year rent-free period.
3	276.00	£127,500	1/4/2020	The landlord carried out £10,000 of improvements for the tenant as an incentive to take the property.
4	70.00	£26,250	1/9/2021	The tenant has replaced the heating system and re-plastered all walls to the retail area at a cost of £10,000.
5	73.00	£35,000	1/8/2019	
6	81.00	£56,000	1/9/2021	New letting on a five-year review pattern.
7	97.00	Owner occupied		
8	94.00	£71,000	1/4/2020	
9	70.00	£50,000	1/5/2021	Property let with a three-month rent-free period.
10	82.00	£66,200	1/8/2021	
11	100.00	£91,500	1/4/2020	
12	200.00	£130,000	1/4/2020	
13	105.00	£81,300	1/3/2021	Tenant has the right to break on review.
14	120.00	£71,000	1/9/2021	Tenant has recently installed a new shop front at a cost of £25,000.

Odd number shops are on the opposite side of the road from the even properties.

For the purposes of the example it is assumed there is no other relevant rental evidence available in the area other than that shown.

All leases are on the basis of a 20-year FRI lease (unless otherwise stated) with a five-year rent review pattern and all rents have been fixed as a result of rent review or lease renewal negotiations.

In undertaking the analysis for the 2023 Revaluation, the valuer should have regard to:

- the AVD of 1 April 2021; and
- the definition of 'rateable value' and its legal interpretation: paragraph 6(1) of Schedule 6 of the Local Government Finance Act 1988.

In order to undertake the revaluation exercise for this particular town, the following approach will be adopted:

- The analysis of the evidence – this analysis will analyse the 'facts' of each letting. It will not at this stage attempt to interpret the information. It will enable such interpretation at a later stage.
- The summarising of the evidence – this stage will summarise the evidence, highlighting the important matters that may have an effect on value.
- Drawing conclusions – this stage will involve the adjustment of the evidence to the terms of rateable value and weighing the evidence in order to arrive at Zone A rents for particular locations in the town centre.
- Preparation of the rating list – this is the valuation stage using the figures in Table 7.8 to arrive at the rateable value of the individual hereditaments.

7.17.2.1 Stage 1

The order of the analysis must reflect the information that is available. In some instances it may not be possible to analyse one rent until other information from the analysis of other properties has been obtained (see Table 7.8).

Table 7.8 Analysis of rents

No. 1		
Date fixed	1/4/2018	
Rent	£65,390	
ITZA	135 m²	
Remarks	Tenant spent £20,000 on fitting out and repairing the property as a condition of the lease	
Rent	£65,390	
Annual equivalent of £20,000 over 5 years (1) @ 6% (2) – 4.21	£4,750	
Rent in terms of rateable value	£70,140	
Rent ITZA	£519.56	
Rent ITZA (before adjustment)	£484.37	

(1) 5 years adopted, as at rent review the rent will reflect the value of the repairs.
(2) 6% adopted based on market evidence for this type of property in the locality.

No. 2		
Date fixed	1/4/2021	
Rent	£45,750	
ITZA	85 m²	
Remarks	2-year rent-free period	
Headline rent		£45,750
YP 3 years def 2 years @ 6% (all risks yield)	2.38	£108,885
Equated rent		
YP 5 years @ 6%	4.21	£25,863
Rent in terms of rateable value		£25,863
Rent ITZA		£304.28
Rent ITZA (before adjustment)		£538.24

(*Continued*)

Table 7.8 (Continued)

In this case the headline rent is in excess of market value. For rating purposes the valuer needs to ascertain the true value of the property. In undertaking the analysis it is considered that rents will have risen to in excess of the rent paid by the time the next rent review takes place.

No. 3

Date fixed	1/4/2020
Rent	£127,500
ITZA	276 m²
Remarks	Landlord carried out £10,000 of improvements

The improvements were carried out by the landlord at the tenant's request and any value will be included in the rent the landlord has charged. No adjustment to the rent is therefore needed.

No. 4

Date fixed	1/4/2021
Rent	£26,250
ITZA	70 m²
Remarks	7-year review pattern

Tenant replaced heating system and re-plastered all walls to the retail area at a cost of £10,000.

The question to be asked is to what extent the £10,000 expenditure will enhance the rental value of the property. It is suggested that the re-plastering would not be fully reflected, say, £8,000 for all the works. It could further be argued that the work is normal repairs imposed on the tenant under both the terms of the lease and the definition of rateable value. In which case, it could be argued that no addition is required to the rent.

Rent		£35,000
Annual equivalent of £8,000 over 20 years @ 6%	11.47	£697
Rent ITRV	£35,697	
Rent ITZA	£509.96	
Rent ITZA (before adjustment)	£500.00	

No. 5

Date fixed	1/8/2019	
Rent	£35,000	
ITZA	73 m²	
Rent		£35,000
Rent ITRV		£35,000
Rent ITZA		£479.45

No adjustments required to the rent passing.

No. 6

Date fixed	1/9/2021	
Rent	£56,000	
ITZA	81 m²	
Remarks	New letting	
Rent		£56,000
Rent ITRV		£56,000

Table 7.8 (Continued)

Rent ITZA		£691.36

No. 7 – owner occupied

No. 8		
Date fixed	1/4/2020	
Rent	£71,000	
ITZA	94 m^2	
Rent		£71,000
Rent ITRV		£71,000
Rent ITZA		£755.00

No adjustments required to the rent passing.

No. 9		
Date fixed	1/5/2021	
Rent	£50,000	
ITZA	70 m^2	
Remarks	3 months rent-free	

As the rent-free period is short, it is proposed that it be ignored for the purpose of analysis on the grounds that it may be a fitting out period. However, if considered appropriate, the incentive could be more fully analysed.

Rent		£50,000
Rent ITRV		£50,000
Rent ITZA		£714.29

No. 10		
Date fixed	1/8/2020	
Rent	£66,200	
ITZA	82 m^2	
Rent		£66,200
Rent ITRV		£66,200
Rent ITZA		£807.32

No adjustments required to the rent passing.

No. 11		
Date fixed	1/4/2020	
Rent	£74,000	
ITZA	100 m^2	
Rent		£74,000
Rent ITRV		£74,000
Rent ITZA		£740.00

No. 12		
Date fixed	1/4/2020	
Rent	£130,000	
ITZA	200 m^2	
Rent		£130,000
Rent ITRV		£130,000
Rent ITZA		£650.00

No. 13		
Date fixed	1/3/2021	
Rent	£81,300	

(Continued)

Table 7.8 (Continued)

ITZA	105 m²	
Remarks	Tenant has right to break on review	
Rent		£81,300
Rent ITRV		£81,300
Rent ITZA		£774.29

No. 14

Date fixed	1/9/2021	
Rent	£71,000	
ITZA	120 m²	
Remarks	Tenant has installed a new shop front at a cost of £25,000	

Consideration needs to be given to whether the shop front adds to the rental value of the premises and, if so, by how much. On the basis that this is a small town centre shopping centre it is suggested that the full amount is taken into consideration and that the life of the shop front is assumed to be 15 years.

Rent		£71,000
Annual equivalent of £25,000 over 15 years @ 6%	9.71	£2,574
Rent ITRV		£73,574
Rent ITZA		£613.12
Rent ITZA (before adjustment)		£591.67

7.17.2.2 Stage 2

7.17.2.3 Conclusions

It is difficult to show with any degree of certainty that rental values have changed over the period when the rents are available.

The rents of numbers 1–5 seem to suggest that this area of the road is less valuable than the area with shops having the higher numbers. The rents for 1–5 range from £462 to £519 but range in dates from 2018 to 2020. In addition, the rents of 2–4 opposite also show a lower level of values than the rest of the road. This would also support the view that the lower numbers are less valuable than the higher numbers.

It is suggested that the value of 1–5 should be £500 Zone A, reflecting the age of the rents and the value of 2–4 should be £525 Zone A based on 2021 rents.

Numbers 9–13 seem to have values between £714 and £774 Zone A, and it is suggested they should be valued at £750 Zone A. It should be noted that the lowest and highest rents were both fixed in 2021, so it would be possible to raise other arguments!

On the opposite side of the road value of shops 6–14 is less clear. It would appear that the value continues to rise from number 4 to its height at number 10, and then falls away to number 14. Looking at numbers 6–8 and the properties

Table 7.9 Summary of the rental evidence

No.	Date	Zone A (ITZA m²)	Unadjusted rent	Adjusted rent ITZA	Remarks
1	1/4/2018	135	484	519	Repairs and improvements spread over 20 years. Repairs and improvements a condition of the grant of the lease.
3	1/4/2020	276	462	462	Landlord carried out £10,000 of improvements.
5	1/8/2020	73	479	479	Owner occupied.
9	1/5/2021	94	714	714	
11	1/4/2020	100	740	740	
13	1/3/2021	105	774	774	
2	1/4/2021	85	291	538	2 years rent-free.
4	1/4/2021	70	500	509	Tenant replaced heating system and re-plastered walls to the retail area at a cost of £10,000.
6	1/9/2021	81	691	691	
8	1/4/2020	94	755	755	
10	1/8/2020	82	807	807	
12	1/4/2020	200	650	650	
14	1/9/2020	120	591	613	New shop front.

opposite it is suggested that £750 Zone A should be adopted for these two properties as well as those on the other side of the road.

For number 10 a value of £810 will be adopted and for numbers 12–14 where the value appears to be decreasing, £650 Zone A will be adopted.

This leaves only number 7 which is owner occupied. It seems there is a general rise in values from the lower numbers on both sides of the road towards the higher numbers. The side road appears to be a 'natural break' and as number 7 falls on the higher value side a figure of £750 may be justified. On the other hand, there may be some element of 'tailing off' and a value below £750 of £650 Zone A has been adopted.

7.17.2.4 Valuation of town centre

In the valuation of the town centre it should be remembered that each property is valued individually but that vacant and to let it is to be expected that each shop in each part of the shopping street will have the same Zone A rent (see Table 7.10).

It should be noted that in the rateable values rounding has taken place and reflects market practice.

Table 7.10 Summary of rental evidence and rateable values

No.	Area	Zone A (ITZA m²)	£ adopted	Rateable value
1	135	£475	£64,125	£64,100
3	276	£475	£131,100	£131,100
5	73	£475	£34,675	£34,700
7	97	£650	£63,050	£63,000
9	94	£750	£70,500	£70,500
11	100	£750	£75,000	£75,000
13	105	£750	£78,750	£78,800
2	85	£525	£44,625	£44,600
4	70	£525	£36,750	£36,800
6	81	£750	£60,750	£60,800
8	94	£750	£70,500	£70,500
10	82	£810	£66,420	£66,400
12	200	£650	£130,000	£130,000
14	120	£650	£78,000	£78,000

7.18 Wales

There are no differences between England and Wales.

8   The contractor's basis

8.1 Introduction

As we have seen in the earlier chapters the basis of valuation for rating is rental value. For the majority of properties the availability of rents will not be a problem, as there will be sufficient rental evidence available on which to base a rating assessment.

However, for some types of property which never let in the market, such as local authority swimming baths, steel works, oil refineries and the like, there will be no rental evidence available. Consequently, the valuer will have to use alternative methods of valuation to ascertain the rental value of the property. One such alternative method of valuation is termed the 'Contractor's Basis (or "Test") of Valuation'.

The contractor's basis of valuation has developed over a substantial period of time. During that period it has been constantly refined to deal with the ever-changing problems that arise from the rating system and the property market. Despite sometimes having been called 'the method of last resort', it has stood the test of time and still provides a highly useful basis of valuation for many types of property that are not normally let in the marketplace, and for the valuation of plant and machinery.

Today, the method is used for the valuation of a wide variety of properties, including schools (public and private), colleges, universities, local authority occupations, airports and many types of larger industrial property such as chemical works, shipyards, steelworks and the like.

8.2 Historical development

It is difficult to trace the exact origins of the contractor's basis of valuation. During the nineteenth century valuers were required to value for rating public utilities such as canals, gas, water and electricity. These properties were not let and so other approaches to finding their value had to be devised.

In one of the early cases, *R* v *Chaplin* (1831) 1 B & Ad 926, which concerned the valuation of a canal, the courts found that the interest paid

DOI: 10.1201/9781003297369-8

on a loan to purchase the canal was equivalent to a rent of the property. In cases that followed, the approach of interest on capital was developed into being one of interest on the cost of constructing the property. Even at this time the courts emphasised the need for care in the adoption of the approach, as cost may not always equal value. This is still true today. By the second half of the nineteenth century the approach was being used for a wider variety of properties, including schools, colleges and universities, together with some of the more specialised industrial property and plant and machinery.

As time went on, the rather simple approach of a percentage of the construction cost was developed to allow for the adjustment of the cost to reflect the disadvantages of an existing property compared to that of a new property. In addition, the value of the land was included as a specific item. One of the major problems involved in the approach was often the discount rate to be used to convert the capital cost to rental value and a whole series of cases went through the courts on this point alone.

8.3 The six stages

A helpful and often quoted definition of the contractor's basis comes from the case of *Dawkins (VO)* v *Royal Leamington Spa BC and Warwickshire CC* [1961] RVR 291:

> *As I understand it, the argument is that the hypothetical tenant has an alternative to leasing the hereditament and paying rent for it; he can build a precisely similar building himself. He could borrow the money, on which he would have to pay interest; or use his own capital on which he would have to forego interest to put up a similar building for his owner/occupation rather than rent it, and he will do that rather than pay what he would regard as an excessive rent – that is, a rent which is greater than the interest he foregoes by using his own capital to build the building himself. The argument is that he will therefore be unwilling to pay more as an annual rent for a hereditament than it would cost him in the way of annual interest on the capital sum necessary to build a similar hereditament. On the other hand, if the annual rent demanded is fixed marginally below what it would cost him in the way of annual interest on the capital sum necessary to build a similar hereditament, it will be in his interest to rent the hereditament rather than build it.*

In effect the method is valuation by analogy. The hypothetical landlord and tenant are seeking to agree a rent but have no comparison or yardstick to guide them. The hypothetical landlord suggests a rent figure but the tenant says this is too high suggesting he or she could leave the negotiations, buy a piece of land and build a similar building and the interest cost in borrowing the money to do this would be less than the landlord is asking. By reference to this imaginary situation and interest cost the hypothetical landlord and tenant reach agreement on the value of the actual hereditament.

The method is normally described as having five stages. This stemmed from the Lands Tribunal decision in *Gilmore (VO) v Baker-Carr* [1961] 1 RVR 598. While this staged approach provides a logical valuation framework and has been approved by the courts, it should not constrain the valuer from modifying the approach where circumstances warrant such modification. A sixth stage was identified in the case of *Imperial College of Science and Technology v Ebdon (VO) and Westminster City Council* [1986] RA 233 concerning the assessment of a university college in central London.

The stages can be summarised as:

Stage 1: Estimation of Cost Construction
Stage 2: Deductions from Cost to Arrive at Effective Capital Value
Stage 3: Estimation of Land Value
Stage 4: Application of the Appropriate Decapitalisation Rate
Stage 5: 'Stand Back and Look'
Stage 6: Effect of Negotiations.

These stages will be examined in more detail later in this chapter.

8.4 Problems with the contractor's basis

Today the format and general application of the approach is well accepted. However, the difficulties that arise in its application can be considerable. The following authorities highlight some of the more difficult issues.

Lord Denman in *R v Mile End Old Town* (1874) 10 QB 208 commented: 'The outlay of capital may furnish no such criterion, since it may have been injudiciously expended, and what was costly may have become worthless with subsequent change.'

Rating Forum, *Contractor's Basis of Valuation for Rating Purposes – A Guidance Note* (RICS Business Services, 1995): 'In the view of the Rating Forum, the further the valuer strays from reality the less weight can be attached to the valuation produced.'

Valuation Office Agency (VOA) Rating Manual Section 4: Part 3 App.1 para 2.1:

Any method of (rating) valuation is only a means to an end, namely to establish the rental value of the hereditament and the rental basis (whether direct or indirect) should be the valuer's preferred approach, providing of course that the rental evidence is sufficiently soundly based, and that where comparison is necessary, it can properly be made having regard to the use and class of the properties involved.

The Rating of Plant and Machinery in Industries Currently Subject to Prescribed Assessment: A Report by the Second Wood Committee (Cm 4283) commented:

> there may be some problems encountered in seeking to apply the contractor's basis to industries where, for historic reasons, capital investment may not have been related to the returns available. As a result, valuations based upon the contractor's basis may be disproportionately high when compared to the industries turnover and profits.

8.5 Approval of the method

As has been highlighted in this book, the basis of rating is the rental value of a hereditament as defined in Schedule 6 of the Local Government Finance Act 1988. A direct rental approach is usually the 'best evidence' of value and to be preferred over other approaches.

However, in two recent cases the rental approach was rejected in favour of the contractor's basis:

In *Turnbull VO* v *Goodwyn School and others* [2014] UKUT 0068 (LC), the Upper Tribunal had to simply decide whether a number of independent schools should be valued on the contractor's basis or whether there was sufficient reliable rental evidence for them to be valued on that basis. The Tribunal found that on the facts of that case (and that case alone) there was not sufficient reliable rental evidence on which to base the valuation and approved the use of the contractor's basis.

Gallagher VO v *Read & Partners and others* [2015] UKUT 0001 (LC) concerned the approach to the valuation of three purpose-built doctors' surgeries where the method of valuation was again at issue – rental or contractor's. The Tribunal rejected the rental basis in favour of the contractors:

> it seems to me that the answer is relatively simple, and in my judgment the VTE's decision was correct. I am not persuaded ... as to the nature of the lease rents or CMRs. They are just not reliable transactions, whether open market, market or not. They are not based upon the vacant and to let principle and do not reflect the value of occupation to the occupier.

8.6 Rejection of the method

Prior to examining the method in detail it is useful to review the cases where the method has been rejected by the Lands Tribunal as being inappropriate.

In *Barclays Bank Plc* v *Gerdes (VO)* [1987] RA 117, the use of the contractor's approach was rejected by the Lands Tribunal in favour of settled assessments of similar types of property.

In a valuation tribunal case, *British Telecommunications Plc* v *Central Valuation Officer* [1998] RVR 86, the contractor's basis was rejected in favour of the receipts and expenditure approach. The approach was criticised for not being able to take into account many of the economic factors that would influence negotiations between the landlord and tenant and which were inherent in valuing a property producing a profit.

In *Hardiman (VO)* v *Crystal Palace Football and Athletic Ltd* (1955) 48 R&IT 91, in the valuation of a football ground, an attempt was made to adopt the contractor's basis. The Tribunal rejected the approach:

> *The hereditament now in question was built over 30 years ago, and economic and other circumstances affecting it and its use are now so different as to render cost, whether past or present, and however adjusted, useless as a guide to its current rental value.*

In *Kidd (Archie) Ltd* v *Sellick (VO)* (1977) 243 EG 135, in valuing an owner-occupied factory premises, the ratepayer adopted a contractor's basis in the absence of rental evidence on the subject property and where there were no rents of comparable properties in the area.

The valuation officer adopted a rental approach having regard to a rather limited amount of evidence spread over a wide area.

The Tribunal preferred the valuation officer's approach. In rejecting the contractor's basis the Tribunal stated:

> *it is true that it starts from a solid foundation of actual costs, although even here it is not entirely clear how the question of site works has been dealt with. Any disagreement as to the adjustment which may be called for to take into account building cost indices is of no real significance, but there remain two imponderables which seem to lack any solid foundation. One is the price of the land and the other is the rate per cent to be applied to the capital cost in order to arrive at the net annual value.*

Hughes (VO) v *York Museums and Galleries Trust* [2017] UKUT 0007 (LC) concerned the valuation of three separate sites comprising The Castle Museum, Heritage Centre and Art Gallery. The Art Gallery dated from 1879, the Castle Museum dated from early 1700s and the Heritage Centre 1200s.

The case raised the issue as to how historic properties, often run at a loss, should be valued – the contractor's basis, the receipts and expenditure basis or the shortened 'profits' basis.

The Tribunal commented:

> *His [the ratepayers] principal objection was … that all new museums were constructed with the assistance of lottery funding, so that no tenant would contemplate spending its own capital on constructing a museum. This seemed to the Tribunal to be an example of the literal approach to the contractor's basis which the Lands Tribunal had warned against in Monsanto. Whether a hypothetical tenant would or would not be prepared to spend its own capital on such*

*a project is nothing to the point. **The theoretical underpinning of the method is that the annualised cost of constructing a fictional alternative building would appear to the parties to be a suitable measure of the rental value of the actual building, and not that the construction of an alternative building would be a course of action which either of them would contemplate in reality.***

It further commented:

> *The quality of the evidence we received on the contractor's basis was disappointing. Despite producing substantial reports neither … provided much explanation of the valuations themselves. Spreadsheets with little or no commentary to explain how rates, figures or allowances had been arrived at were of limited assistance and necessitated detailed questioning from the Tribunal, which should not have been required. We would have been helped by much more cooperation between the experts and by a more detailed statement explaining where they agreed and disagreed and why.*
>
> .....
>
> *In the event, these deficiencies in the evidence have not proved critical because **we are satisfied that the use of the contractor's basis for the purpose of these valuations is not appropriate**. At the most fundamental level we can see no justification in the case of historic buildings used as museums or visitor attractions for the assumption underlying the contractor's basis that notional costs of construction bear some consistent relationship to rental value. Such buildings are often, as the evidence demonstrates, inherently unprofitable yet would be very expensive to construct either in their original form or as a modern substitute. The statutory rating hypothesis requires the assumption of a letting of expensive and inefficient historic buildings on terms which the evidence suggests could not be achieved in reality; it does not require that the commercial viability of the enterprise or the risks of assuming responsibility for the maintenance of such premises be ignored. Nor does the statutory hypothesis permit the making of an assumption, without evidence, that the sort of tenant who might take a tenancy of an historic building for use as a museum would have external sources of revenue sufficient to enable the payment of a commercial rent where the purpose for which they wish to occupy the premises themselves is likely to be loss making.*

A subsequent case of *Hughes (VO)* v *Exeter City Council* [2020] UKUT 0007 (LC) considered the valuation of a museum and art gallery. The Tribunal again rejected the use of the contractor's basis.

> *There is no general legal principle in England and Wales which requires the R & E or CB method to be used in any particular case, and certainly none which determines the outcome of the present dispute. In particular, there is no legal principle which would preclude the use of the R & E method in the present case*

and would require the CB to be used. Likewise, there is no legal principle or valuation practice which would preclude the modification of the R & E method for properties of the unusual kind we have in this appeal, e.g. by use of an appropriate overbid or uplift, or a revenue-based method (e.g. percentage of gross receipts), or perhaps a percentage or amount related to outgoings, to reflect a socio-economic or cultural motivation to occupy, so long as all relevant considerations are taken into account and weighed. However, we have previously explained why the Tribunal has not been asked to approach this appeal in that manner.

Even where a non-profit making hereditament is occupied for very important socio-economic or cultural purposes, the CB method may be inappropriate (because, for example, it would provide an unrealistically high valuation which cannot reliably be adjusted at stage 5), or the valuer's judgment is that the property has a nil value (see e.g. Morecambe, Hingley and Hoare (VO) *v* National Trust *[1998] R.A. 391; [1997] 2 E.G.L.R 229 (CA),). The property may have been constructed many years ago and there may be no reason to think that any party bidding for the hypothetical letting would be prepared to construct the building or any equivalent now.*

8.6.1 Reasons for rejection

Although *Garton* v *Hunter (VO)* [1969] 2 QB 37 established the principle that all evidence is admissible, the issue to be decided in any valuation is the weight that should be given to that evidence. Analysis of legal precedent would indicate that:

- The contractor's basis was often rejected as insufficient information was provided in order to test the valuation approach properly; for example, there was no detailed breakdown of costings.
- There was too much scope for interpretation in the figures that went to make up the valuation.
- The method was often rejected in favour of more 'direct' evidence, either of rents or the analysis of comparable assessments which could more reliably form the basis of the assessment, that is, following the principles laid down in *Lotus & Delta Ltd* v *Culverwell (VO) and Leicester City Council* (1976) 239 EG 287.
- The method may not arrive at the appropriate value in the situation where it can be shown that cost does not equate to annual value.

8.7 The six stages in detail

We will now look at each of the stages of a contractor's basis valuation in more detail.

8.7.1 Stage 1: Estimation of construction cost

The first stage of a contractor's basis of valuation is to ascertain the cost of constructing or replacing the hereditament. The cost to be ascertained is the total cost which must include the cost, not just of the buildings, but of all the ancillary accommodation, car parks, fencing, external lighting, plant and machinery, site works, services and other associated features of the property. It will include the costs associated with the design of the building, all fees associated with the construction and any other incidental cost.

Two alternative approaches to the estimation of cost can be adopted depending on the circumstances of each case. As originally developed, the method involved costing an identical replacement building though utilising modern materials and techniques where old materials and techniques were either no longer available or practised. This method is termed the 'replacement building approach'.

While this approach may be suitable for many types of building, it became apparent that many, usually older buildings if they were to be rebuilt, would not be rebuilt as originally constructed, but rather a modern substitute would be built which would be more suitable for modern needs. This led to what is termed the 'substituted building approach'.

8.7.1.1 Replacement building approach

The replacement building approach follows from the long accepted principle of *rebus sic stantibus*, that it is the actual building that is being valued. The approach assumes that a precisely similar building is being constructed. However, the valuer is permitted to envisage modern construction techniques being adopted in constructing the replacement building. For example, intricate stonework may be factory machined rather than being handcrafted by a mason.

8.7.1.2 Substituted building approach

It should be noted that the substituted building approach was originally considered to be a different and separate method of valuation to that of the contractor's basis (*Oxford University* v *Mayor and City of Oxford* No 1 [1902] Ryde and Konstam's Rating Appeals 87). It was only in later years that it was accepted as an alternative approach in stage 1:

> *A method by which the antagonistic opinions as to the value of these buildings were brought into some relation with actualities, was adopted by Mr Ryan for the assessment committee, and with slight variation by Mr Eve for the University. This basis I may call for brevity the 'substituted building' basis. The method pursued by Mr Ryan was to measure up an existing building whose assessment was in debate, to ascertain the purpose for which it was used, and then to estimate the cost of erecting an imaginary*

building in its place suitable for that purpose and worthy in his view of the dignity of a university, but curtailed of unnecessary spaces and of orna- ment referable rather to the memorial than to the practical character of the fabric. Now Mr Eve for the University adopted a substantially similar method, but the buildings which the University would in his view erect if compelled to abandon those now existing were buildings plainer in char- acter than those estimated for by Mr Ryan, though fitted for the purpose for which they were to be used. I adopt this method as one which appears capable of leading to a just and reasonable conclusion as to the rental which it is my duty to ascertain.

The substituted building approach envisages that the original building would not be rebuilt in exactly the same form, but rather a modern substitute would replace it. That substitute would design out much of any excessive size, height, embellishment and so on that is no longer required.

It is usual to exclude from the cost the replacement of architectural embel- lishment or excessive ornamentation which adds greatly to the cost but not sig- nificantly to the value, it being assumed that the hypothetical tenant in carrying out the valuation exercise would exclude such costs. In *Magdalen Jesus and Keble Colleges, Oxford* v *Howard (VO) and Oxford Corporation* (1959) 5 RRC 122, the substituted building was described as imaginary, 'but suitable for its purpose and worthy of the dignity of a university, but curtailed of unnecessary spaces and of ornament referable rather to the memorial than to the practical character of the fabric'.

The main problem of this approach lies in deciding what the 'substitute' that has to be valued is. Different approaches have been taken in different cases. The Rating Forum suggests that the building should be of the same dimensions as the one to be valued. Adopting any different size would be a further departure from reality than is necessary. However, other arguments and approaches are also possible.

8.7.1.3 Definition and ascertainment of substituted building

Where a substituted building approach is adopted, the valuer is faced with a number of problems in deciding what is envisaged in the substituted building. The following issues need to be considered:

- What is the substituted building envisaged – how does it differ from the existing structure?
- How has the substitute been arrived at – how can the valuer justify the substitute?
- What is its specification, size and so on?

Clearly, the choice of substituted building can have a considerable effect on the resultant value.

Generally, the substituted building should be of similar floor area to that of the building being considered, but it may be appropriate to estimate the cost of a substitute building that is not necessarily the same size and shape but has the same functional capacity as the actual building.

This simple substitute building approach appears to be derived from *Shrewsbury School (Governors)* v *Shrewsbury Borough Council and Hudd (VO)* [1966] RA 439 where the Lands Tribunal criticised the valuation officer's use of modern building costs derived from local authority schools applied to the actual building's gross internal areas (GIA).

The Tribunal considered that modern buildings on the existing GIA would allow space for twice as many pupils as were accommodated in the actual school buildings. This was presumably due to modern internal layout and was certainly evident by the occupation levels of local authority schools compared to the public school. The Tribunal was unable to accept that Shrewsbury would not use the substituted modern buildings, as put forward by the valuation officer, to their full capacity.

The Tribunal mentioned the case of *Magdalen Jesus and Keble Colleges, Oxford* v *Howard (VO) and Oxford Corporation* (1959) 5 RRC 122, where the substituted building used was one *contracted* in height and width but giving the same accommodation and as such could be 'readily appreciated' or understood. The valuation officer's approach was criticised in the *Shrewsbury* case because the apparent difference in capacity between the actual and substituted buildings did not make the substituted buildings 'easy to visualise'. The Tribunal appeared to be suggesting that functional capacity should also be taken into account so that the substituted buildings could readily be compared to (or visualised as similar to) the actual buildings.

A tenant arguing for a lower rent on the basis of the cost of constructing its own alternative would certainly not envisage building more than was required and would use both modern materials and a modern and efficient layout. This, though, should not be taken too far, as the Lands Tribunal for Scotland found in *Shell UK Exploration and Production Ltd* v *Grampian Assessor* [2000] RA 295 concerning a gas terminal where it criticised the ratepayers for costing an imaginary building designed merely to achieve low costs and not a practical substitute for the actual building. It is the actual building that is being valued. Similarly, the Upper Tribunal, in *British Car Auctions (T/A Blackbushe Airport Ltd)* v *Hazell (VO)* [2014] UKUT 0164 (LC); [2015] RA 108 considering an airport, was not happy with the ratepayer's modern substitute which reduced the size of the runway as it regarded it as important that the modern substitute should reflect the use of the actual property and should be able to do the same basic job. It was not an opportunity to adopt a 'new business model'. In this case the nature of the substituted hereditament and in particular the dimension of the runway were in dispute:

At the first stage of the contractor's test, when the parties are contemplating the construction of a modern substitute, it is important that the modern substitute chosen must reflect the use of the actual hereditament that has to be valued. The modern substitute chosen should be able to do the same basic job that the actual hereditament does. The choice of a modern substitute is not the opportunity to adopt a new business model … in our judgment the exercise must always start by identifying a modern substitute that stays as close to the use of the actual hereditament as it can. If such a modern substitute can be accurately identified then, in principle, the application of the test should be based upon it even if, for sound reasons, substantial end allowances become necessary.

In *Semlogistics Milford Haven Ltd* v *Webb (VO)* [2018] UKUT 0019 (LC) the Upper Tribunal considered in deciding on the assessment of an oil storage depot, originally built as an oil refinery, that there were exceptional circumstances fully justifying consideration of a modern equivalent particularly as the actual hereditament had been designed as a refinery. A tenant could argue with some force that if it was building an oil terminal it would have a substantially different site area and layout. It was satisfied the ratepayers' design provided a sound basis not least because the VO's expert had described the design as very reasonable and 'eminently sensible.'

8.7.1.4 Choice of approach

The method to be used in any particular case depends on the circumstances, particularly the age of the buildings. In the *Imperial College of Science and Technology* case, the parties agreed estimated replacement costs of the University's buildings as the:

Cost of construction including fees and site works of a modern building in substitution for each building comprising the hereditament on a site having no features which would significantly affect construction costs, construction being completed at April 1 1973. The said substituted buildings would:

(a) Have the same GIA as the buildings comprising the hereditament; and
(b) be of such standard, finish and specification and have such services as would be required to enable the uses carried on in the existing building at the dates of the proposals to be carried on satisfactorily in all respects.

This approach appears to be essentially the substituted building method.

The amount of any deductions under stage 2 will vary depending on the approach to building cost used.

In general terms, the replacement building approach is adopted for modern buildings and for those hereditaments where the design, nature and character of the hereditament, however outdated, is such that it is part of the essential characteristics and desirability of the hereditament. Many buildings are designed in such a way as to convey their importance, whether this is by the size of the structure, the materials, the design or the ornamentation, and it would simply not be acceptable if a modern and simplistic design was adopted. For example, the Houses of Parliament would not have the same element of authority if situated in a modern multi-storey office building, though they might satisfy the same function. The same would apply to many older universities, town halls and other public buildings.

Consequently, the replacement building approach should normally be adopted for those hereditaments that would be replaced virtually unchanged from their current form, other than the use of modern materials.

8.7.1.5 *Ascertainment of cost*

Having decided to value the building on the replacement or the substituted building approach, the next step is to calculate the cost of construction. Where such a calculation is being based on actual construction costs then, in order to undertake the costing, two matters need to be determined:

- the date at which the cost is to be ascertained and
- the type and size of contract to be employed.

8.7.1.6 *The date at which the cost is to be ascertained*

Prior to 1990 and the change in legislation brought about by the Local Government Finance Act 1988, the date of valuation was the date at which a rating list came into force. This was confirmed by the House of Lords in *K Shoe Shops Ltd* v *Hardy (VO) and Westminster City Council* [1983] 1 WLR 1273.

The Local Government Finance Act 1988 introduced the concept of the antecedent valuation date (AVD), and it was generally accepted that this fixed not only the valuation date for a rating list but also the date at which costs were to be ascertained for the contractor's basis of valuation. However, some valuers argue that the date at which costs should be ascertained is not the AVD but some time prior to that date. This being on the basis that construction would need to commence some time prior to the AVD for the property to be completed by that date. Typically, it is argued that a construction period of two years should be adopted. Support for such an assumption can be found in the Royal Institution of Chartered Surveyors' (RICS) *Appraisal and Valuation Manual* (*'The Red Book'*) when undertaking valuations on a depreciated replacement cost (DRC) approach.

This argument has not generally been tested before the Lands Tribunal, but in recent cases ratepayers have come to accept that the correct date for ascertainment of costs is indeed the AVD on the basis that what is really being valued is the actual hereditament at the AVD and not what it might have cost had construction started two years before. This does not, however, rule out a challenge to that assumption sometime in the future.

8.7.1.7 *Adjustment of the contract price to date of valuation*

Where an actual contract price is available, it may need to be adjusted due to the following factors:

- The need to adjust the contract price to the date of valuation. It is usual for the contract price to be assumed to reflect costs at the mid-point of the construction period (or tender date to completion) and to adjust it on this basis.
- Where actual costs are available but need to be adjusted to take into account the date of the costs compared to the AVD, it may be necessary to use one of the published indices. Care needs to be taken in deciding which index to adopt. In *Hailbury Investments Ltd* v *Westminster City Council* [1986] RA 187, the Tribunal was critical of the expert evidence because the reason for the choice of the index and its basis could not be explained. In *Willacre Ltd* v *Bond (VO)* [1986] 1 EGLR 224, the Lands Tribunal warned that fine distinctions between the use of three different building cost indices could lead to an 'illusory sense of accuracy'.
- Changed market conditions as at the AVD.
- Changes in physical factors between the contract date and the AVD or date of valuation. The hereditament may have changed from that which was originally constructed.
- The incurring of expenditure on non-rateable plant and machinery. In simple terms, the cost of construction may include an element for items which should not form part of the rating assessment. For example, the construction cost of a factory may include the cost of buying and installing some of the non-rateable production machinery.
- The incurring of unremunerative expenditure. For example, in *Coppin (VO)* v *East Midlands Airport Joint Committee* (1970) 221 EG 794, the ratepayer had incurred additional expenditure by building a runway that was larger than required by the types of aircraft then using the airport. Until such time as larger aircraft were introduced, the ratepayers could obtain no benefit from that expenditure. They had made the decision to incur the expense in advance of requirements in order to minimise any disruption to services in future years. In this case an allowance was made for this factor.
- The incurring of expenditure outside the hereditament. In some instances the cost may cover the cost of constructing more than one hereditament and may need to be apportioned between two or more hereditaments.

8.7.1.8 *What should be included in 'costs'?*

Stage 1 costs should include all those items necessary to build the buildings and their associated external environment. This will include the following:

- cost of constructing the buildings;
- cost of providing services to the site;

- cost of onsite facilities (car parking, landscaping, fencing and so on);
- planning and building regulation application costs;
- contingencies;
- fees.

Abnormal costs, such as special foundations, should be excluded because they would not increase the premises' rental value compared to premises which did not need such foundations. However, care should be taken not to 'double count' by using a land value that reflects subsoil difficulties.

8.7.1.9 Grant and financial aid

The valuation problems associated with grant aid and other forms of financial assistance have increased in recent years, with changes in grant structures and new forms of funding becoming available. This has meant that many more properties are being developed with some form of financial assistance, whether by way of a grant or by some other form of financial assistance, such as lottery funding. Many of these projects would not otherwise have been viable. The issue of how, if at all, that aid should be taken into account in the rating hypothesis has been a matter of concern for many years.

In *Willacre Ltd* v *Bond (VO)* [1986] 1 EGLR 224, the Lands Tribunal discussed the effect of a Sports Council grant of £90,000 on a contractor's basis valuation of an indoor tennis centre. The valuation officer regarded the payment as irrelevant and would 'form no part of the hypothetical tenancy negotiations'. The ratepayers considered that as the grant was an actual amount paid it should be taken into account. The Lands Tribunal said:

> *I have come to the conclusion that … the figure of £90,000 should be deducted from the agreed construction costs. The valuation for rating purposes is a theoretical exercise which is deemed to have taken place prior to the negotiations between a hypothetical landlord and a hypothetical tenant. In this case the actual occupier has received a grant and I consider it most probable that the hypothetical tenant in his place could reasonably expect such a grant and would have it in mind when considering what rent he could afford to pay. Thus the amount of the grant is, in my opinion, to be taken into account in estimating the deemed actual capital cost.*

The Rating Forum originally suggested that the most appropriate place to consider this matter was at stage 5. However, in the Bisham Abbey case, *Allen (VO)* v *English Sports Council/Sports Council Trust Company* [2009] UKUT (LC) 187, the Lands Tribunal said:

> *In its decision the VT said that a hereditament could be built, because of the availability of grant aid, to a level, standard or size of facilities for which there was little or no demand; so that a contractor's basis valuation would produce*

a value that was more than the market could or would bear. It was on the basis of this reasoning that it made a stage 5 reduction.

We do not think that this reasoning is correct. It is not clear to what particular market the VT was referring. It was not suggested that there was any hypothetical tenant other than SE or its land-owning component, SCTC, or any potential hereditament for it to occupy other than the subject hereditament. As in the case of many hereditaments valued on the contractor's basis, there is no market in the general sense. That is the reason this method of valuation is used ... The contractor's basis seeks to determine the annual value of the occupier's occupation of the hereditament by reference to the effective capital value of the buildings upon it, reaching by this means the rent that the tenant and the landlord would agree upon in the hypothetical transaction. Reference to the rent that the market (as something distinct from the occupier of this hereditament) would bear as the basis for making a stage 5 deduction appears to us erroneous. There can be no justification for reducing the value so obtained on the basis that it is more than the market could or would bear, because there is no market in the general sense.

Of course, if the buildings have been built to a quality or size that go beyond what the tenant would see fit to pay for, that should be reflected at stage 1 of the valuation in determining what the substitute buildings would consist of. It is agreed ... that a deduction should be made for the additional cost of design elements that did not add value to the hereditament. But apart from this the facts are emphatically against any suggestion that the newly constructed buildings are excessive. They were built to the requirements of SE, and the grant was awarded under statutory procedures designed to ensure that the works proposed justify the award of the grant.

It rejected the ratepayer's argument that a 94% grant should be deducted from the construction cost of a sports facility on three main grounds:

* Rating was a tax on the value of the occupier's occupation. The House of Lords decision in *Metropolitan Water Board* v *Chertsey Assessment Committee* [1916] 1 AC 337 had rejected the idea that a gift or present from a 'generous benefactor' reduced the value of an occupation. A government grant was no different and did not reduce the value of the occupation to the actual occupier.
* It was 'a necessary assumption that the benefaction is to be made available on an annual basis, for example, by deed of covenant, in which case it is quite clearly available to pay a rent', following *Leicester City Council* v *Nuffield Nursing Homes Trust* [1979] RA 299; 252 EG 385; [1980] JPL 120.
* Matters bearing on the amount that the tenant would be prepared to pay are, following *Williams (VO)* v *Cardiff City Council* (1973) 226 EG 613 subsumed in the prescribed decapitalisation rate and cannot form the basis for any adjustment of the rateable value.

However, more recently, in *Hughes (VO)* v *Exeter City Council* [2020] UKUT 0007 (LC), the Upper Tribunal reconsidered its earlier decision in the 'Bisham Abbey' case, and wholly rejected the then President's view that most matters that might have been considered at stage 5 were wrapped up in the statutory decapitalisation rate and very much reaffirmed the use and importance of stage 5 in enabling valuers to properly value. It preferred the earlier analysis in the 'Eastbourne' case. 'We entirely agree with the analysis in Eastbourne of the relationship between stages 4 and 5 following the introduction of prescribed decapitalisation rates. We are unable to agree with the contrary view taken by the Tribunal in (Bisham Abbey).' And 'We consider that from now on the Tribunal's decision in *Eastbourne Borough Council* v *Allen*, and not its decision in *Allen* v *English Sports Council*, should be treated as stating the law on this issue accurately, unless and until a superior court should decide otherwise.'

In Eastbourne it was explained the figure arrived at by the end of stage 4 of the Contractor's Basis is generally a 'ceiling value', and in negotiations with the landlord the tenant would generally not pay more than this amount. At stage 5 such deductions may be made as appear to the valuer to be justified in order to arrive at the appropriate rateable value. It is at stage 5 that the valuer must use his or her valuation judgment to reflect all the matters relevant to the hypothetical letting which have not so far been taken into account in the method. For example, a valuer or the Tribunal might be satisfied that a substantial reduction should be made at stage 5 because the hypothetical tenant genuinely could not afford, or justifiably would not be prepared to pay, the figure resulting from stage 4.

Unfortunately, of course, the prescription of the decapitalisation rates somewhat restricts valuers arriving at true values (if it is considered the statutory decapitalisation rate would not otherwise be the chosen rate to achieve a proper valuation), though, of course, somewhat paradoxically, the statutory decapitalisation rate is deemed 'correct' by the legislation and therefore a valuation arrived at using it cannot be 'wrong' simply because of the rate.

As can be seen, the Bisham Abbey and Exeter Museum cases provide two differing views on what is encompassed within the statutory decapitalisation rates and what is not covered by the rates. This question may need a decision by a superior court to settle which of the two views is correct.

The Upper Tribunal in the Exeter case emphasised again the importance of standing back and looking at whether the result of using the contractor's method properly represented rental value in the market.

8.7.2 *Stage 2: Estimation of adjusted replacement cost*

Stage 1 estimates the cost of the provision of a new building, but the property to be valued may not be new and some adjustment may be required to reflect its deficiencies. Stage 2 seeks to adjust cost to reflect the differences in the characteristics of the 'new' costed building compared with the actual hereditament that is the property really being valued. Often this stage is referred to as 'adjustments for age and obsolescence', but such a term does not fully reflect the true nature or purpose of the adjustment.

For a long time it has been accepted that just because a certain amount of cost has been incurred it does not necessarily mean that such expenditure is reflected in the value of the property. Lord Denman in the *Mile End Old Town* case commented: 'The outlay of capital may furnish no such criterion, since it may have been injudiciously expended, and what was costly may have become worthless with subsequent change.'

In *God's Port Housing Association* v *Davey (VO)* (1958) 51 R&IT 651, the actual cost of providing the hereditament was disregarded as it was found to be in excess of that for which the hypothetical tenant could have provided a similar hereditament.

The Wood Committee also stated:

Moreover, there may be some problems encountered in seeking to apply the contractor's basis to industries where, for historic reasons, capital investment may not have been related to the returns available. As a result, valuations based on the contractor's basis may be disproportionately high when compared to the industries' turnover and profits.

The receipts and expenditure method is also more likely to reflect the impact of the regulatory regime under which most industries operate and changes in competition.

Consequently, at this stage great care is required in order to arrive at the appropriate rateable value. However, in general terms, stage 2 allowances tend to be those related to some form of obsolescence which may take a number of different forms, as follows:

- 'Physical obsolescence' relates to the deterioration of the buildings or other parts of the property through the wear and tear of the components. Although age is not in itself justification for an allowance, the tenant will reflect the prospect of maintenance costs due to deterioration over time by a reduction in the rental bid.
- 'Functional obsolescence' covers the problems which may be present in the design of the property which could be deficient by comparison with current requirements, e.g. excessive ceiling heights; inappropriate layout; inadequate load bearing of floors; inferior heating and ventilation, etc.
- 'Technical obsolescence', which may be regarded as an extension of physical obsolescence, arises where current technology has so changed that the actual plant to be valued or the building housing the equipment has become significantly redundant or economically outmoded.
- Obsolescence will also apply equally to the plant and machinery, site works and all other constituent parts of the hereditament and not solely to the actual buildings. In certain instances it may also apply to the land as well as the buildings.

Where the substituted building approach has been adopted at stage 1 and allowances have already been made in the substitute for excess accommodation and so on, it is inappropriate to make further allowance for these factors in stage 2, otherwise there would be double counting.

While stage 2 is often considered to include 'age and obsolescence', it can be inappropriate to suggest that a building is to some extent obsolete merely because it is old. In many instances older buildings are just as suitable for their purpose as newer ones or even better. For example, schools built in the 1960s tend to be built to a higher standard than those built in the 1970s and are better suited to current requirements. Each building must be considered on its own merits.

8.7.2.1 *Means of adjustment*

This is one of the most subjective and difficult parts of the whole valuation process and the amount of adjustment is often down to the experience of the valuer. For some types of property, scales may have been agreed between the VOA and representatives of particular industries. This will then provide a framework by which the adjustments can be made, but it does not prevent disputes arising, as such scales cannot cover every eventuality – each property must be valued on its own merits.

Where scales are not available then some attempt to compare the efficiency of the building to a modern building may be a useful starting point.

In the case of *Sheerness Steel Co. Plc* v *Maudling (VO)* [1986] RA 45, the difficulty of quantifying the adjustment for a steel works was highlighted. The Tribunal commented:

> *The degree to which excess capacity and obsolescence affects rental value of the hereditament is not easy to measure. It is helpful to compare the capital cost of a substitute hereditament of the same size, also the running costs, if possible distinguishing between savings due to obsolescence or reduced capacity, but essentially identifying those savings affected by rateable and non rateable plant. In the alternative, it would also be helpful to know the measure of savings that could be affected at the existing hereditament by modernisation of non rateable plant without alteration of the hereditament.*

In that case an allowance of 35% was given for excess capacity of the buildings, plant and land and an allowance of 15% for obsolescence.

A number of cases have considered the question of excess capacity and under-utilisation. In *Semlogistics Milford Haven Ltd* v *Webb (VO)* [2018] the Upper Tribunal was content that it had been established in cases that, as a matter of principle, a reduction could properly be made when using the Contractor's basis to reflect under-utilisation. In Civil Aviation Authority v Assessor for Strathclyde a 70% allowance to ECV was made to reflect the limited use which could be made of the airport. Again, in another Scottish case, *UKAEA* v *Assessor for Highland and Western Isles* [2006] RA 153, the Lands Tribunal for Scotland said: '*In principle over-capacity on a long-term basis should be reflected in an appropriate allowance.*'

There is no difficulty, of course, where the hereditament is simply too large and a smaller substitute can be envisaged, but there is some difficulty with the concept when, on occasion, the full capacity is needed. Nonetheless, in the Semlogistics case the Upper Tribunal said:

> *There is no rule of law or valuation principle to prohibit from being considered as a topic of possible relevance, when assessing effective capital value, the question of whether the capacity of the hereditament is excessive having regard to prospective usefulness of that capacity assessed at the AVD, and taking into account all relevant matters including the state of the market and the future prospects for business at the hereditament.*

The Tribunal saw the excess capacity as a bargaining counter in the hypothetical negotiations.

8.7.3 Stage 3: The value of the land

At stage 3 of the valuation an amount has to be added to the cost for the value of the land comprising the hereditament. This should be the value of the land for the actual 'mode or category of use', that is, *rebus sic stantibus*. Thus, any development potential in the site for other uses must be ignored.

As Sir William Fitz-Gerald commented in the decision in *Dawkins (VO)* v *Royal Leamington Spa BC and Warwickshire CC* [1961] RVR 291: 'To assess land as a school as if it were land used for some other purpose would be a violation of the principle of *rebus sic stantibus.*'

The site is assumed to be cleared of buildings and available for development for the particular purpose. To the cost of the land must be added the cost of providing all site works and so on which have not been taken into account in the cost of construction. Traditionally such costs will be taken into account at stage 1.

The land value in a contractor's basis valuation is of considerable importance. Given that building costs do not vary greatly across the country, and as traditionally nationwide decapitalisation rates have been adopted, it is this element which gives identical hereditaments (say) in London and Liverpool different values.

While *rebus sic stantibus* clearly indicates that the land has to be valued for its existing use, this is not of great assistance when there are no sales of land for the particular existing use or the actual site has not been recently acquired when the purchase price may form good evidence. However, the following two cases do provide some assistance.

In *Dawkins (VO)* v *Royal Leamington Spa BC and Warwickshire CC* [1961] RVR 291, the valuation officer argued that the site value should be 'the prevailing use value of land in the locality in which the school is situated', i.e. in the particular case residential land. In cross-examination it was put to him that this was contrary to *rebus sic stantibus* and the land should be valued for school use and nothing else.

> *He replied to the contrary. Envisage, he said, a piece of land subject to a covenant restricting its use to the site of a school. Assume that both vendor and purchaser recognise that the prevailing use of similar land is residential, and they both know the price which the land in question will fetch for that use. The vendor would be unwilling to sell below the residential price and would hold on rather than do so. The purchaser would be unwilling to pay a high residential price. In the outcome, they would strike a bargain and the purchase price would be the value of the land for residential purposes slightly deflated. That, he said, was the basis he had adopted.*

The Tribunal accepted the ratepayer's argument that the land must be valued for use as a school because of the principle of *rebus sic stantibus* but accepted the valuation officer's approach that 'the open market price of land for use as a school would be (in this case) near to its value for residential purposes'.

In *Downing College etc, Cambridge* v *Cambridge CC and Allsop (VO)* (1968) 208 EG 805, the Lands Tribunal considered that the site of King's College, which is on the opposite side of the road to shops and offices, 'should be valued in the light of residential values'. The Tribunal accepted that the college did not gain by its potentially valuable commercial frontage and if having to build afresh would use the cheaper residential land behind such frontages.

Thus, in using adjacent land values slightly depressed as suggested in *Dawkins*, it is important to consider whether the property is situated in that location because it actually needs to be there or is only there for historical reasons and could (or nearly could) be situated as advantageously in a lower land value area.

It is probably a reasonable first presumption for recently purchased sites that the cost and location of the site represents value.

Land which is surplus, under-utilised or retained for future expansion must also be valued *rebus sic stantibus*. If the site is largely unused it may have some amenity value or it may be used for car parking and valued as such. Its potential cannot, because of *rebus sic stantibus*, be considered in assessing the rateable value.

Adjustments may be necessary to the land value in the same way as the buildings. Where the buildings have deductions from their cost (stage 2), for example for obsolescence, bad planning, etc., it may be appropriate to apply the same allowance to the land in order to value it *rebus sic stantibus* encumbered by the actual buildings. However, as stated by the Lands Tribunal in *Imperial College of Science and Technology* v *Ebdon (VO) and Westminster City Council*

[1986] RA 233, the amount of the deduction for the land does not necessarily follow from the buildings and will depend on the facts in each case.

A poor layout of buildings may cover a greater area of land than an efficiently designed layout plan would. In this circumstance a reasonable figure for the land may be obtained by applying a value only to the smaller area of land.

8.7.4 Stage 4: Decapitalisation

8.7.4.1 The choice of decapitalisation rate

The choice of a suitable decapitalisation rate has always been a most important matter as a mere 1% difference in opinion could have a 20% or more effect on value. The approach to deciding on a rate has been the subject of much discussion and debate over the years. In *Imperial College of Science and Technology v Ebdon (VO) and Westminster City Council* [1986] RA 233, expert evidence was given on behalf of the various parties as to the appropriate rate for the local 1973 list for a university. The case resulted in criticism from various quarters that it had taken some 11 years for the question of what should be the decapitalisation rate for universities in the 1973 lists to be decided. To remove the prospect of potentially long periods of uncertainty for those classes of hereditament usually valued on the contractor's basis, the Government decided to designate an appropriate rate by statutory instrument.

With effect from 1 April 1990, the decapitalisation rate has been prescribed by statutory instrument, and as a result the valuer has no choice in what rate to apply. While the prescription of the rate has many advantages, not least the reduction in the amount of litigation going to the courts on whether the 'correct' rate has been adopted, it also has many disadvantages. The valuer's freedom to arrive at the required rateable value is restricted where part of the valuation process is prescribed by law.

The Non-Domestic Rating (Miscellaneous Provisions) (No. 2) Regulations 1989 (Amendment) (England) Regulations 2016 No, 777 and The Non-Domestic Rating (Miscellaneous Provisions) (No. 2) (Amendment) (Wales) Regulations 2021 No. 1141 (W. 278) prescribe the decapitalisation rates to be adopted for the purposes of the 2023 rating lists. In England the rates are 4.4% for all types of property other than health, education and defence properties where 2.6% is adopted which is the same as applied for the 2017 revaluation, but in Wales the rates are 3.4% and 1.9%, respectively. In addition, public conveniences in Wales are also subject to the lower rate.

The statutory rates apply where:

2(1F) Paragraph (2F) of this regulation applies in relation to a hereditament shown in a non-domestic rating list compiled for a billing authority in England on or after 1 April 2017 the rateable value of which is being ascertained using the contractor's basis of valuation.

Regulation (2F) talks about the 'appropriate rate' and gives the percentages mentioned above and then the regulations define the 'appropriate rate' as follows:

2(2F)(3) 'the appropriate rate' means the percentage rate applicable in relation to the notional cost of constructing or providing the hereditament or any part of it for the purpose of estimating the rent at which it might reasonably be expected to let from year to year;

Interestingly, the statutory rates apply where the contractor's basis is not only being used to ascertain the rateable value of the hereditament but also any part of it. In *Bunyan (VO) v Acenden Ltd* [2023] UKUT 17 (LC) the Upper Tribunal said this regulation only applied where the contractor's basis was being used and not where ratepayer's expenditure on fitting out an office building was being amortised to an annual figure as this was not using the contractor's basis for part of the hereditament. It saw this as a 'similar exercise, but not the same exercise'. It gave an example of an engine test bed at an aero engineering complex which might be valued on the contractor's basis, whereas the rest of the hereditament comprising offices and factory would be valued on a rentals comparison method. The engine test bed could become a separate hereditament if separately occupied so that was a good example of part of a hereditament being valued on the contractor's basis.

The statutory decapitalisation rates have been strongly criticised as reducing the contractor's method, at least partly, to a formula and potentially resulting in some hereditaments being valued contrary to the statutory basis of the rent which might reasonably be expected. It will be appreciated that only if the statutory decapitalisation rate is in accordance with whatever would be the 'true' rate will the resulting figure equate to actual rental value.

It has been suggested that where the statutory decapitalisation rate produces (because it is the 'wrong' rate) an incorrect answer, this can be rectified at stage 5 – the 'Stand back and look' stage. This would appear to be an erroneous view because if the valuer seeks at stage 5 to alter the decapitalisation rate to say 3% or 7% on the grounds that the statutory rate produces the wrong answer, the effect of the Regulation will be to immediately restore the statutory rate.

Despite the prescription of the statutory decapitalisation rate, the theory behind the rate is still the same post-1990 as it was pre-1990.

The theory behind the rate of decapitalisation was not fixed but rather was continually evolving and developing in order to arrive at a rent based on the statutory definition of rateable value.

The classic explanation given in *Dawkins (VO) v Royal Leamington Spa BC and Warwickshire CC* [1961] RVR 291, quoted at the beginning of this chapter, could also be called the 'forgone interest' approach as it envisages the hypothetical tenant:

can borrow money, on which he would have to pay interest; or he could use his own capital on which he would have to forgo interest and put up a similar building for his own occupation rather than pay rent for it, and will do that rather than pay what he regards as an excessive rent.

This explanation needed further modification as per Lord Denning in *Williams:*

To that statement, however, I would make this qualification. The annual rent must not be fixed so as to be only 'marginally below' the interest charge. It must be fixed much below it, and for this reason: by paying the interest charge on capital cost, he gets not only the use of the buildings for life, but he also gets title to it, together with any appreciation in its value due to inflation: whereas, by paying the annual rent he only gets the use of the buildings from year to year – without any title to it whatsoever – and without any benefit from inflation.

In *Oxford University* v *Mayor and City of Oxford* No. 1 [1902] Ryde and Konstam's Appeal 87, the Recorder stated:

I am of opinion that 3% is under all the circumstances of the present case the proper rate of interest. The University was proved to be able to raise money at that rate, and that being so, I can conceive no adequate cause which should induce such a body to charge themselves with 5%.

Here, back at the turn of the century, the rate adopted was the rate at which the university could borrow money, i.e. 3%. The university, through all its connections and so on, had the ability to borrow money at favourable rates.

More recently the rate at which money can be borrowed is only one of the factors taken into account.

In *Westminster City Council* v *American School in London* [1980] 2 EGLR 168, it was found that:

In the passage quoted from the Solicitor General's opening speech in the Leamington case, there is a reference to the possibility of the hypothetical tenant using his own capital, 'on which he would have to forgo interest to put up a similar building'. None of the valuers in the present case obtained any help from looking at the matter in this way. Indeed it would seem that the interest which the hypothetical tenant would seem likely to forgo must depend very largely on his personal circumstances, and inevitably provide a less reliable means of arriving at an annual value for rating purposes.

Over the years the base rate of interest on which to found a decapitalisation rate has changed. It is suggested that at the present time regard must be had first to base rate, i.e. the rate from which all other rates of interest derive. However, this rate may vary on a weekly basis, or even more frequently in times of economic crisis. Consequently, the rate must be looked at over a period of time. A tenant taking an annual tenancy does have a prospect of continuance and will take a longer-term view of interest rates. How long a view should be taken will depend on the type of property and the nature of the occupation.

In recent times it has been argued that the real rate of interest should be looked at rather than base rate. The real rate of interest is taken to mean the true rate of interest obtained after excluding the effects of inflation. Such an approach was adopted in *Westminster City Council* v *American School in London* [1980] 2 EGLR 168 and has further been adopted in recent Scottish cases.

How one arrives at the real rate of interest is a matter of debate but will involve the deduction of inflation from the base rate.

It may be appropriate, where a long-term view of interest rates is adopted, to give some form of weighting to the figures to allow for the time period.

Having arrived at the 'starting rate', a number of adjustments needed to be carried out:

- First, the borrower's premium. Borrowers do not simply pay base rate or the real rate, but an amount over and above this figure dependent upon their status, security and the purpose for which the money is required. Generally a borrower's premium of between 1% and 5% is common.
- Second, an adjustment must be made for the 'Cardiff' factors mentioned above, i.e. to reflect the fact that the tenant is just a tenant, and does not get the benefits associated with the ownership of the property, such as security, increase in capital value and so on.
- Finally, consideration should be given to the commercial nature of the use of the property. Where a property is used for commercial purposes no reduction in the rate obtained is appropriate.

However, if the property is used for non-commercial purposes, then following Lord Denning in *Williams*, the landlord would consider:

> *First of all, the hypothetical landlord would be anxious not to leave the premises empty, bringing in no return on his capital. He would be content to accept much less than 6% which the landlord of commercial premises would expect. He would be providing premises for the public good, namely, for educational purposes, and would be disposed to accept such a rent as the hypothetical tenant could afford to pay without crippling him or his activities, especially if he was a non-profit making body. But on the other hand, the hypothetical tenant might be very anxious to get possession of the premises. If he was under a statutory compulsion to establish such a college, he would feel that he must have them; not perhaps at any cost, because if the rent were too high he would look elsewhere, but at such a reasonable rent as he could persuade the hypothetical landlord to accept. If he felt a moral compulsion to establish such a college out of his general public duty, he would probably pay as much as if he were under a statutory duty. In other circumstances, he might wish to have the premises*

as being desirable but not as being compulsory or as a matter of urgency. He would probably then be inclined to hold back and pay less than a tenant who felt a pressing need. In all cases he would of course have to consider the means at his disposal. If he was a government establishment, such as a local authority, supported entirely by government grants and rates, he would count the cost, but not so anxiously as a non-governmental establishment under a pressing necessity to make both ends meet.

Thus, a number of additional matters needed to be taken into account in arriving at the decapitalisation rate prior to the introduction of statutory prescription:

- The amount of competition for the property. Where there were a number of possible tenants for the property, all possible tenants were to be considered as being in the market and in competition with each other to secure the tenancy. Consequently, the possibility of obtaining a reduced rate would be lessened either wholly or partly.
- The nature of the occupier. Where a property was occupied by a charity or a similar type of organisation with limited funds available, the ability to pay and to make ends meet was to be considered.
- The reason for occupation. The reason for the occupation of the property, whether for a statutory, social or other purpose, was to be taken into consideration in deciding whether a lower rate was appropriate.

8.7.5 Stage 5: Adjusting the rental value – 'stand back and look'

This stage is effectively divided into two main parts. The first part is concerned with the consideration of reflecting any factors that would affect the rental value of the property (as opposed to the capital value/replacement cost) and the second part is to 'stand back and look' at the result and try to evaluate whether it fairly represents the rental value of the property on statutory terms.

The Rating Forum suggests that the effect of grants, if any, is more properly to be considered at this stage rather than at stage 1. For certain cultural and recreational types of property which have been grant funded, it is recognised that some allowance should be given at this stage to have regard to the fact that the actual property was constructed with the aid of grants. The amount of the allowance will vary depending on the amount of grant in relation to the cost of construction but can vary from 0% up to 50% in the case where nearly the whole of the construction was grant funded.

In *Allen (VO)* v *English Sports Council/Sports Council Trust Company* [2009] UKUT (LC) 187, the issue was the extent to which, if at all, an allowance should be made to reflect the fact that 93.9% of the construction cost was grant funded. The Tribunal found that no allowance should be made at this stage for the grant.

8.7.5.1   Adjustment for factors affecting rental value rather than capital value

Stages 1–3 are concerned with cost and the adjustment of costs. Stage 5 adjustments are usually those that affect the rental value of the property rather than the cost of construction. Such items will include poor layout of the actual building on the site, differences in levels between buildings and so on.

In addition it may also be appropriate to make allowances at this stage to recognise that cost does not equal value.

The type and degree of adjustment will vary from property to property, but examples of adjustments made are given from the following cases.

Eton College (Provost and Fellows) v *Lane (VO) and Eton Urban District Council* [1971] RA 186 – a 10% allowance was made for occupational disadvantages.

Imperial College of Science and Technology v *Ebdon (VO) and Westminster City Council* [1986] RA 233 – a 7.5% allowance was made for difficulties of goods access, access from one building to another and piecemeal development.

Leeds University v *Leeds City Council and Burge* [1962] 2 RVR 311 – a 12.5% allowance was made for the disability of roads dividing the hereditament.

Imperial College of Science and Technology v *Ebdon (VO) and Westminster City Council* [1986] RA 233 – a 6% allowance was made for the cramped site, lack of recreational facilities, different floor levels and piecemeal development.

Aberdeen University v *Grampian Regional Assessor* [1990] RA 27 – a 5% allowance was made for the over-provision of space on the basis that where there was only one possible tenant for the property, the landlord would have to accept that the property was larger than normally required and be prepared to make an allowance for this fact. A further allowance for the dispersed location of the building within the university was disallowed as the Tribunal was not convinced that this was a real disadvantage for this type of property.

8.7.5.2   Stand back and look

Having arrived at the rateable value, the valuer needs to stand back and take a hard look at the resultant figure in order to be certain that it is appropriate as the rental or rateable value. It must be remembered that throughout the valuation a large number of valuation components have been ascertained:

- the replacement or substituted building;
- building costs;
- ancillary costs;
- adjustment of costs at stage 2;
- value of land;
- adjustment of the value of the land.

At each of the above valuation steps a value judgement has had to be made. A small error or variation at any one of these steps may mean that the resultant valuation does not properly reflect the rental value of the hereditament. Where this occurs, the valuer will have to re-examine all the component parts of the valuation and make the necessary adjustments.

In many cases the 'stand back' approach can be undertaken by analysing the resultant value to a rent per m² which can then be compared with assessments of properties with some similarity. For example, a police station or fire station could be compared to offices in order to ensure the value did not seem anomalous when compared to an alternative use.

Other unit comparators can be adopted for different types of property. For example, price per pupil for schools and universities was adopted in *Downing College etc, Cambridge* v *Cambridge CC and Allsop (VO)* (1968) 208 EG 805. An allowance of around 15% was made for Churchill College having been analysed to a price per student (£117) that was found to be too high when compared to other colleges within the university (£66–£69).

There will be many instances where such a comparison is not possible, for example an airport or chemical works. Here the valuers must use their own professional judgement. In reviewing the appropriateness of the figure, the Lands Tribunal has often considered the tenant's ability to pay the rent determined and often made appropriate allowances. In *Shrewsbury School (Governors)* v *Shrewsbury Borough Council and Hudd (VO)* [1966] RA 439, a 5% allowance was given, as on review of the assessment the Tribunal considered that the hypothetical tenant could not afford to pay the rent.

8.7.6 Stage 6

This was suggested in *Imperial College of Science and Technology* v *Ebdon (VO) and Westminster City Council* [1986] RA 233. In that decision the member, Mr Mallett, said that:

> *I think it logical first to determine the annual equivalent of the likely cost to the hypothetical tenant on the assumption that he sought to provide his own premises, then to consider if this figure is likely to be pushed up or down in the negotiations between the hypothetical landlord and a hypothetical tenant having regard to the relative bargaining strengths of the parties.*

He further went on:

> *In the present case there is only one tenant and although the buildings and the site have advantages and disadvantages they do fulfil the tenant's needs. The location is most desirable even if the site has certain disadvantages, but the tenant is not irrevocably tied to either the site or the location. The tenant is supported by government grants both as to his capital expenditure and annual running cost. I think the weight of evidence is in favour of the view that the*

rent arrived at by the contractor's test based upon a decapitalisation rate of 3.5% would not be varied.

This stage represents the 'higgling of the market' and would also reflect the natural rounding up or down of the valuation arrived at under stage 5.

It is to be noted that no subsequent case has adopted a stage 6 and the Joint Professional Institutions Rating Valuation Forum guidance note, second edition, says that 'if stages 1 to 5 are properly applied then it should not be necessary to have a stage 6'.

8.7.7 *Method of last resort*

The method has been referred to by the Tribunal in several cases as a method of last resort. However, in *Eton College (Provost and Fellows) v Lane (VO) and Eton Urban District Council* [1971] RA 186, the Tribunal said that with the refinements the method had undergone, the time had come to say a good word for the method:

> *Providing a valuer using the approach is sufficiently experienced and is aware of what he is doing, and knows just how he is using his particular variance of the method, and provided he constantly keeps in mind what he is comparing with what, we are satisfied that the contractors basis provides a valuation instrument at least as precise as any other approach. For this kind of case (a public school) it is, almost certainly, the best substantive method that has been devised so far.*

Following the Court of Appeal decision in *Garton v Hunter (VO)* [1969] 2 QB 37, it is not correct to say that any one method of valuation should be used for any type of hereditament since evidence of all methods is admissible, the goodness or badness of the evidence going to weight and not to admissibility:

> *Summarising then: We do not look on any of these tests (methods) as being either a 'right' method or a 'wrong' method of valuation; all three are means to the same end; all three are legitimate ways of seeking to arrive at a rental figure that would correspond with an actual market rent on the statutory hypothesis and if they are properly applied all the tests should in fact point to the same answer; but the greater the margin for error in any particular test, the less is the weight that can be attached to it.*

That properly applied, all tests should point to the same answer, as illustrated in *Willacre Ltd v Bond (VO)* [1986] 1 EGLR 224 when the Lands Tribunal gave a decision based on both the contractors and comparative or rentals approach (however, note the problem discussed under stage 4 of the statutory decapitalisation rate, potentially producing a 'wrong' answer).

In *Barclays Bank Plc* v *Gerdes (VO)* [1987] RA 117, which concerned the valuation of a purpose-built high security warehouse or cash centre, the valuation officer argued for a contractor's basis valuation. The ratepayer's valuer argued for a comparative approach valuation on the basis of the assessments of other cash centres. The Tribunal noted that there was a wide disparity between the valuations and this indicated 'that either one method or the other is inappropriate in the present circumstance or that a method had been wrongly applied or perhaps neither method has produced the right answer'.

The Tribunal also noted that given the availability of close comparables, the comparative method gave 'only a small margin for possible error', whereas the 'same cannot be said for the contractor's basis of valuation' given the various estimates, adjustments and selection of an appropriate decapitalisation rate that are required.

The contractor's basis, while appearing fairly straightforward at first sight, has become more complex as the method has been refined and valuers have explored the implications of judicial decisions. Despite the complications and the various approaches within the method which have been and are used, it is a useful and possibly essential valuation tool which is quite widely used, particularly in Scotland where the 'Contractor's Principle' has found particular favour.

8.8 Illustrative valuation

Prepare a fully annotated assessment of a fire station for entry in the local 2017 rating list.

The property was built in 1930 and is situated close to the industrial area of the town. The property is of ornate brick construction and the accommodation comprises:

Fire station garage	350 m²
Stores	100 m²
First floor	
Offices	100 m²
Yard – concreted	2,000 m²

The Fire Authority considers that 20% of the floor space in the garage area is unusable due to the design of the building in relation to modern requirements. A modern fire station would cost £1,600 per m² to construct as at 1 April 2015.

8.8.1 Illustrative approach

The appropriate method for the valuation of the fire station would be the contractor's basis of valuation.

In undertaking this valuation one must decide on which of the two bases should the cost be calculated:

- replacement building approach;
- substituted building approach.

In this case it is considered that the substituted building approach should be adopted as a replacement building would not be constructed but rather a modern substitute taking into account the changing requirement for this type of building would be constructed.

It is stated that 20% of the garage floor space is wasted due to the design of the building. This would indicate that when envisaging the substituted building, an area less than the current 400 m² would be contemplated. However, the building envisaged should not be the absolute minimum required but rather that which would be built by a prudent landlord, i.e. it would allow for some expansion.

Stage 1
Basis: simple substituted building 350 m²

Garage	350 m²	@	£1,600	£560,000
Stores	100 m²	@	£1,600	£160,000
Offices	100 m²	@	£1,600	£160,000
Yard	2,000 m²	@	£100	£200,000
				£1,080,000

Stage 2
Less age and obsolescence allowance 40% £432,000

£648,000

Stage 3
Add value of land as fire station
Say 3,000 m² @ £800,000 per ha £240,000

£888,000

Stage 4
Decapitalise 4.4% £39,072

Stage 5
Any allowances that should be reflected in rental value £–

Stand back – is it reasonable
Say, rateable value £39,000

8.8.1.1 Notes to valuation

The current six-stage approach is derived from *Gilmore (VO)* v *Baker-Carr* [1961] 1 RVR 598 previously referred to.

A sixth or negotiation stage was suggested in *Imperial College of Science and Technology* v *Ebdon (VO) and Westminster City Council* [1986] RA 233 but reference to it has been deliberately omitted from The Contractor's Basis of Valuation for Rating Purposes – A guidance Note (April 2017) by the Joint Professional Institutions' Rating Valuation Forum (JPIRVF).

- Stage 1 – ascertain cost of either a replacement or a substituted building together with the cost of providing external features.
- Stage 2 – where appropriate, adjust the cost to have regard to the deficiencies of the actual hereditament being valued as opposed to a modern one.
- Stage 3 – add the value of the land based on its current use, *rebus sic stantibus*.
- Stage 4 – apply the statutory decapitalisation rate.
- Stage 5 – where appropriate, adjust the value to reflect any matters that would affect rental value as opposed to capital value.
- Stand back – see whether the 'answer' is appropriate!

8.9 Summary

The method is used for the valuation of those properties where there is a general absence of rental information and the use of the receipts and expenditure approach is not appropriate.

It may also be used as a check on valuations undertaken using other approaches to test the reliability of those other valuations.

The method is based on the assumption that a tenant could, in theory, build an alternative property and would be prepared to pay a rent that would equate to the interest on the cost of construction.

The method is based on five valuation stages as laid down in *Gilmore (VO)* v *Baker-Carr* [1961] 1 RVR 598.

The use of a prescribed decapitalisation rate at stage 4 is a statutory requirement.

The application of the method can be highly complex and a great deal of thought and consideration needs to be given to the valuations if the valuer is to arrive at the true rental value of the property on the statutory basis.

8.10 Recent Tribunal decisions

Allen (VO) v *Tyne & Wear Archives and Museums* [2022] UKUT 206 (LC) UTLC

Belfast International Airport v Commissioner of Valuation [2017] VR/9/2016 – application of method

Celsa Steel (UK) Ltd v *Webb (VO)* [2017] UKUT 0133 (LC) – application of
   method; stage 5 allowances
Hobbs v *Gidman (VO)* [2017] UKUT 0063 (LC) – method rejected
Hughes (VO) v *Exeter City Council* [2020] UKUT 0007 (LC)) – method
   rejected
Hughes (VO) v *York Museums and Galleries Trust* [2017] UKUT 0007 (LC)) –
   method rejected
Semlogistics Milford Haven Ltd v *Webb (VO)* [2018] UKUT 0019 (LC)) –
   application of method

8.11 Wales

The Non-Domestic Rating (Miscellaneous Provisions) (No. 2) (Amendment)
(Wales) Regulations 2021 No. 1141 (W. 278) prescribe the decapitalisation
rates to be adopted for the purposes of the 2023 rating lists.
   The rates are:

- in the case of a defence hereditament, an educational hereditament, a health
   care hereditament or a hereditament which is wholly a public convenience, 1.9%;
- for all other properties, 3.4%.e different from those applied in England.

Decapitalisation rates (1990–2023)

List	Normal	Education and health (and defence from 2000)
1990	6.00%	4.00%
1995	5.50%	3.67%
2000	5.50%	3.67%
2005	5.00%	3.33% (Wales 3.3%)
2010	5.00% (Wales 4.5%)	3.33% (Wales 2.97% – also applies to public conveniences)
2017	4.4% (Wales 3.8%)	2.6% (Wales 2.1% – also applies to public conveniences)
2023	4.4% (Wales 3.4%)	2.6% (Wales 1.9% – also applies to public conveniences)

9 Receipts and expenditure method

9.1 Introduction

An alternative approach to the valuation of property where there is no, or only limited, rental evidence is the use of what is termed the 'receipts and expenditure method'. This approach is used for the valuation of properties such as hotels, sports and leisure centres, safari parks, race courses and race tracks.

The approach may also be used as a check on other methods of valuation in order to ensure the correct rateable value is determined.

9.2 Historical development

The application of the method can be traced back alongside that of the contractor's basis. The receipts and expenditure approach, then termed the 'profit's method', also owed its development to the valuation of statutory undertakings or public utilities – railways, canals, dock ports and gas, electric and water undertakings. The method was used for the valuation of the productive parts of the undertakings, while the contractor's basis was used for the unproductive parts. Many of the leading cases were concerned with public utilities, especially water undertakings.

While originating in the realms of public utilities, the method began to be applied to other types of hereditament. The rationale for this development was:

- that there was no, or unsuitable, rental evidence on which to base the assessment, and that the hypothetical tenant would have regard to the profits in determining what rent to pay;
- that there was some element of monopoly value associated with the hereditament which enabled the landlord to charge a higher rent for the property by reason of its profits.

In the development of the approach it has long been accepted that profits are to be those of the hypothetical tenant, not the actual occupier. Where the actual occupier, due to particular personal skills, is outperforming what would be expected from a typical occupier or hypothetical tenant, then some allowance

DOI: 10.1201/9781003297369-9

will have to be made for the personal skill of the occupier. The converse will also apply where it can be shown that the property is not doing as well as its potential would indicate.

The name change from 'profits' to 'receipts and expenditure' has been increasingly accepted and has been encouraged by the publication of the Rating Forum Practice Note. Profits, as such, have never been rateable but rather suitably adjusted accounts have been used to ascertain how much a tenant could afford to pay in rent.

The valuation rationale of the approach was encapsulated in more recent times by the Court of Appeal in *Hoare (VO) v National Trust [1998]* RA 391; [1997] 2 EGLR 229 (CA):

Sometimes in the case of properties which are rarely, if ever, let it is appropriate to arrive at the annual value by a method of valuation known as the profits basis. This is a somewhat confusing name since profits as such are not rated and are not rateable. But the broad theory is that where a property can be used so as to yield profits then the hypothetical tenants would be prepared to pay a rent for the use of that property in order to be able to make those profits and that the level of rent would reflect the level of anticipated profits.

9.3 Overview of the approach

The approach to the valuation is not as prescribed as that of the contractor's basis of valuation and follows a fairly consistent and common sense approach:

Net receipts	£
Less	
Working expenses	£
Divisible balance (or net profit)	£
Less	
Tenant's share	£
Rent	£

The method does involve a large amount of calculation. Net receipts and working expenses have to be calculated. This involves totalling sums in accounts and usually making adjustments for items in the accounts which, while properly in a balance sheet, are not relevant or appropriate for the valuation. The simplest example would be a mortgage payment which is an expense of the business but would not be an expense for a rating valuation. The hypothetical tenant, not being an owner, would not pay a mortgage but instead would pay a rent – which is of course the object of the valuation exercise. In the past, receipts and expenditure valuation calculations were very time consuming and therefore tended to be avoided wherever possible. Nowadays, spreadsheets lend themselves ideally to receipts and expenditure valuations allowing 'what if' calculations to be performed easily and quickly. This should encourage a more widespread use of the method.

The detailed application of the approach will be considered later. For specific types of properties valuation approaches have often been agreed by the Valuation Office Agency (VOA) and with trade associations or similar.

For many properties that do not usually let in the market and in the absence of rental or other evidence the valuer is faced with the choice of having to adopt either the receipts and expenditure basis or the contractor's basis. It was suggested in *Garton* v *Hunter (VO)* that all methods 'if properly applied' should produce the same answer – in practice that may not be the case. Over the years there have been a whole series of cases where the choice of the method of valuation has been at issue. Since the last edition three particular cases have been heard by the Upper Tribunal in respect of heritage buildings used as museums. These have involved buildings that are hundreds of years old; where the actual occupiers may not charge for admission and they may make a substantial loss. The question arose – how should they be valued? The valuation officers preferring to base valuations on the contractor's basis and the ratepayers upon the receipts and expenditure basis. These are:

Allen (VO) v *Tyne & Wear Archives and Museums* [2022] UKUT 206 (LC) UTLC
Hughes (VO) v *Exeter City Council* [2020] UKUT 0007 (LC)
Hughes (VO) v *York Museums and Galleries Trust [2017]* UKUT 0007 (LC)

They are discussed both in this and the Contractor's Basis chapter (Chapter 8).

9.3.1 Rejection of the method

Prior to examining the application of the method in detail, it is useful to review some of the instances where it has been found that the method is inapplicable and why.

In *Edinburgh Parish Council* v *Edinburgh Magistrates* (1912) SC 793, the Court concluded that:

> *It may well be that the practice of valuing cemetery companies by reference to the profits basis is so well settled that it would be difficult to upset it even if a company made a loss in any particular year or years, but it seems to me that to say that a burial ground must be valued on the basis of profits when it is quite clear on the facts that it never has been and never will be operated with any idea of making a profit is really a contradiction in terms.*

In *Warwickshire County Cricket Ground* v *Rennick (VO)* [1959] R&IT 787, the method was rejected as the property was making a loss, and the approach would have resulted in a nil assessment which did not reflect the true benefit the occupiers were receiving from its occupation.

In *Fir Mill Ltd* v *Royton UDC and Jones (VO)* [1960] R&IT 389, it was found that:

> *In the light of this finding, evidence of the profits of individual companies in the cotton trade is of no assistance. Although in the past rating authorities in Lancashire have undoubtedly paid attention to the state of the cotton industry as a whole, in our opinion that is relevant only so far as it may have affected the rental value of industrial properties generally in the area.*

In *Melville and Rees (VO)* v *Airedale and Wharfedale Joint Crematorium Committee* (1963) 3 RVR 201, the approach was rejected in favour of the contractor's basis as no accounts were available for the property and to estimate them would be unreasonable.

9.3.2 Reasons for rejection

The reasons why the courts and Tribunal have rejected the adoption of the method are varied but include the following:

- Rental evidence is available on which to base the assessment. Good quality rental evidence will carry far more weight than any other means of assessing the rental value. Consequently, especially in retail property where ratepayers have often argued that they should be valued on a receipts and expenditure basis, this has been rejected on the grounds that direct rental evidence is available on which to base the assessment. This does not mean that regard cannot be had to the accounts to demonstrate a trend or some other matter.
- The property does not make a profit, and a strict application of the approach would lead to a nil assessment, when it is clear that the occupier receives some substantial benefit from its occupation.
- There is no intention to make a profit and consequently to rely on accounts where there is no intention of making a profit would be inappropriate.
- Insufficient information with regard to the accounts is available. However, in a number of instances the courts have been happy to accept that where accounts are not available it is reasonable to estimate the income and expenditure just as the hypothetical tenant would do.

9.4 Use of accounts as a check for other approaches

In the rating sphere, the reference to accounts is an integral part of the receipts and expenditure method of valuation. However, reference to the accounts of the business may also be used for other purposes such as to prove or disprove a valuation undertaken on a different basis whether rental, contractor's or by comparison:

- *Clark* v *Alderbury Union* (1880) 6 QBD 138 – evidence of the trade that was done in refreshment rooms at Salisbury Railway Station was admitted to show that the tenant could not afford to pay the agreed rent.
- *Trocadero (Swanage) Ltd* v *Perrins (VO)* (1958) 51 R&IT 140 – the Tribunal found that although the trading accounts of a restaurant could not be accepted as a basis for its assessment, they did confirm conclusions drawn from other sources showing it was considerably over-assessed on the valuation officer's basis of values per sq ft of floor space.
- *Durose Cafe & Garage Ltd* v *Hampshire (VO)* (1959) 52 R&IT 228 – with regard to the valuation of a transport hostel, cafe and garage, it was held that while the ratepayers were at liberty to put in accounts, as supporting evidence, it was not necessary to have recourse to a profits valuation.
- *Carpanini* v *Fardo (VO)* (1958) 51 R&IT 723 – it was found that rental value could not be directly deduced from the accounts, but they might be accepted in confirmation of a claim for a reduction in rental value by the closing of a railway station to passenger traffic.

9.5 The approach in detail

In this section we will look at the approach in more detail, beginning by considering some general matters, then examining what should be included in receipts and working expenses and discussing the approaches to the tenant's share.

9.5.1 Disclosure of accounts

Ideally, the method requires the valuer to have access to the accounts of a business or sufficiently similar information.

Until relatively recent times there has been no statutory requirement for a ratepayer to disclose accounts for the purpose of determining the ratepayer's rating assessment. Whether ratepayers wished to disclose accounts was a matter for them. Most ratepayers were often more than willing to produce their accounts, especially where it supported a point that they were making.

The ability of a valuation officer to call for rent returns where trade information must be disclosed is of more recent origin. What information can reasonably be requested in a rent return was considered in *Watney Mann Ltd* v *Langley (VO)* [1966] 1 QB 457. Following that decision, the valuation officer may ask for all the information necessary for him or her to undertake the valuation. This usually will fall short of requesting actual accounts but rather requires full details of the receipts and expenditure associated with the property. Often, the VOA will have agreed with the appropriate trade body the exact nature of information to be requested and supplied.

9.5.2 Period of accounts

The method supposes that a prospective tenant, in considering a rental bid, will have regard to the profit that can be made by occupying the hereditament. The prospective tenant will therefore wish to see the most recently available full year's accounts in order to judge the potential for profit in the next and subsequent years. Given that trade tends to fluctuate, a prudent prospective tenant will also wish to examine earlier accounts. These will enable the tenant to discern any trends and to take into account fluctuations in trade in order to form a judgement as to the likely appearance of future accounts and the reasonably maintainable profit.

Traditionally, it is normal to look at the last three years' accounts so that any trends can be discerned. This does not mean that the accounts should be averaged, but rather the valuation is based on the most appropriate set of accounts for the relevant valuation date but reflecting normal fluctuations in income and expenses. It must be emphasised that the use of a number of years' accounts is only to ensure that any trends in the accounts can be properly identified. What has to be ascertained is how the hypothetical tenants for the hereditament, and in particular the hypothetical tenant, would use the accounts to determine their rental bid.

What is sought is the rental bid of the hypothetical tenant for the year after the antecedent valuation date (AVD) based on the evidence of the last three or more years' accounts up to the AVD and having regard to expectations of future trends at the AVD.

Where additional information or accounts become available after the AVD, then that information should not be used as a basis for the valuation as it would not have been available at the AVD. However, regard may be had to that information to confirm or support a trend which had previously been identified.

9.5.3 Identification of the hereditament

Properties valued on the receipts and expenditure basis can often be complex in character and can involve parts of the property being let or licensed to other occupiers. It is therefore important to correctly identify the hereditament being valued and what parts, if any, are to be separately assessed. Furthermore, even where no part of the property is sublet, it may be located in such a way that consideration must be given to whether the property comprises a single assessment or whether there should be more than one assessment. The normal rules will have to be considered in arriving at the appropriate number of assessments.

It is essential that the identity of the hereditament is correctly identified early on. Failure to do so will mean that it is not possible to adjust the accounts to reflect the actual hereditament.

Cases may also arise where on first appearance it is considered that there are a number of separately rateable hereditaments, but further investigation reveals that they are really occupied by the same body trading under different names. In these circumstances accounts may have to be consolidated.

9.5.4 *Adjustment of accounts to date of valuation*

Where accounts are for a period prior to the date of valuation then they will need to be projected forward to the relevant date. Undertaking such a projection is difficult and can be subject to much debate and speculation between the parties.

The simplest form of adjustment that has been accepted by the courts has been the subjective adjustment of receipts based on the experience of the management and the valuers with regard to their knowledge of the nature of the business.

Expenditure has also often been adjusted by reference to the Retail Price Index (RPI), although this is a rather rough and ready approach.

9.5.5 *Adjustment of accounts*

The use of the accounts is to help inform what the hypothetical tenant would be prepared to pay in rent for the hereditament. It is necessary to ensure that the accounts fully reflect the potential of the property or otherwise. In some cases the property may not be run at its optimum and some form of adjustment of the accounts may be necessary to properly reflect its potential to the hypothetical tenant. In these cases the adjustment of the income, either upwards or downwards, to reflect the true potential of the property can be difficult to prove and such an adjustment may rely heavily on the experience of the parties.

9.5.6 *Restriction on making a profit*

One of the potential problems with the use of the method is that the profit-earning capacity of a property may be restricted by statute or some other form of quasi-statutory control. It is generally accepted that, where there is a restriction on the ability to make a profit from a particular hereditament, that fact should be reflected in the rating assessment.

In *London County Council* v *Erith and West Ham (Church Wardens and Overseers)* [1893] AC 552, the House of Lords said: 'any restrictions which the law has imposed upon the profit earning capacity of the undertaking must of course be considered'.

Furthermore, in *Port of London Authority* v *Orsett Union AC* [1920] AC 273:

> *The actual hereditament of which the hypothetical tenant is to be determined must be the particular hereditament as it stands, with all its privileges, opportunities and disabilities created or imposed either by its natural position or by*

the artificial conditions of an Act of Parliament. The character and the extent of the various deductions from the gross revenue must be fixed in relation to the conditions.

In Sculcoates Union v Hull Docks [1899] 1 QB 667, a railway company had the statutory right to run toll-free over the railway lines which belonged to the dock company. It was held that the statutory restriction on the charging of tolls must be taken into account in arriving at the rating assessment.

In the above cases the courts held that the statutory restrictions should be taken into account in arriving at the assessment. Where, however, some other form of restriction on the profit-earning capacity is imposed, it is argued that these would have to be ignored as vacant and to let they would not be imposed by the hypothetical landlord as they would be inconsistent with the statutory basis of valuation.

In some instances the maximum level of charges may also be subject to statutory limitation, but the occupier may decide to charge a lower amount. While it may be arguable that the hypothetical tenant would charge the maximum permitted, the courts have looked at the business rationale for the level for charge to determine what would be charged by the hypothetical tenant. There may be sound business reasons for a lower charge.

9.6 Gross receipts

This will comprise all the income from all sources providing it arises directly from the occupation of the hereditament.

In *Surrey County Valuation Committee* v *Chessington Zoo Ltd* [1950] 1 KB 640, income from animals let out to other concerns outside the hereditament was included in revenue for the purpose of assessment of a zoo, since normally the animals had to be kept in cages at Chessington Zoo.

In *Rank Organisation Ltd* v *Priest (VO)* [1966] RA 540, in the valuation of a cinema, it was held that the receipts should include box office takings, sales of ices and chocolates, screen advertising, car park takings and other sundry items.

9.6.1 *Receipts must be attributable to the hereditament*

All receipts should be directly attributable to the occupation of the actual hereditament being valued. Consequently, any income obtained from outside the hereditament will need to be excluded. For example, income from a separately assessed car park would have to be excluded, as would rents from separately assessed properties.

It can often be a difficult decision as to where to draw the line between receipts directly derived from the hereditament and receipts derived from the wider nature of the business.

9.6.2 Rents received

Rents of properties let out and separately assessed should be excluded. Where rents are received but are for the use of parts of the hereditament, then these rents or payments should be included.

9.6.3 Value Added Tax

Generally the figures will exclude VAT unless the business is unregistered (the current registration level is £85,000 (same since 2017).

9.6.4 Reserve funds

Money transferred from reserve funds should not be taken as income as these sums will have been taken into account in previous years. Money transferred to reserve funds is not deductible from gross receipts being a form of profit distribution.

9.6.5 Grants and subsidies

Generally, revenue grants and subsidies are taken as a normal receipt, unless there are indications to the contrary. However, problems arise when the grant is 'one off' or the amount of grant either is variable in amount or cannot be guaranteed to be payable every year. In many instances the amount of grant, if any, may be dependent upon the amount of funding that the allocating body receives in its own right and which may vary from one year to another, as, for example, the Arts Council.

In respect of football grounds, the income from the sale of TV rights can vary considerably and may well depend on such factors as the success or otherwise of a particular club or even who the other team is for any particular match, especially with an overseas opponent.

In deciding the approach to be adopted, one must not lose sight of what the valuer is trying to determine – the rental value of the property. In practical terms, the hypothetical tenant would take a view as to the likelihood of receiving a grant or some form of subsidy, the likely amount and whether it would vary from one year to another in order to arrive at a rental bid.

Capital grants relating to capital projects should be ignored. However, at times it may be necessary to examine in more detail the exact nature of the grant as it may include an element of both revenue grant and capital grant for the construction, extension or refurbishment of the property.

9.6.6 Donations received

Donations can take the form of a direct financial contribution to support a particular undertaking or it may be in the form of the supply of voluntary labour or services.

This can give rise to many problems and the issue has been considered by the courts on a number of occasions. Where it can be shown that the hypothetical tenant could normally expect to receive such donations then their value should be reflected in the gross receipts. A similar position would apply to the use of voluntary labour.

In *Winchester and Alton Railway Ltd* v *Whyment (VO)* [1981] RA 258, the amount of voluntary donations ranged between £1,400 and £1,800 over a number of years.

The ratepayers argued that as there was no legal entitlement to donations and, as they did not arise from the working of the hereditament, a potential tenant would disregard the source of income as being uncertain and transitory. Mr Whyment was of the view that the hypothetical tenant would have similar characteristics to the company and would expect donations to be the same or similar having regard to the figure for donations shown in the accounts for the last two years:

> *Despite the contentions of the ratepayers, it seems to me that the donations received are directly connected with the ratepayers' occupation of the hereditament. A collection is made by placing money boxes on the hereditaments and some of the donations received from other sources are specifically for the purchase of items of equipment to be used on the hereditaments. All donations are treated as income in the accounts. Therefore, I find that it is correct to treat donations as income in the valuation, but the amount may be questioned. The accounts for the year ending 31 October 1978 are accepted as the proper starting point of the valuation. There is no evidence that by the 3 March 1979 there was any indication of a change in the receipts and expenditures in general or in receipts from donations in particular. Therefore, I accept the figure of £1,830 as a proper one to add to the income in respect of donations.*

In *Bluebell Railway Ltd* v *Ball (VO)* [1984] RA 113, it was noted:

> *The only item of dispute here is the donations. As I have indicated, many of these are earmarked for particular projects, but I am of the opinion that the company can properly look forward to continuing receipts under this heading, and that such donations would be available to meet the demand for rent, bearing in mind that the occupation of the hereditament is a prerequisite before any rolling stock can be operated or maintained. I think the approach of the valuation officer in disregarding part of the very large sum subscribed as sundry donations for 1978 is right and I adopt his figure of £146,837 as gross receipts.*
>
> *I turn next to the question of voluntary labour. I regard this in very much the same light as the donations. The willingness of the members of the society to devote their time, skill and energy to help in the running of the undertaking is an established fact and without it the operations could not be continued in*

their present form. These are services, however, which are an essential feature of the company and, in my opinion, the company in agreeing to pay a rent would reasonably anticipate that services on this scale would continue to be provided. They do not represent an appropriate charge against income and in fact no payments are made.

9.6.7 Sponsorship received

One of the issues that has become more relevant in recent times is that of sponsorship, with many sports clubs receiving considerable sums of money in respect of sponsorship. Income from sponsorship would normally be reflected as part of the income. In some cases the extent of the sponsorship may be wider than just the hereditament being considered and some adjustments will have to be made.

The main benefit the sponsor receives is usually some form of advertising. This can take many different forms from naming the ground after the sponsor (the Autoglass Stadium, for example), advertising around the ground and advertising on shirts. In other instances the sponsor may provide equipment, such as cars, suitably endorsed with its name.

One of the problems with sponsorship is that it can be fairly transitory in nature – only available while the club is doing well and should the club not do so well and be relegated, then the level of sponsorship may fall substantially. In many sporting areas the sponsorship can relate to the sport as a whole or a particular event – NatWest and Benson & Hedges Cricket, cigarette companies and Formula 1, for example. Again, sponsors may pull out of events for a wide range of reasons, sometimes at very short notice if they feel that their continued association with an event or sport is having a detrimental impact on their business.

One of the problems with sponsorship is whether changes in levels of sponsorship during the life of a rating list can be taken into account as a Material Change of Circumstances (MCC) or has one to imagine the typical level of sponsorship taking the circumstances as at the AVD, this being an economic factor rather than a physical one.

9.6.8 TV rights

Some sports receive considerable sums from being televised and the recent deals to broadcast some of the major sporting events show just how lucrative this is. Generally the sums of money for these rights are received by the sports organisations and distributed to the clubs in whatever is considered to be an appropriate manner.

As with other forms of grants and sponsorship it will be necessary to determine whether the hypothetical tenant would be likely to receive such income and its amount on a year-by-year basis. Where such income is likely then an appropriate amount should be reflected in the receipts of the business.

9.7 Expenditure

9.7.1 Cost of purchases

The cost of purchases needs to be allowed for in the approach. Care needs to be taken with regard to this item for a number of reasons:

- The costs may not be those that would be available to the hypothetical tenant in that the current occupier may be able to obtain discounts that would not be available to the hypothetical tenant. For example, the actual occupier may be a company running a chain of properties and can secure additional discounts for bulk purchasing. However, if the likely hypothetical tenant would be a chain then the discounted figures can be adopted.
- The costs of purchases may include an amount for items used other than on the property and will need to be adjusted accordingly.
- The level of purchases can vary substantially from year to year and may have a substantial impact on the profitability or otherwise of the undertaking. Levels of stock may be used to artificially manipulate the profits for other reasons such as tax liability. Adjustments may be required to be made for the level of purchases in any one year.
- Purchases should only relate to those required for the hereditament and any that relate to uses outside the hereditament, such as those for personal use of the proprietor, should be excluded.
- Attention needs to be given to the general stock levels associated with the hereditament as it is possible for the operator to manipulate the levels of stock and purchases to show different trading outcomes.
- Changes in stock need to be allowed for. If the stock is lower at the end of the year then in effect the amount of 'purchases' is higher because some of the stock has been consumed. The difference should be added to the 'purchases'. Conversely, if the stock is higher at the end of the year then the increase should be deducted from the purchases as it is not a purchase that has been used in earning the gross receipts. The holding of the stock is taken into account later in the valuation as part of the tenant's capital.

9.7.2 Working expenses

Typically, working expenses will include wages, insurance, professional fees, upkeep of a motor car, telephones and postage. The object of adjusting and examining the accounts is to determine the reasonably maintainable profit; therefore, if any item is abnormally high or low in any one year of accounts compared to others it may need adjusting to a 'normal' figure.

9.7.3 Rent

Any rent appearing in accounts should be treated as a deduction, as rent is the item that has to be ascertained. Usually a rent figure appearing in the accounts will be a ground rent or a rent for a small part of the hereditament.

9.7.4 Head office expenses

These can be a somewhat arbitrary figure and may be used as a means of transferring expenses from one property to another, often for taxation purposes. There is nothing in principle that prevents head office expenses being allowed as an expense.

It is necessary to ascertain the exact nature of the services provided by the head office and to form a view as to whether these are reasonable. In *London Cemetery Co.* v *St Pancras Assessment Committee* [1936] 24 R&IT 208, the Court accepted the need for the head office to be elsewhere and allowed, as expenses, deductions for the expenses incurred by the head office. However, it did reduce some expenses on the basis that it owned more than one property and the expenses must be properly apportioned.

9.7.5 Directors' fees and salaries

Whether directors' fees should or should not be allowed as a deduction from working expenses, and where deductible and how much, has long been an issue for debate. The principle would seem to have been established that directors' fees and expenses should be allowed as a deduction so long as the payments are directly related to the activity carried on in the hereditament. In many cases if there had not been directors actively involved in running the business then additional staff would have had to be employed.

How much should be allowed is really a matter of valuation opinion. The amount should reflect the amount of work undertaken by the directors concerned. In some cases the courts have allowed only a proportion of the fees while in other cases the full fees have been allowed. In recent years it has become widely accepted practice to allow 50% of the actual fees on the basis that the directors will spend part of the time actively engaged in running the business and the remainder of the time promoting the landlord's hereditament.

9.7.6 Sinking fund

Sinking funds can be established for a number of different purposes. The most common type of sinking fund encountered is one where the fund is established for the purpose of setting aside sums of money for the eventual payment of a large item of repair. In such cases the sinking fund should be allowed as an expense. Care needs to be taken not to 'double count' and cover the same element of repair in an allowance for 'repairs'.

The amount of the sinking fund should be the amount that would be required to be set aside at the relevant date, having regard to the then current state of repair of the item or items under consideration.

9.7.7 Payments of a capital nature

Expenditure on capital items such as furniture, motor cars, non-rateable chattels, etc., should not be included as working expenses. The hypothetical landlord is deemed to provide the hereditament and the hypothetical tenant to provide tenant's capital. Items such as motor cars will be provided out of tenant's capital and interest on this capital is allowed later as part of the tenant's share. Sometimes capital expenditure will be shown in accounts which are payment for an improvement to the hereditament. These are properly expenses to be borne by the landlord who provides the hereditament rather than the tenant, e.g. the cost of sinking a mine (*R* v *Attwood* (1827) 6 B&C 277).

9.7.8 Rates

Rates are a deductible expense and are taken as the actual amount that was payable for the forthcoming year as at the AVD. For the 2023 Revaluation this will be that payable for the 2020/2021 rate year and will be a known figure.

This flows from the change to having an AVD from the 1990 Revaluation onwards. It is well accepted that the level of rates is that which would properly have been paid at the AVD. In *Thomason* v *Rowland (VO)* [1995] RA 255, the President of the Lands Tribunal said, 'I believe that the rates in fact paid should be shown as a working expense, as the valuation officer asserted.'

The case involved a receipts and expenditure valuation and the question was whether the rates to be deducted were those payable at the AVD or those at compilation following the revaluation. The decision was that it was the level at the AVD.

9.7.9 Mortgage payments

Mortgage payments are not allowed as this would not be consistent with the rating hypothesis. They are payments made to cover interest on acquiring or building the property and are therefore expenses a landlord should bear.

9.7.10 Interest payments

Interest payments on the cost of the tenant's chattels and other borrowing needs of the tenant are not treated as a deduction. This is because they will be taken into account when considering the tenant's share later on in the valuation.

While normally allowed, care needs to be taken to ensure that bank charges and bank interest charges are typical for running the type of business.

9.7.11 Bad debts

Bad debts and refunds should be allowed as long as they are reasonable and not abnormal for that type of business.

9.7.12 Levies

Any statutory levy is to be considered as an allowable expense as it would be incurred by any hypothetical tenant taking on the hereditament. In the case of *Rank Organisation Ltd* v *Billet (VO)* [1958] R&IT 650, it was held that a voluntary levy should not be allowed as an expense of the business.

9.7.13 Repairs

Repairs are allowable but will often require adjustment for the periodic nature of many items of repair. Exceptional repair expenditure in one year will not be borne in another year. Repair expenditure is required to be evened out.

This item may raise the somewhat interesting question 'what standard of repair is envisaged?' as was highlighted in *Brighton Marine Palace & Pier Co.* v *Rees (VO)* [1961] 9 RRC 77. Depending on the type of property being considered, it may be appropriate not only to allow for general repairs but also to make provision of a sinking fund for major items of both repair and renewal.

9.7.14 Repairs to tenant's chattels

Generally, repairs to the tenant's chattels are allowable but again with the proviso that periodic fluctuations are evened out.

9.7.15 Depreciation to tenant's chattels

A number of different approaches need to be considered including whether a sinking fund should be set up for the renewal of chattels or whether an annual amount should be allowed for depreciation.

In *Thomason* v *Rowland (VO)* [1995] RA 255, a sinking fund was adopted using the receipts and expenditure basis to allow for the replacement of the tenant's equipment which was considered to have a life of ten years.

In *Bluebell Railway Ltd* v *Ball (VO)* [1984] RA 113 a sinking fund was also considered:

> *The next area of disagreement is the question of depreciation and sinking fund allowances. At this stage I think there was some confusion in the minds of both valuers. First of all they both accepted the figure of £40,632 as being the expenditure on maintenance, and also included a figure in respect of the bridge survey. The figure appearing in the company's accounts for depreciation was*

£6,763 under three headings: a) freehold land and buildings, sheds and railway track; b) equipment and fixtures and fittings; and c) rolling stock. The valuation officer adopted this total figure in his valuation. Mr Appleby also based his figure on the accounts, but doubled the allowance in respect of the rolling stock. He also included a figure, as I have already explained, as a sinking fund for replacement of tenants rolling stock etc.

Neither approach is, in my judgment, correct. The tenant is entitled to consider as a charge against his income the amount which he sets aside on maintenance and for depreciation of his own equipment. The sinking fund provision is to meet the statutory requirement in the definition of net annual value that the hypothetical tenant is to be deemed to maintain the hereditament in a state to command the rent. It therefore applies only to the landlord's hereditament, that is to say, to the land, buildings and fixed plant. On the facts of this case I am prepared to adopt the figure of £6,763 for depreciation which, together with the amount spent on maintenance, seems to me an adequate provision for the tenant's own chattels. It is the figure adopted by Mr Ball, and although it includes £2,427 in respect of what is clearly part of the landlord's hereditament, I accept the evidence of Mr Appleby that something more than the figure in the accounts is appropriate for some of the rolling stock. As to the sinking fund, this represents a charge which the hypothetical tenant must accept before paying any rent, but it does not apply of course to the engines and the coaches which form the greater part of Mr Appleby's calculations.

9.8 Tenant's share

Once the allowable expenses have been deducted from the income of the property the sum remaining, the net profit, is termed the divisible balance for rating.

This represents the amount of money that is available for division between the tenant as a return or inducement for undertaking the business and the landlord in the form of rent.

A number of different methods have been adopted over the years in order to arrive at the tenant's share. Each method has its own advantages and disadvantages and may be more appropriate or inappropriate in some circumstances.

As the Upper Tribunal put it in *Beaconside Country House and Cottages* v *Gidman (VO)* [2016] UKUT 0497 (LC):

It is necessary, in every case, once the Fair Maintainable Trade and the working expenses have been established, to 'stand back and look' and to ask what the hypothetical landlord and tenant would agree would be a sufficient return to reward those efforts. The correct answer is unlikely to be the same in every case.

The House of Lords in *Railway Assessment Authority* v *Southern Railway Co.* (1936) 24 R&IT 52 established that the inducement to the tenant to take the hereditament, the tenant's share, must be a proper and sufficient inducement and not merely an arbitrary fraction of the divisible balance. It was therefore a first charge on the divisible balance and fell to be deducted first even if this left nothing for rent, the logic of this being that if the inducement to the prospective tenant is insufficient then he or she will not take a tenancy.

As shown by the Court of Appeal *in Sandown Park Ltd* v *Esher UDC & Castle (VO)* [1955] 1 WLR 747; 47 R&IT 351; 25 DRA 320 (CA); reversing (1952) 46 R&IT 70; 23 DRA 66 the division should not be seen as a 'joint adventure' between landlord and tenant. The tenant's share is a first charge on the divisible balance though as Somervell LJ put it, 'One imagines possible tenants offering different rents. The landlord is entitled to, and will, refuse any offer which is based on the tenant's retaining more than sufficient to induce a reasonable tenant to take the premises.'

The general principle governing the apportioning of the tenant's share is that the greater the competition for the property, the higher the rent and the lower the profit margin.

9.8.1 What is included?

The tenant's share is normally considered to include the following items:

- A remuneration for the tenant. This will reflect the nature of the business and the time, effort and experience required to carry out the business. In some businesses, a proprietor's salary may already have been allowed for in the accounts under a general heading of wages and salaries. This will need to be deducted as the tenant's share is calculated as the final element of a receipts and expenditure valuation, and it is confusing to have elements of tenant's share, e.g. a salary or own consumption at various places in the valuation calculation.
- Pure interest on capital employed in the business. The capital referred to is the capital a tenant will need to provide by way of chattels, cars, a money float, etc. Most properties valued on this basis require a considerable amount of capital to be invested by the tenant. For example, a hotel needs to be fully fitted out, furnished and equipped by the tenant.
- A risk premium that reflects a sum of money a tenant would require to take on the running of the business. The amount a tenant would require as an inducement to take on the business will vary with the nature of the business concerned. It reflects the risk the tenant is taking in employing capital in the business rather than simply earning interest on it in the bank.

The recognised alternative approaches to the tenant's share are:

- a percentage of the tenant's capital;
- a percentage of gross receipts or turnover;
- a percentage of the divisible balance;

- a direct estimate of the likely remuneration the tenant will require or accept;
- the use of individual figures for remuneration, risk and interest on capital.

9.8.2 Percentage of tenant's capital

The approach adopts a straight percentage of the tenant's capital. The approach is often used for small scale undertakings. Problems associated with the approach are in ascertaining:

- the amount of capital that the tenant has invested in the undertaking and
- the percentage rate to be applied.

The Royal Institution of Chartered Surveyors (RICS) Rating Forum suggested that when adopting this approach, the return to be adopted could be derived from:

- the discount rate used in a discounted cash flow valuation for valuation and appraisal purposes;
- return on capital employed – but note that this can be calculated in a number of different ways and each method has its own advantages and disadvantages;
- target return on capital employed;
- weighted average cost of capital – this approach has been adopted for public utilities (*China Light & Power Co.* v *Commissioner for Rating* [1996] RA 475 and *British Telecommunications Plc* v *Central Valuation Officer* [1998] RVR 86) and is perhaps too complicated for normal use;
- the Rating Forum's guidance note on 'The Receipts and Expenditure Method of Valuation for Non-Domestic Rating', at paragraph 5.51, states that 'when considering the individual elements of the tenant's share, interest on the tenant's capital may be found by having regard to the yield obtainable from low-risk investments'.

9.8.3 Percentage of gross receipts

The choice of the appropriate percentage is difficult and often subjective. It needs to reflect the amount of capital invested and the nature of the risk associated with the type of venture being undertaken.

9.8.4 Percentage of divisible balance

This is probably the most widely used approach, especially for smaller hereditaments, yet it tends to lack sophistication and evaluation of the rationale for adoption. There is too often a tendency to adopt a 50% approach without taking a more objective view.

9.8.5 A spot figure

This tends to be an approach of the last resort, although it may be appropriate depending on the knowledge and skill of the valuer in relation to the property.

9.8.6 Individual breakdown

In this approach the valuer seeks to reflect the individual items that go to make up the tenant's share. The approach has the advantage that individually the items are easier to ascertain and to justify.

Whatever approach is adopted, it is necessary to ensure that there is no double counting, especially in those instances where the proprietor's remuneration is taken out in the expenses of the business.

9.8.7 Choice of method

The original approaches were to take a percentage either on capital or on gross receipts. In *Underdown (VO)* v *Clacton Pier Co. Ltd* [1958] RVR 460, the Tribunal said:

> *The weight of text book authority and decided cases seem to me to indicate that prima facie the method that one follows in most cases is tenant's capital, except where there would be an insufficiency of capital to give you a reasonable figure; an illustration of this latter is the case of water undertakings.*

The capital involved in a water undertaking was often comparatively small compared to, for example, a gas undertaking, because the operation mostly involved landlord's capital (the hereditament) rather than capital brought along by an occupier. In order to achieve a realistic result a very high percentage on the tenant's capital would have had to be used and therefore a percentage on gross receipts was preferred.

The modern view appears to be that the choice of method is for the valuer (or Tribunal) to select depending on which he or she (or the Tribunal) considers would give the best indication as to the allowance the hypothetical tenant would require. The Tribunal has indicated that it is wary of too mathematical an approach and has shown a preference for the direct estimate or spot figure. In *Garton* v *Hunter (VO)* [1969] 2 QB 37, the Tribunal said on remission from the Court of Appeal:

> *In our view a prospective tenant would quantify his requisite share in terms of an amount, in keeping with both the gross receipts and the divisible balance, rather than in terms of a percentage.*

9.8.8 Criticisms of the approaches

The existence of various approaches to the tenant's share indicates that no one method can be regarded as better than the others. In practice, one approach is normally used for a particular class of hereditament due to convention and, presumably, because it has been found the most satisfactory for the class.

The approach by taking a percentage of tenant's capital can really be used only where the amount of tenant's capital is large and the tenant would be particularly concerned to achieve a reasonable return on it. Even so, the approach does not provide a link between the tenant's remuneration (or share) and the profitability of the enterprise because the tenant's share depends on the capital and not on net profits.

The use of a percentage on divisible balance has been criticised as relying too heavily on conventional percentages and mathematical calculation.

The robust approach of taking a lump sum suffers from the defect of having no point of reference but is a matter of opinion (albeit expert opinion) only.

9.8.9 Examples of tenant's shares in decided cases

There are several cases which serve as examples when considering tenant's shares:

- *St Albans City Council* v *St Albans Waterworks Co. and Clare (VO)* (1954) 47 R&IT 191 – 8% of gross receipts adopted as the tenant's share.
- *Brighton Marine Palace & Pier Co.* v *Rees (VO)* [1961] 9 RRC 77 – 15% of the tenant's capital adopted for a pier.
- *Vaughan (VO)* v *Great Yarmouth Seashore Caravans Ltd* [1960] R&IT 522 – 27.5% of gross receipts adopted for a holiday caravan camp based on evidence of earlier settlements.
- *Ebbutt (VO)* v *Nottingham Colwick Estates Light Rail Co.* (1960) 53 R&IT 537 – the case concerned a small light railway serving an industrial estate and the tribunal said:

  For this particular undertaking, which is a small one, I feel that a tenant would not be satisfied with 10% of the gross receipts and I find that the tenant's share should be £900, which is about 12.5% of the gross receipts or 33% of the net receipts (i.e. divisible balance).

- *LE Walwin and Partners Ltd* v *Baird (VO)* [1968] 14 RRC 265; [1968] JPL 463; 206 EG 1149; [1968] RA 338 – 50% of the divisible balance adopted for a hereditament consisting of 60 bathing hut sites, a car park and premises.
- *Garton* v *Hunter (VO)* [1969] 2 QB 37 – the case concerned a holiday caravan site. The Tribunal adopted a spot figure. 'On the evidence before us

we consider the hypothetical tenant would have based his rental bid on the expectation of receiving £10,000 as his share.' Interestingly, this is equivalent to about 32% of gross receipts or 50% of the divisible balance.

9.9 Review – stand back and look

Just as with the contractor's basis, it is necessary to take a critical look at the resultant figure to ensure that it properly reflects the true rental value of the hereditament. Often, some form of comparative analysis is possible to check whether the valuation is within the expected range for the type of property.

9.10 Local authority overbid

The concept of a local authority 'overbid' was raised in a number of cases in the mid-1950s. The argument behind the overbid is that where the local authority is a potential tenant for the property they would, in some circumstances, be prepared to bid a higher amount in rent than would normally be shown by a strict application of the receipts and expenditure approach. Often, a strict analysis of the accounts would reveal the authority is either making a loss or just covering its costs.

The rationale for the overbid is that the local authority would receive something other than the financial benefit of the occupation of the subject hereditament – this is normally considered to be of benefit to its population and the town in general. As it was receiving these additional benefits it would, in theory at least, be prepared to pay a higher rent than that indicated by the analysis of the accounts.

In *Morecambe & Heysham Borough Council* v *Robinson (VO)* [1961] WLR 373, it was noted:

The fundamental matter of principle in this case is whether there is a point in the scale of proved profits, below which the actual figures cease to be the basis of assessment because they are clearly less than the true value to the tenant. It is my view that there is, and that it has been reached in this case. Mr Stiles' figure in valuation C must be amended by increasing the administrative expenditure and the tenant's share, and the wages increase is too little. The result would be to bring the rateable value below £1,000, and still lower, if the wages and upkeep of gardens and shelters are included. I am quite satisfied from a study of the literature in this case that the value of the foreshore and promenade amenities to the town of Morecambe is such that the corporation would be prepared to suffer a moderate loss, rather than permit this hereditament to remain unlet and unmanaged. I am quite sure that in hypothetical negotiations between a landlord and a tenant, this fact would have its effect upon the rental value agreed.

In *Oswestry Borough Council* v *Plumpton (VO)* (1961) 180 EG 487, it was noted:

> *I think that a person looking at the market from an investment angle would not bid above £1,000, the figure brought out by a valuation on the basis of receipts and expenses, and that whilst it is impossible to be in any degree precise, taking everything into account I feel that the limit to which the Corporation would be willing to go would be an overbid of 25%, giving a rent of £1,250. The rateable value will, therefore, be £1,250.*

Where the receipts and expenditure method is applied to properties occupied by local authorities, then regard should be had not only to the results of the valuation, but also to any additional benefits such a body may receive from its occupation. These benefits could include:

- an increase in status of the town by the presence of the subject property – market, leisure facility, etc.
- promoting and enhancing the business in the town.

Obviously, difficulties arise in defining what the benefits actually are and what they are worth. In some instances it could be argued that, where it is difficult to define the benefits and then to quantify them, an alternative form of valuation should be used, normally the contractor's basis. This approach should overcome the problems. In many instances, however, the adoption of such an approach may well create its own problems. A number of cases considered a local authority overbid:

- *Taunton Borough Council* v *Sture (VO)* [1958] 51 R&IT 749 – a 50% overbid was made to reflect the substantial benefits that a market brings to the town – the benefits being in terms of the status of the town as well as supporting its economic wellbeing.
- *Morecambe & Heysham Borough Council* v *Robinson (VO)* [1961] WLR 373 – the assessment of an esplanade was increased from between £600 and £700 to £1,000 to reflect the added value the amenity gave to the town. The Court of Appeal upheld this decision.
- *Hereford City Council* v *Taylor (VO)* [1960] DRA 611 – a 35% overbid was made for the benefits to the local authority of a market. The Tribunal considered that the local authority would wish to see the continuance of the market and its benefit to the community and would wish to control the market itself, thus ensuring the proper conduct of the market.
- *Oswestry Borough Council* v *Plumpton (VO)* (1961) 180 EG 487 – a 25% overbid was made to allow for the substantial benefit that the local authority receives from the presence of a market in the town.

In recent times there has been a tendency for the term 'overbid' not to be used and the term 'socio-economic benefits' to be used instead, this term encompassing the other, non-financial, advantages that the authority obtains from the occupation of its property. In essence they are the same, but perhaps the latter term more accurately describes what is being valued. This has been particularly the case since 1990, when arguments have arisen over the valuation of a range of local authority leisure properties where the motive of occupation and the true benefits received have been in dispute, for example in *Eastbourne Borough Council and Wealden Borough Council* v *Allen (VO)* [2001] RA 273.

In *Allen (VO)* v *Tyne & Wear Archives and Museum* [2022] UKUT 206 (LC) UTLC an attempt was made to quantify the 'socio economic benefits' arising from a museum, but the Upper Tribunal did not consider the valuation officer had established a method for translating the social value to the public into value to the local authority itself. See section 9.11 Loss-making properties.

In a local Valuation Tribunal case, *Cardiff City Council* v *Clement (VO)* (1996), a profits basis was adopted for the valuation of St David's Hall, Cardiff. In the valuation an additional amount was included under receipts for the 'socio-economic benefits' the council received. The Valuation Tribunal approved the addition for these factors but did not accept that a further addition should be made to reflect the local authority overbid, presumably on the grounds that this would have been a double counting of the benefits. This was the first recorded case where an attempt was made to quantify the value of the benefits and add them to the receipts of the property.

In most cases the approach adopted has been to add a percentage to the rental value found by the application of the profits basis to arrive at a value that reflects the 'overbid'.

The use of the 'overbid' adjustment seems to have fallen into disuse following the post-1990 revaluations, although it may still be a valid means of solving some of the differences in values arising out of the use of the contractor's basis and receipts and expenditure basis of valuation for the same property.

9.11 Loss-making properties

Where a property does not make a profit, a number of arguments have been put forward with regard to the valuation approach to be adopted.

In *Kingston Union AC* v *Metropolitan Water Board* [1926] AC 331, it was noted:

> *It is the occupation of the hereditament in respect of which the hereditament is assessed. Should that occupation be absolutely worthless to everybody, it will have no rateable value, but, on the other hand, if the occupation of it may be very valuable to a hypothetical tenant though that occupation does not secure to him any profit, or bring in to him any income, it may be assessed as by this section provided. No statement of the law on this point could be more unsound and misleading than that which I rather thought had been put forward by the appellants' counsel in argument in this appeal namely, 'No income, no taxable value, no profits, no taxable value'.*

In *Jones* v *Mersey Docks and Harbour Board* (1865) 11 HLCas 443, the following was discussed:

> *The learned judge, in my opinion, did not and could not have meant that it is essential to rateability that a particular occupier of the land can make a pecuniary profit by the use to which he is putting it. It is I think rateable whenever its occupation is of value.*

The first argument is that the property is not capable of beneficial occupation and therefore the occupier cannot be held to be in rateable occupation. In the majority of cases this argument will fail as it can usually be demonstrated that the occupier is receiving some other type of benefit rather than a financial return. This could be in the form of complying with some form of statutory requirement or providing a social service. Since 1990, the term 'socio-economic benefits' has often been adopted to encompass these non-financial benefits.

On the basis that there is beneficial occupation, the next issue is usually what method of valuation should be adopted. If the property is not making a profit, then it would seem inappropriate to adopt the receipts and expenditure basis, as that assumes that there will be some element of profit which can be used to reward the tenant and to pay the landlord in the form of rent. Consequently, it is often argued at this stage that some alternative approach to the valuation is required – usually the contractor's basis.

The courts have generally held that where the property is running at a loss then it is inappropriate to apply the receipts and expenditure method. However, in a number of cases they have accepted that, just because the property is making a loss, there is no reason to automatically reject the method. They have also accepted that a suitable adjustment of the accounts may be sufficient to indicate an appropriate assessment because to use an alternative method may result in an inappropriate level of assessment.

In *Hoare (VO)* v *National Trust [1998]* RA 391; [1997] 2 EGLR 229 (CA), the expenses in the upkeep of Castle Drogo, a stately home, were in excess of its income. The Court, in summary, found that while the National Trust, as the only possible tenant, would take on the considerable repairing liability it would not, in addition, be prepared to pay a rent.

In *Allen (VO)* v *Tyne & Wear Archives and Museum* [2022] UKUT 206 (LC) UTLC an attempt was made to quantify the 'socio-economic benefits' arising from the property.

> *Of course, the local authority funds its museum in order to generate socio-economic value, but there is no methodology available to translate that social value to the public into value to the local authority itself. This is more than a theoretical point; it is fundamental to the rating hypothesis.*
>
> *… there is no methodology available to translate that value to the local authority into willingness to pay any rent at all, let alone how much, on the part of the local authority …*

… It is not obvious to us at all that where rent is paid for a museum that operates at a deficit the reason for that rent is an overbid based on socio-economic value. There may be all manner of reasons why that rent is paid and we are not going to speculate.

The Tribunal determined that the hereditament should be assessed at a nominal value of £10.

There have also been many cases where the property has been making a loss and yet it has been agreed that there should be some assessment as the occupation is of value, though not financially rewarding.

9.12 Valuation by taking a percentage of gross receipts

In recent years there has been an increasing trend to assess and agree assessments of more and more classes of hereditament, by reference to a percentage of gross income. This has been undertaken rather than using a full receipts and expenditure approach.

This approach appears to be comparatively recent and originally used as a check on cinema valuations where the usual practice was to value on a full receipts and expenditure approach. In *Rank Organisation Ltd* v *Billet (VO)* [1958] R&IT 650, this approach was preferred to that of comparison on a value per seat basis:

I prefer the method of arriving at the gross value by means of a percentage on estimated normal receipts, for this method of valuation enables the relationship between capacity receipts and the estimated receipts to be ascertained, and inferences to be drawn therefrom as to whether or not a cinema can be expected to be profitable.

In *Rank Organisation Ltd* v *Priest (VO)* [1966] RA 540, both parties adopted the approach described by the Tribunal as: 'really a shortened and simplified way of making a profits valuation without actually going into all the details such a valuation requires …'

Later, in *Associated Cinematograph Theatres Ltd* v *Holyoak (VO)* [1969] RA 305, the Tribunal stated that 'there can now be no doubt that a proper method of assessing a cinema is by reference to its gross receipts'.

The sort of properties now regularly considered by reference to receipts include bingo halls, bowling alleys, cinemas, football grounds, hotels, pubs, roadside restaurants, stately homes, theme parks and zoos.

The description of the approach of taking a percentage of gross receipts as really being a shortened form of the receipts and expenditure method is somewhat misleading. It is really just a method of comparison in situations where a comparison by reference to gross receipts is more reliable than a comparison on a value per m² basis. If a rent is analysed to show, say, 10% of the gross receipts of a property, then it may be that applying 10% to the gross receipts of a similar property will give a reliable guide to its rent. However, if the percentage is derived from rents then the approach is really a rental comparison method of valuation and not a shortened receipts and expenditure method. If valuations prepared on a full receipts and expenditure approach are analysed to percentages of gross receipts and those percentages are used to value other properties, then it is a shortened receipts and expenditure approach.

A difficulty with using a percentage of gross receipts is that the origin of the percentage can become lost, with valuers applying it with little knowledge of its origin or the weight of the evidence establishing it. Worse still it may have no basis, but merely have become an adopted practice. In *Hoare (VO)* v *National Trust* [1998] RA 391; [1997] 2 EGLR 229 (CA), the Court of Appeal criticised the valuation officer's use of a percentage of receipts as 'completely arbitrary with no reference point in the real world'. The cases concerned the assessment of two stately homes which were open to the public to view. The parties had agreed that the National Trust was the only likely occupier or hypothetical tenant, and the Court found for a nil assessment on the basis that the Trust's policy was only to occupy freehold and therefore it would not have offered a rent or overbid over a nil figure. The Lands Tribunal had accepted the valuation officer's approach of valuing on the basis of 3% of gross receipts, but the Court of Appeal rejected this. The Court did not understand why the valuation started from a turnover figure rather than a net profit figure and said:

> *Moreover the amount of the percentage reduction seems to me equally arbitrary. The resulting valuations give a wholly misleading picture of scientific rigour. One suspects that what the valuer does is to use his evaluation of all the facts of the case and arrive at an intuitive figure and then build a theoretical structure to justify it. I cannot see any rational hypothetical tenant, who (unlike the Trust) is prepared to make an overbid, using that theoretical structure to arrive at the amount of his overbid in his negotiations with the hypothetical landlord. Nor can I see the hypothetical landlord having such calculations in mind.*

In a number of recent cases the Upper Tribunal has criticised the use of the 'shortened method'. It has, though, by no means rejected it out of hand. It described it in one case as an 'approach which seeks to determine the rent at which a hereditament would be expected to let by basing the assessment on a percentage of turnover, rather than a full appraisal of both receipts and expenditure', and suggested:

> *Where, in respect of a particular mode of occupation, a consistent relationship can be demonstrated between the turnover of businesses of that type and the level of profit they generate, a shortened approach can be useful –* Hughes (VO) *v* York Museums and Galleries Trust *[2017] UKUT 0007 (LC) UKUT*.

However, in a succession of cases it was clearly unhappy with its use as a valuation approach or check:

- It was unhappy at the idea of 'distilling all of the attributes, trading prospects and locational aspects of a property into a single numeric factor', i.e. the percentage of gross receipts – BNPPDS Ltd & BNPPDS Ltd (Jersey) as trustees for Blackrock UK Property Fund 2022 LC-2021-102.
- It considered looking at just gross receipts diminished the role of costs as 'no reasonable commercial operator determines the true income potential of a site without reference to expenditure' – BNPPDS Ltd & BNPPDS Ltd (Jersey) as trustees for Blackrock UK Property Fund 2022 LC-2021-102.
- By its nature the shortened method 'cannot fully take account of the different operational costs of different hereditaments'.
- It was 'only appropriate after a rigorous analysis of the available rental evidence, or a significant number of assessments agreed on the full receipts and expenditure basis' – *Allen (VO)* v *Tyne & Wear Archives and Museums* [2022] UKUT 206 (LC).
- Its use was 'ill-advised in the absence of rental evidence or full R & E valuations on which to base the selection of the appropriate percentage' – *Fryer* v *Cox* (VO) 2022 LC-2021-181.

Nonetheless, the approach has a number of benefits:

- One advantage is the comparative ease of obtaining gross receipts. Net receipts tend to be rather more difficult to secure.
- Another is consistency. It is easy to compare one property's assessment with another.
- Some properties have a percentage of gross receipts as the rent review method showing it is used in the market.
- Variations in the quality of similar properties can be accounted for by varying the percentage.

9.13 Illustrative valuation

Describe how you would value for rating purposes the small hotel described below for inclusion in the local 2023 rating list.

The Golden Lion Hotel is situated in a small market town and has 12 double bedrooms and six singles, all with their own bathrooms. The property has a thriving restaurant which is open to both residents and non-residents. For simplicity only the 2020/2021 accounts are given in Table 9.1, but these are in line with earlier accounts.

There are a number of different approaches that could be used to value this type of property (See Table 9.2):

Table 9.1 2020/21 accounts

Receipts	
Restaurant	£1,200,000
Bedrooms	£900,000
Bar sales	£750,000
Function room hire	£61,500
Total income	£2,911,500
Expenditure	
Wages and salaries	£900,000
Purchases	£750,000
Heating and lighting	£271,500
Rates	£150,000
Telephone and stationery	£30,600
Insurance – building	£27,000
Insurance – contents	£31,500
Repairs – building	£60,000
Repairs – tenant's chattels	£120,000
Hire purchase	£61,500

Table 9.2 Valuation of hotel

Receipts	£	
Restaurant	£1,200,000	
Bedrooms	£900,000	
Bar sales	£750,000	
Function room hire	£61,500	
Total income	**£2,911,500**	
Expenditure		
Wages and salaries	£900,000	
Purchases	£750,000	
Heating and lighting	£271,500	
Rates	£150,000	
Telephone and stationery	£30,600	
Insurance – building	£27,000	
Insurance – contents	£31,500	
Repairs – building	£60,000	
Repairs – tenant's chattels	£120,000	
Hire purchase	Not allowed	
Mortgage interest	Not allowed	
Total expenditure	**£2,340,600**	
Gross receipts	£2,911,500	
Less		
Expenditure	£2,340,600	
Divisible balance	£570,900	
Less tenant's share		
i. Remuneration	£100,000	
ii. Interest on capital employed		
Stock	£80,000	
Fixtures	£200,000	
Cash	£25,000	
	£305,000	
@ 7%		£21,350
iii. Risk		£55,000
Rent		**£394,550**

1 full receipts and expenditure approach;
2 shortened receipts and expenditure approach.

Based on an assessment of £394,550 the assessment could be analysed to a price per bedroom as can be seen in Table 9.3. (The ratio would have been determined from the relativities between single and double rooms as indicated in the analysed accounts.)

Therefore, a double bedroom would analyse at a rental value of £23,485 and a single bedroom at £18,788. This with other comparables could then be used as a basis for the valuation of similar properties.

Table 9.3 Analysis of assessment for shortened receipts and expenditure basis

12 doubles	@ 1.00
6 singles	@ 0.80

9.14 Recent tribunal cases

Fryer v *Cox VO* [2022] UKUT 0229 (LC)

Allen (VO) v *Tyne & Wear Archives and Museums* [2022] UKUT 206 (LC) UTLC

Hughes (VO) v *Exeter City Council* [2020] UKUT 0007 (LC)*Hughes (VO)* v *York Museums and Galleries Trust [2017]* UKUT 0007 (LC))

9.15 Wales

There are no differences between England and Wales.

10 The valuation of shops

10.1 Introduction

Retail is the largest class of non-domestic hereditaments. There are some 510,226 shops, supermarkets, banks, department stores and other retail units in England and Wales in the 2023 draft rating lists with a total rateable value of £14,756,178,476.

As a generality, rental evidence is plentiful for shops and other retail premises and valuations are undertaken by comparison using the rental method of valuation. When using a comparative approach an analysis scheme is needed which best fits the class of property. For most shops the zoning approach to analysis has been found most suitable. Zoning is not a method of valuation but a means of comparing one shop with another for analysis and valuation. Over the years shopkeepers realised that, in most cases, they made a greater proportion of sales from goods displayed at the front rather than the rear of shops. They have therefore paid less additional rent for extra area to the rear of shops compared to what they have been prepared to pay for extra area at the front.

Valuers have attempted to interpret this in a formal way by attributing a higher value per m^2 to area at the front of a shop compared to the area at the back. They have usually done this by dividing the shop into a number of zones parallel to the frontage and treating each zone as being worth half that of the preceding zone. The usual depth of these zones is 6.1 m, which is the metric equivalent of 20 ft. Commonly, three zones are taken, A, B and C, with any additional depth being described as the Remainder rather than Zone D. Other parts of the shop are usually allocated values in proportion to the Zone A value.

10.2 Zoning

Zoning takes as its basic assumption that the most valuable part of a shop is the area closest to the frontage and areas of sales space further back into the shop are less valuable. It has therefore been accepted that devaluing rents on an overall basis per m^2 will not give a good common unit of comparison for comparing shops of different depths. Zoning is a method of analysis that takes depth into account. For large shops it is nowadays normal to devalue their

DOI: 10.1201/9781003297369-10

rents on an overall basis rather than a zoning approach, but it remains appropriate for the small and medium-sized shops.

Two zones and a remainder were often used in the past, but nowadays it is usual to take three or four 6.1 m depth zones and a remainder, and to halve back the values for each successive zone.

10.2.1 Example 1

A lock up shop has been let for £33,000 per annum (pa). It has a depth of 20.9 m and a width of 6 m. Analyse the rent of a zoning approach.

Zone A	6×6.1	=	36.6	@ £500 per m^2	=	£18,300
Zone B	6×6.1	=	36.6	@ £250 per m^2	=	£9,150
Zone C	6×6.1	=	36.6	@ £125 per m^2	=	£4,575
Remainder	6×2.6	=	15.6	@ £62.5 per m^2	=	£975
						£33,000

Halving back is used in the above example and by trial and error it would be possible to work out the Zone A value of £500 together with the Zone B, Zone C and remainder values of £250, £125 and £62.50. An easier and quicker approach to analysis is to convert the areas of the various zones into the equivalent Zone A area. As values are 'halved back' the value of each square metre of Zone B will be half that of Zone A. Zone B can therefore be shown in terms of Zone A (ITZA) as equal to half the area of Zone A. In the above example this would show as:

Zone A	6×6.1	=	36.6	@ A/1	=	36.60
Zone B	6×6.1	=	36.6	@ A/2	=	18.30
Zone C	6×6.1	=	36.6	@ A/4	=	9.15
Remainder	6×2.6	=	15.6	@ A/8	=	1.95
						66.0 m^2 ITZA

£33,000/66.0 m^2 ITZA = £500 per m^2 Zone A

This provides a means of quickly and easily analysing shop rents if the area ITZA is known. This enables comparison of different shops' rents to establish a level of value. In *Sainsbury (J)* v *Wood VO* [1980] RA 204, it was said that zoning was devised 'to apply the evidence of rents to shops of varying depths to arrive at a pattern of values to be attributed to other, broadly similar, shops where no rental evidence was available'.

Ancillary areas such as rear or first-floor stores are usually viewed as having a value per m^2 as a given fraction of Zone A. If a scheme of analysis treats ancillary stores as having a value of a twelfth of Zone A, a rear store of 24 m^2 will have an area ITZA of 2 m^2.

10.2.2 Example 2

A lock up shop has been let for £50,000. It has a depth of 26.4 m and a width of 7 m. There is an ancillary store with dimensions of 3 m × 4 m and a staff room of 2 m × 3 m. The usual analysis scheme for this locality adopts a tenth of Zone A for staff rooms and a twelfth for store rooms.

Zone A	7 × 6.1	=	42.7	@ A/1	=	42.70	
Zone B	7 × 6.1	=	42.7	@ A/2	=	21.35	
Zone C	7 × 6.1	=	42.7	@ A/4	=	10.67	
Remainder	7 × 8	=	56	@ A/8	=	7.00	
Staff room	2 × 3	=	6	@ A/8	=	0.75	
Store	3 × 4	=	12	@ A/12	=	1.00	
						83.47 m^2 ITZA	

£50,000/83.47 m^2 ITZA = £599 per m^2 Zone A

Note that A/8 and A/12 are based on the analysis of rents in the local area. This will vary from one location to another and will also depend on the quality of the accommodation.

10.3 Numbers of zones and zone depths

As mentioned, it is the usual practice across the country to adopt an analysis scheme of three 6.1 m zones and a remainder (6.1 m being the metric equivalent of 20 ft). While it is an attractive idea to adopt a more truly metric zone depth and use a round number such as 7 m or 10 m zones, this has not been done due to the amount of work involved in re-calculating the areas of all the shops in the country. In some areas different zoning patterns are used. Shops may be zoned back in 6.1 m zones without a remainder. In Oxford Street a 9.1 m, 9.1 m, 9.1 m Remainder pattern is used.

Other more esoteric approaches to zoning exist, such as natural zoning and logarithmic zoning. This diversity indicates there are no hard and fast rules and valuers are at liberty, in analysing rents, to adopt the approach they consider correct. However, there are two important cautions to doing this:

- It is essential to value as you devalue. Analysed evidence on one zoning scheme cannot be applied to areas ITZA calculated on a different basis in order to reach a valuation. Clearly, the wrong answer will be achieved if rents ITZA analysed on an analysis scheme of 9.1, 9.1, 9.1 Remainder are applied to a shop whose area has been analysed on a 6.1, 6.1, 6.1 Remainder basis.
- To apply a different zoning approach or change some aspect of an existing scheme such as the fraction of Zone A used for ancillaries requires

re-calculating the area ITZA for all the rented comparables as well as the properties to be valued so they are all analysed on the same basis. This can be very time consuming.

10.4 Ancillaries

As mentioned, ancillary areas are normally analysed and valued as a fraction of Zone A values. Sometimes these fractions are referred to as 'X factors' with X being the denominator of the fraction A/X.

While the use of fractions of Zone A is well established as an approach, it is perhaps clearer to use percentages, i.e. 25% or 10% of the Zone A value for the analysis and valuation of ancillaries.

The fractions for ancillary areas depend on the use, quality and position of the various ancillary parts. While the selection of suitable factors is a matter for the valuer's judgement backed by an analysis of the evidence, the maxim of 'as you devalue so must you value' also applies. It is not enough merely to consider that a particular part, say a first floor, has been taken at too high a figure by the valuation officer in a valuation if the analysis of comparable shop rents has also been made on this basis. A re-analysis of all the comparables is necessary to satisfy the principle of comparing like with like. If a higher 'X factor' for first floors is adopted, say A/14 instead of A/10, the effect on the analysis will be to increase the Zone A value while decreasing the value attributable to first floors.

10.5 Plant and machinery

Certain plant and machinery which is found in retail premises may be rateable (see Chapter 13 The Valuation of Plant and Machinery). Many items which are rateable are often reflected in the unit value for the accommodation.

The most common items are sprinklers, air conditioning plant and racking, though not all systems of air conditioning or racking will be rateable.

10.6 Quantity allowances

Zoning takes account of changes in value due to depth, and does, therefore, incorporate some allowance for size but is not primarily designed to cope with size as such. For shops size can be seen as having three different components:

- Size in depth – the depth of a shop may be more than is normal in a locality. By adopting zones analysed and valued at half of the preceding zone, zoning may take account of this size in depth, as well as allowing for depth as such, by using extra zones or changing the remainder's 'X factor'.
- Size in width – the width of a shop also affects its size but in its basic form zoning makes no allowance for this variable.

- Size in different levels – the existence of upper and lower floors are also part of the size of a shop. The use of varying X factors allows the valuer to take the relative values of different floors into account.

It is often considered that values for property drop with size: an allowance for buying in bulk in effect. However, such a quantum allowance should not be regarded as automatic and is dependent on the demand for a particular type and size of shop in a locality. In one town there may be a shortage of larger units and demand may even be such as to give them a premium value!

In the case of *Trevail (VO)* v *C and A Modes Ltd* [1967] RA 124, the Lands Tribunal established a three-stage test for quantum allowances:

- no presumption in favour of quantum;
- only comparatively slender evidence is required to establish that the market still allows for quantum in any location;
- once this is done positive evidence is required in order to displace the assumption of a quantity allowance.

These tests can be regarded as being as relevant today as in the 1960s. Each case needs to be decided on its own merits, and what may apply in one locality may not in another. It is a matter of evidence which exists in the locality. In *Trevail (VO)* v *C and A Modes Ltd* [1967] RA 124, 5% was allowed for department and walk round stores, but in *FW Woolworth and Co Ltd* v *Peak (VO)* (1967) 14 RRC 5, the Lands Tribunal held that the evidence showed that a reverse quantum allowance was appropriate, showing an additional value for larger shops.

As the Lands Tribunal said in Trevail there is no automatic presumption of quantum. It needs to be demonstrated by evidence. As the Upper Tribunal said in *Harrods Ltd* v *Baker (VO)* [2007] RA 247:

> *There is thus no inference to be drawn that simply because of its very considerable size Harrods should be valued at a lower rate per square metre than the comparables. If there is to be an allowance for quantum, it must be sought elsewhere.*

Reverse quantum, or an increase in value for size, may result from a shortage of large units in a locality. This might be where there is no room to increase the number of large shops, perhaps in a historic town. National multiples may be keen to be represented but do not wish to trade in a standard sized unit which might suit a local trader or a specialist multiple but is generally too small for the national multiples. In such cases the national multiples may be prepared to pay a premium rent in competition in order to secure occupation. A triple unit in a local parade may command a premium rent rather than a discount for quantity as it is the only unit suitable for use as the (lucrative) local 24-hour supermarket.

Quantum allowances should not be seen as applying just to the very largest of shops. It is quite possible to find evidence in some localities showing a rent less than the total for two single shops is paid for a double shop unit. In any particular case it is for the valuer to carefully examine the market and the evidence and to decide whether it shows quantum.

10.7 Allowances for shape

The shape of a shop will substantially affect the rent the unit will achieve. Unusual shapes causing disabilities can take a number of forms.

10.7.1 Frontage to depth ratio

The zoning approach is designed to allow for depth in a shop, but makes no allowance for a shop having an excessive frontage and therefore excessive Zone A in relation to its depth. If a shop has a long frontage, but a depth of little more than 6.1 m, virtually all of it will be Zone A. If the standard Zone A value derived from the evidence of shops with a normal ratio of frontage to depth is used it is likely the shop will be overvalued.

An allowance will need to be made as the basic zoning method does not cope with this problem. This can be done by an end allowance, or by some recalculating of adjoining shops onto an overall value per m² basis and making a direct comparison adjusted for the advantage of a long and prominent frontage.

The typical ratio of frontage to depth for standard shops is 1:2.5 or 1:3 on sales space.

The factor was particularly noted by the Lands Tribunal in *WH Smith* v *Clee (VO)* [1978] RA 93. The Tribunal adopted a 14% allowance where the ratio was 1:0.34 (137 ft frontage to 46 ft depth) describing the allowance as an 'allowance for shape'.

10.7.2 Unusual shape

In practice a great variety of shop shapes are met and the unusual ones are not always old shops or amalgamations of various shops. Modern developments can have some quite unusual shapes. The shops may have:

- an irregular shape;
- masked areas;
- split levels.

Apart from causing difficulties in the measurement of the unit, these factors may affect value. For smaller shops, irregular shape due to intrusions, such as staircases into the sales area or pillars, can cause difficulties with shop fitting and display. Areas of sales space that are masked may not be noticed by shoppers with a consequent loss of sales, or may require the inconvenient placing or the extra placing of staff to deter shop-lifting in areas masked from what would normally be the counter or till. As always it is important to consider rental evidence. Do the rents show allowances for masking, etc., or as in older centres, are the features common to most of the comparables?

For larger stores an unusual shape may have little effect on value. Staff in such stores will be spread more widely and the effect of shape will be less noticeable to customers given the greater space. They will be just as prepared to walk into the rear arm of an 'L'-shaped large shop as they would have been to walk into the deep remainder of a similar but regularly shaped shop.

If an allowance is justified, it can be considered by either reducing the £ per m² applicable to the masked area or adopting an end allowance for shape as a percentage of sales area value.

Irregular shape such as a frontage not at right angles to the shop's sides can present a problem in knowing where the zones should start and what shape they should be. Usually if the frontage is at an angle, the zones will run parallel to that angle through the shop.

Split levels in the sales area running from front to back or across the sales area are also likely to be a disadvantage as they divide the shop area, can create display difficulties and result in unusable space where the steps or ramps are sited. On the other hand, shop fitters sometimes put in raised floors to parts of sales areas to create interest. Again, in approaching the valuation a change in £ per m², X factor or an end allowance can be made. In *Lonhro Textiles* v *Hanson (VO)* (1984) 271 EG 788, the Lands Tribunal adopted a fraction of approximately A/8 instead of A/4 for the rear lower ground-floor sales area which had an unusual shape (widening to the rear) and was 1.2 m below the front ground-floor sales area.

10.7.3 *Return frontages*

Return frontages vary from those of a few metres onto narrow side streets, to a corner shop in a modern shopping mall with full frontages to both malls. The critical factors suggesting additional value are the size and visibility of the return frontage. The addition, if any, will logically bear a relationship to the main frontage. Ideally, analysis of rental evidence for shops with return frontages compared to those without should yield evidence of value. This is usually analysed as a value per metre run, a spot figure or a percentage of the sales area value.

Difficulties have arisen where a shop has frontages to two good shopping streets. Arguments and interesting zoning calculations have been advanced seeking to zone from both frontages or seeking to zone from one or other frontage.

In *WH Smith* v *Clee (VO)* [1978] RA 93, the Lands Tribunal was content to accept as a check valuation the ratepayer's approach of zoning from the shorter, more valuable, frontage and then making a 30% addition for the return frontage. In *Time Products* v *Smith (VO)* [1984] RVR 106, the Lands Tribunal, again as a check, zoned from the shorter, equally valuable, frontage and made a 10% addition for the return frontage.

Some shops lie between two shopping streets and have access from both of them. The usual approach is to zone from both frontages. Zoning from the inferior entrance ceases when it is less in value than the value attributed when zoning from the primary frontage.

10.7.4 Non-ground-floor sales and ancillaries

The usual practice of ascribing fractions of Zone A value to non-ground-floor sales and ancillaries to indicate their value relative to the rest of the hereditament was described earlier. Ideally, analysis should be undertaken by deducting the actual value of the ancillary parts from the rents before zoning the ground-floor sales area rather than by applying fractions to them.

The fractions used are often pre-determined, representing the valuer's opinions or what has been used in the past for previous rating valuations or schemes of valuation for earlier rating lists. However, in theory there should be for any locality a pattern of relativities for basements, first-floor sales, storage, offices, etc., which best fits the evidence and is therefore likely to represent market rental value.

The value of the sales space on first or basement floors will vary depending on ease of access via stairs, escalators or lifts and where these are situated – in Zone A or at the back of the shop. The Lands Tribunal has also considered the situation where the ground-floor sales space in a shop is a very small proportion of the sales space, perhaps little more than an entrance. In *Westminster City Council* v *Burton Group and Saul (VO)* (1986) 277 EG 757, the Tribunal adopted A/10 for first-floor sales, where there was eight times more sales space on the first floor compared to the ground floor of a shop in the West One Shopping complex on Oxford Street at Bond Street Underground Station. The Tribunal noted, in reaching its decision, that first-floor sales were usually valued at between A/8 and A/10 in Oxford Street. In *Reject Shop Ltd* v *Neale (VO)* [1985] 2 EGLR 219, the Tribunal adopted A/7 for first-floor sales but, noting the very small ground-floor part (35.6 m^2 on the ground floor compared to 583.8 m^2 on the first floor), made a 15% allowance to reflect its view that first-floor sales space is more valuable when occupied together with ground-floor sales space.

10.7.5 Zoning criticism

The zoning approach to analysis and valuation has come under increasing criticism and questions are raised about whether, in the second decade of the twenty-first century and with all the changes in retail stemming from online shopping, COVID-19 and the way shopkeepers set out their shops, zoning is no longer appropriate. Certainly, it is not an approach upper-most in retailers minds when deciding to rent a shop, the retailer being more likely to consider expected turnover and what the operation can afford to pay. It is, though, still used for rent reviews and, of course, Rating where demonstrable uniformity is important.

In the past zoning was certainly used to enable the valuation of large shops by comparison with the zoned analysis of rents on smaller shops but nowadays large shops are looked at on an overall basis and rents of large shops analysed on an overall basis. Should therefore unit and small shops be looked at overall? Is it merely habit that retains the use of Zoning?

The difficulty of a simple overall approach is it completely ignores frontage whereas it is clear frontage is important to retailers as it gives 'impact' and draws shoppers in. The frontage is, indeed, somewhat advertising. If advertising is not just about letting consumers know about new products or products that are new to them but also about reinforcing 'brand awareness', a shop front very much has that function as well as the local draw into that particular shop. It is obvious that a shop which is narrow and deep will not be as valuable as a shop that is a regular shape with a good frontage, sensible depth and the same square area of retail space. Overall approaches would not deal with the value of frontage, and that is something zoning does take very much into account. Perhaps an approach that takes into account both area and frontage is needed but while zoning with its 'halving back' can be criticised, it does not seem an overall approach, as such, will give a sound comparative analysis one shop to another.

10.7.6 Natural zoning

The approach to zoning explained in section 10.2, using pre-determined zone depths and halving back of values, is not the only way of analysing shop rents. Another approach is called natural zoning and involves determining the appropriate zone depths and relative values by an examination of the normal shop depths in the locality. Zone A will be the depth of the shallowest shops in a locality. Zone B will be the difference between their depth and the next size of shop and so on. Halving back does not apply automatically to natural zoning. The relationship between zones A, B and C and the Remainder is determined by the difference in rents between the shallower and deeper shops.

10.7.7 Example – natural zoning

In a locality there are three rented shops. Shop 1 is let for £21,000 and has a depth of 7 m and a width of 6 m, Shop 3 is let for £36,000 and has a depth of 17 m and a width of 6 m and Shop 5 is let for £30,000 and has a depth of 12 m and a width of 6 m. Analyse the rents on a natural zoning approach.

In this example Shop 1 has the smallest depth at 7 m. The whole of the 7 m depth can be termed Zone A and the rent therefore analyses at £500 per m² for Zone A. Shop 5 is the next deepest with a depth of 12 m. The evidence for Shop 1 indicates that the first 7 m of depth, i.e. Zone A, has a value of £21,000 leaving £9,000 for the 5 m. This 5 m can be termed Zone B and the £9,000 analyses at £300 per m². The deepest shop is Shop 3 with a 17 m depth. The evidence for Shops 1 and 5 shows the first 12 m has a value of £30,000 leaving £6,000 for the final 5 m. This 5 m analyses at £200 per m². This evidence produces a zoning pattern of:

Zone A	6 × 7	=	42 m²	@ £500 per m²	A/1
Zone B	6 × 5	=	30 m²	@ £300 per m²	A/1.667
Zone C	6 × 6.1	=	36.6 m²	@ £200 per m²	A/2.5

The shops without rents can then be valued using the evidence from the above analysis. Shop 2 has a depth of 15 m and a width of 6 m. Shop 4 has a depth of 10 m and a width of 6 m. Shop 6 has a depth of 16 m and a width of 8 m.

Shop 2
Zone A	6 × 7	=	42	@ 500 per m² =	21,000
Zone B	6 × 5	=	30	@ £300 per m² =	9,000
Zone C	6 × 3	=	18	@ £200 per m² =	3,600
					£33,600

Shop 4
Zone A	6 × 7	=	42	@ £500 per m² =	21,000
Zone B	6 × 3	=	18	@ £300 per m² =	5,400
					£26,400

Shop 6
Zone A	8 × 7	=	56	@ £500 per m² =	28,000
Zone B	8 × 5	=	40	@ £300 per m² =	12,000
Zone C	8 × 4	=	32	@ £200 per m² =	6,400
					£46,400

Natural zoning has a theoretical purity in that the depth of the zones and the relative value of each zone is derived from the evidence in the locality, unlike the normal method of zoning where pre-conceived or standard zone depths are used together with halving back. An arbitrary zoning pattern of 6.1, 6.1, 6.1 Remainder with halving back may not provide the best fit to the evidence available in a particular locality, whereas natural zoning is tailored to the actual evidence in that locality.

However, the normal Arbitrary or Arithmetical Zoning approach is almost invariably used because it is simpler and quicker to undertake and has a long acceptance of usage. Further, except in perhaps some modern shopping developments, it is usual to find a much greater variety of shop depths than were included in the example which, on a natural zoning approach, would logically result in a large number of comparatively shallow zones.

Other approaches exist. The idea of halving back, while universally followed nowadays, has been questioned in the past with 'thirding back' suggested as an alternative. A more radical solution to avoid the dramatic change in value at the zone margin is to introduce a much larger number of zones or to envisage, instead of the value steps, a steady reduction of value using a value curve. It is of course possible to ignore the idea of the front part of a shop being more valuable than the rear and adopt an overall approach. In a modern centre with little difference in depth from one shop to another, this may be a reasonable approach. Certainly an overall approach is used for very large shops where it has been found that it is not possible to relate the rents of small shops and very large shops on a zoning approach and, indeed, it is difficult to find a connection between the two.

10.8 Turnover rents

Rents derived from a percentage of gross receipts are a comparatively recent idea for shops in Britain and were possibly first used for the Eldon Square shopping centre development in Newcastle.

Turnover rents are usually set at a fixed percentage of a trader's turnover for an individual shop but subject to a minimum base rent. Until a certain set turnover is reached only the base rent is paid. When the turnover exceeds the set figure the rent is calculated using the turnover and the percentage specified in the lease. For example, if the percentage for the trader has been set in the lease at 10% and the base rent set at £30,000 then the additional turnover rent becomes payable when the turnover exceeds £300,000. If in a particular year the turnover for the shop is £350,000 then the total rent payable will be £35,000. If the turnover is only £250,000 then the base rent of £30,000 remains payable.

There are advantages and disadvantages for landlords and tenants in this type of arrangement. A landlord can expect with turnover rents to maintain a constant 'full' income rather than having to wait until each rent review for rents returns to full rental value. Because the leases provide for disclosure of turnover to the landlord – indeed the tills of the shops may be connected to the landlord's computer – the landlord can better determine the optimum tenant mix for the success of the centre; there is a more obvious common aim between landlord and tenant in maximising turnover and making the centre a success, and tenants have some relief in rent because the rent can drop back down to the base rent if trade is poor for a time.

The base rent is usually set at around 75–80% of rack rent. The usual policy is for the landlord to let a few units on a normal basis in order to determine the level of rack rent for a shopping centre, although the base rent can be determined on the basis of previous years' trading.

The proportion of turnover which traders can pay is determined by profit margins. These vary considerably between trades and, indeed, can vary for different retailers in the same trade. Food sales from supermarkets trade on large volumes but narrow margins, whereas jewellery is very much the reverse.

The clauses in a lease for properties let on turnover rents can be complex, with clauses requiring a minimum level of performance, clauses requiring the notional turnover to be increased if the shop is closed for a number of days, restrictions on assignment to similar trades only, etc.

The typical percentages of turnover used for rents are:

Jeweller	10–11%
Ladies' fashions	8–10%
Men's fashions	8–9%
Radio and electrical	7%
Catering	7%
Shoes	8–10%
Records	9%
Sports goods	9%
Greengrocer	6%
Butcher	4–5%
Baker	5%

10.9 Superstores and hypermarkets

These types of store tend to be valued on what is termed the overall approach. This consists of applying a single rate per m^2 to the whole of the sales area of the property and lesser figures to the ancillary accommodation. An alternative approach is to analyse and value on a completely overall approach, with all areas being analysed and valued to a single £ per m^2 gross internal area (GIA). The unit price will be determined by a number of factors including the quality of the accommodation, competition and catchment area. Many stores are rented and valuations can be prepared from an analysis of rents. It is important to be aware of whether the rent is for the fitted out or unfitted unit.

Commonly, large supermarkets are classified as:

Large food stores 750–2,500 m^2 GIA
Superstores 2,500–5,000 m^2 GIA
Hypermarkets 5,000+ m^2 GIA

There have been some cases of this type of store being valued on a zoning basis, but in the main this has been for specific purposes connected to the individual property.

10.9.1 Typical valuation approach

Ground external storage	9.60	@ £63.75	£612
Mezzanine internal storage	15.70	@ £25.50	£400
All main areas	6,808.40	@ £255.00	£1,736,142
Car wash (Class A+)	1.00	@ £10,952.00	£10,952
Petrol forecourt and shop			£134,084
			£1,882,190
Say			**£1,880,000**

10.9.2 Valuation issues

- Quantity allowance – the basic price for the sales area will usually reflect any allowance for quantity where appropriate.
- Concessions – a common feature of these properties is for a part of the store to be 'let out' to a separate company selling a complementary product or service. This can range from cafes, hairdressing and video rentals to banks and building societies. Where this occurs it is important to identify carefully the unit of assessment to be valued (see *Peter Arnold Ltd* v *Riley (VO)* [1986] RA 172 – valuation of unit within a superstore).
- Petrol stations – many stores have associated petrol filling stations. The location of the petrol station may result in it being identified as a separate hereditament.
- Customer parking – the whole ethos of such stores means that a large car park either on site or within close proximity is essential. Often the car park will have a capacity of between 600 and 1,000 cars depending on the size of the store. Ideally parking should be on the same level as the store with no steps, etc., for trolleys to contend with.
- Identity of the hereditament – as with retail warehouses, it may be difficult in certain circumstances to identify the hereditament or hereditaments and the rateable occupier.

10.9.3 Cases

- *Norwich City Council* v *Sainsbury (J) Ltd and Yelland (VO)* [1984] RA 29
- *Sainsbury (J) Ltd* v *Wood (VO)* [1980] RA 204
- *Watts (VO)* v *Royal Arsenal Co-operative Society* [1984] RVR 237.

10.10 Retail warehouses

This type of store is also usually valued on an overall approach, either treating different parts of the accommodation as having different relative values or using a single figure overall to GIA. As opposed to superstores and hypermarkets these warehouses are often grouped on estates with other similar uses and will usually be situated on main roads. The estates may also contain leisure and other facilities.

10.10.1 *Typical valuation approaches*

Ground-floor sales	2,997.21	@ £185.00	£554,484
Air conditioning	2,997.20	@ £7.00	£20,980
Mezzanine floor sales	1,920	@ £30.00	£57,623
			£633,087
Say			**£630,000**

10.10.2 *Valuation of a retail warehouse*

Ground-floor sales	9,240.00	@ £120.00	£1,108,800
Entrance	235.00	@ £120.00	£28,200
Ground-floor office	747.40	@ £120.00	£89,688
First-floor office	339.40	@ £120.00	£40,728
Ground floor covered sales	220.00	@ £33.00	£7,260
Outdoor display	1,345.00	@ £18.00	£24,210
Canopy	1,496.00	@ £24.00	£35,904
Builder's yard	399.00	@ £24.00	£9,576
Greenhouse	310.00	@ £30.00	£9,300
			£1,353,666
Say			**£1,353,650**

The assessment reflects the presence of car parking to the front and side of the property and a service yard to the rear.

10.10.3 *Valuation issues*

- Car parking – car parking is essential to this type of property. In many instances such stores may be situated on retail parks which share a common car park. In these instances it may be difficult to identify the rateable occupier of the car park (if there is one) but the assessment of the units will reflect the availability of such parking facilities.
- External sales area – many DIY-type stores now require external sales areas for garden centres, sales of flagging, stonework, etc.
- Service yards – well laid out service yards are essential to this type of property. It is a common requirement for loading bays to have adjustable ramps for ease of unloading pallets, etc.
- Concessions – a common feature in these types of properties is for part of the store to be 'let out' to a separate company selling a complementary product or service. This can range from double glazing or central heating

to other similar products. Where this occurs it is important to identify carefully the unit of assessment to be valued.

10.10.4 Cases

- *Norwich City Council* v *Sainsbury (J) Ltd and Yelland (VO)* [1984] RA 29
- *Sainsbury (J) Ltd* v *Wood (VO)* [1980] RA 204
- *Watts (VO)* v *Royal Arsenal Co-operative Society* [1984] RVR 237.
- *Go Outdoors Ltd* v *Lacy (VO)* [2019] UKUT 0051 (LC)

10.10.5 Department stores

In *John Lewis & Co Ltd* v *Goodwin (VO) and Westminster City Council* (1979) 252 EG 499, the Lands Tribunal decided on a fairly simple definition of a department store as a 'large shop or store, having a substantial sales area above the first floor' drawing a distinction between department stores 'and (possibly equally large) shops or stores with any floor areas above the first floor being used for non-sales'. Generally, department stores will have a GIA in excess of 9,250 m^2.

The valuation approach for department stores has exercised the minds of valuers for many years, with one early textbook even suggesting the use of the contractor's basis. In *Trevail (VO)* v *C and A Modes Ltd* [1967] RA 124, the Lands Tribunal commented, in valuing a department store and a large variety store, that the problem of quantity allowances did raise doubt as to 'whether in cases such as this the zoning method is not being stretched a good deal beyond its capabilities'. Zoning is now regarded as an unsatisfactory approach to analysing department store rents, although it may still have some use as a secondary check. The approaches used are either to analyse rents and undertake valuations on an overall approach applying a single value per m^2 to the overall area reflecting all the different advantages and disadvantages, or to attribute different factors to the different floors.

Due to a number of sale and leaseback transactions in recent years there are perhaps a surprising number of rented department stores to give evidence of rents. Analysis by overall area, including staff and storage accommodation, is undertaken using GIA and department stores are compared having regard to their advantages and disadvantages. Many stores are the result of merging units over the years and may not be as conveniently set out when compared to purpose built stores, often having changes in level and unsatisfactorily located stairs and escalators. With increasing offsite storage, it is usual to find that sales space accounts for 70–75% of GIA. If it is less than this then the proportion of storage is unusual and this would be a disadvantage.

In *Harrods Ltd* v *Baker (VO)* [2007] RA 247, the Lands Tribunal considered the valuation of the largest department store in the United Kingdom. The Tribunal, following the parties, used an overall approach to its valuation derived from the rent of another department store.

Basic value 140,133.9 m² @ £135	£18,918,077
Less lack of air conditioning	–£161,154
Less exceptional maintenance costs	–£1,586,000
Less exceptional delivery costs	–£586,225
	£16,584,698
Say rateable value	**£16,575,000**

10.10.6 Cases

* *Trevail (VO)* v *C and A Modes Ltd* [1967] RA 124
* *John Lewis & Co Ltd* v *Goodwin (VO) and Westminster City Council* (1979) 252 EG 499
* *Harrods Ltd* v *Baker (VO)* [2007] RA 247.
* *Ballcroft Estates Ltd* v *Virk (VO)* [2022] UKUT 153 (LC)

10.11 Kiosks

The zoning approach does not seem to work well with the very largest of shops and the same is true for the smallest – kiosks. With a kiosk all the area is in Zone A but it is likely that applying the local Zone A rate to the kiosk will undervalue the unit. There appears to be a basic value for just being able to trade in a locality. This is known as 'foot in' value. In the case of *Suri* v *Masters (VO)* (1982) RA 271 262 EG 1308, the Lands Tribunal considered the most important factor in valuing kiosks was their location rather than their size. It also regarded frontage as more important than depth and size, commenting on the very narrow frontage (0.7 m) of the appeal hereditament.

10.11.1 Typical valuation approach

Table 10.1 Valuation of kiosk

Description	Area	£/m²	£
Sales area	2.20	£7,700	£20,020
Say			**£20,000**

10.11.2 Cases

* *Suri* v *Masters (VO)* (1982) RA 271 262 EG 1308
* *Weston Super Mare Borough Council* v *Escott VO* [1954] R&IT 23.

11 The valuation of offices

11.1 Introduction

Offices can range from a single room above a shop through an office building within a factory complex to a city centre office block. An occupier's requirement for the location of the office space and the facilities that he or she requires from that accommodation will vary according to the category for which the office is ultimately used.

Offices may broadly be classified into a number of different categories:

1. those designed primarily to provide office accommodation and that are generally a purpose designed building;
2. those where the public may gain access, such as banks and building societies;
3. those that are ancillary to alternative uses, such as offices in industrial complexes (this category is not considered in this chapter);
4. offices above shops and other properties.

11.2 Basis of measurement

The Royal Institution of Chartered Surveyors (RICS) recommends that offices should be measured to IPMS All Buildings Standard following the international standard and the RICS *Property Measurement* 3rd Edition 2023. However, it is normal practice for rating to measure to net internal area (NIA), having regard to the old definition in the RICS *Code of Measuring Practice*, sixth edition.

High Tech office like Planning Class E(g) hereditaments are likely to be compared to industrial and warehouse hereditaments as well as offices and should be measured to gross internal area (GIA) in addition to NIA to allow comparison to take place.

11.3 Survey requirements

While carrying out the survey special attention should be given to the following features.

DOI: 10.1201/9781003297369-11

11.3.1 *External*

- general description – a general overview of the property;
- construction – details of the type of construction and its main features;
- age – plus details of when any extensions or refurbishment took place;
- type of office – purpose built, hi-tech, converted, above shops;
- location – general local description, central business district, peripheral, out of town, business park, etc.;
- access – general access to the property, off side street, visibility of access, etc.;
- transport facilities – public transport, cars, delivery vehicles, accessibility for pedestrians.

11.3.2 *Internal*

- entrance – sole or shared, quality of entrance;
- walls – structural or non-structural, type and quality of finish;
- floors – solid, timber, raised, channelled, floor loading;
- ceilings – finish, suspended, floor to ceiling height;
- windows – construction, glazing (single, double, tinted, etc.);
- lighting – natural, type and quality of artificial lighting;
- heating – type of fuel, type of system (radiators, ducts, under floor), extent of heating;
- air conditioning – type (variable air volume, fan coil, etc.) and whether it provides conditioning cleaning, cooling, humidification, etc., and extent of provision;
- toilets – whether communal to the building or exclusive, quality of finish, whether sufficient for requirements.

11.4 Categorisation

As mentioned, there are many types of office from the temporary hut on a building site, through converted offices above shops and rather basic 1960s buildings to ultra-modern 'green' office towers.

For new offices a broad system of categorising the level of finish has come into use though there is not an industry standard classification in the industry, but the following gives a general guide:

11.4.1 *Shell and core fit-out*

This is rather as the term describes – the structure and external cladding is provided but no fit out to the interior space. There are merely screeded concrete floors but with no lighting or facilities. If there are common parts to be shared between tenants these will have been completed and finished to a high standard ready for occupation. The lifts, lavatories, staircases and building reception will be complete.

11.4.2 Category A fit-out – 'Cat A'

Cat A suggests a fairly standard open-plan layout ready for a tenant to complete the space with its own Cat B fit-out. Cat A typically includes:

- raised access floors;
- suspended ceilings;
- basic mechanical and electrical services including lighting;
- fire detection services and smoke alarms;
- heating, cooling and ventilation services;
- basic internal finishes.

A variation is Cat A+ fit-out (or 'Plug and Play' space). This is undertaken by landlords as a way of attracting tenants by largely finishing the offices creating a functional office rather than leaving the tenant to spend on fitting out. It has become a popular type of fit-out with landlords who are offering flexible space under built to lease programmes.

11.4.3 Category B fit-out – Cat B

Cat B fit-out suggests office space ready to walk into and use. If undertaken specifically for a tenant (Tailored Cat B) it may involve individual branding and specific needs. Cat B typically Includes:

- fully-fitted kitchens and non-communal staff facilities;
- partitioning, including directors and senior executive's offices, meeting rooms and patch rooms;
- full electrical and other sockets;
- IT installation and infrastructure.

11.5 Unit of assessment

The unit of assessment is usually quite clear and self-evident. Sometimes an occupier may occupy space on more than one floor of an office building and where the two or more occupations are contiguous vertically and/or horizontally, it is likely there will be a single hereditament.

The rules covered in Chapter 3 on 'The Hereditament' should also be fully considered in determining the number of hereditaments.

11.6 Valuation approach

For most offices, the rental approach is the most commonly adopted approach as there is a substantial amount of rental evidence available in most parts of the country. Only in very exceptional circumstances would a different approach to the valuation be adopted. The usual unit of comparison is a straight rent per m².

For offices situated in what are, in effect, shop units in retail locations, comparisons to similar properties may be on a price per m² in terms of Zone A (ITZA) or on an overall rate per m².

11.7 Rental analysis

While the subject of rental analysis has been covered in an earlier chapter, it is important to remember that the rent passing may require adjustment before it is suitable for rating purposes.

Typically, adjustments may be needed for:

- rent-free periods;
- capital contributions to the fitting out;
- tenant's improvements;
- non-standard lease terms.

11.8 Illustrative valuation

A single owner-occupied office block has a total NIA of 625 m². There are separately defined areas created by structural partitions providing a reception, kitchen and stores. The offices are situated on three floors without a lift, and there are 15 car spaces. A typical valuation assuming a value of £100 per m² for offices.

	m²	£m²	£
Ground floor			
Offices	200	100.00	20,000
Reception	30	50.00	1,500
Stores	20	20.00	400
First-floor offices	250	100.00	25,000
Second floor			
Offices	100	90.00	9,000
Kitchen	25	45.00	1,125
Car spaces	15	75.00	1,125
Rateable value			£58,150

In the above valuation account has been taken in the valuation of those areas that are separated from the office accommodation by structural partitions. The stores, kitchen and reception area may well have poor natural light, restricted head room and an inferior finish compared with the office accommodation.

Consequently, they are valued on a lower basis. In addition, regard has been had to the disability of the absence of a lift on the second floor by reducing the rate per m^2 on this accommodation.

11.9 Factors affecting value

The most important evidence for any rating assessment will be the actual rent agreed on the property, particularly if both it is close to the valuation date and the lease terms are similar to the basis of rateable value. If the actual rent on a property is of no assistance then the rent passing on similar or identical properties may be used to arrive at the rental value of that property.

The actual rent will reflect all the disadvantages and advantages of the office location and quality of accommodation.

The valuation of offices has at times caused problems for valuers due to a change in market conditions around the time of the revaluation. The valuation must reflect market conditions at the antecedent valuation date (AVD) and changes in the state of the economy can be reflected only at the next revaluation.

11.9.1 Quantity allowance

Problems concerning whether or not an end allowance should be made on account of the difference in size between the property valued and others in the locality are more commonly associated with retail users than office occupiers.

The concept, however, of quantity allowances is the same irrespective of category of use, i.e. a hypothetical tenant occupying, by way of example, 10,000 m^2 of office accommodation may be able to secure, in negotiations with the hypothetical landlord, a lower rental per m^2 than a tenant occupying, for example, 1,000 m^2 of identical accommodation.

11.9.2 Fit out

In *Bunyan (VO)* v *Acenden Ltd* [2023] UKUT 17 (LC) the Upper Tribunal considered whether fitting out works to an office block increased the rental value. The principal contention of the respondent ratepayer was that the additional 'Category B' works carried out by a tenant to render a high-quality office building capable of beneficial occupation made no difference to the building's rateable value. The RV could, therefore, be ascertained substantially by reference to the rent and other financial terms agreed when it was let in its 'Category A' condition.

For the appellant VO, the proposition that the expenditure of such a substantial sum of money to make the building fit for occupation had added not a single penny to its RV was illogical and impossible of acceptance.

The Tribunal preferred the VO's evidence on the willingness of the market to pay more if offered in the fitted out, 'Category B' condition. It was 'consistent' with the Tribunal members' own expectations of how a rational market would behave and reflected the approach taken by the Lands Tribunal in earlier shop fitting cases. It was satisfied the ratepayer's 'Category B' fit out was, in the main, sufficiently generic and unexceptional and would have had 'general market' appeal.

The approach to moving from a cost figure to an annual sum to add to a rent excluding fitting out is discussed in Chapter 7.

12 The valuation of factories, workshops and warehouses

12.1 Introduction

Properties within this category cover a wide variety of property types from small nursery units to distribution warehouses, and from Victorian mills to the Class B1 'high-tech' building.

The Use Classes Order 1987 (SI 1987/764), as amended, highlights the great range of property that is covered within this general classification:

Part B

B2 General industrial – Use for industrial process other than one falling within class E(g) (previously class B1) (excluding incineration purposes, chemical treatment or landfill or hazardous waste)

B8 Storage or distribution – This class includes open air storage.

A significant reform of the Use Classes Order came into force on 21 April 2021 somewhat simplifying the classification within industrial and storage.

12.2 Survey and basis of measurement

Almost invariably, industrial and warehouse buildings will be measured on a gross internal area (GIA) basis as defined in the Royal Institution of Chartered Surveyors (RICS) *Code of Measurement Practice*. Units in high specification almost office like 'hi-tech' Class E(g) buildings will usually be measured on a net internal area (NIA) basis.

The value of these types of properties can be influenced by a large number of different factors associated with the location, design and construction of the property as well as its condition. It is therefore important to note all these matters when undertaking the inspection of the property.

In most cases, other than for properties on industrial estates that comprise single buildings, the hereditament is likely to comprise a range of different buildings used for different purposes such as production, services, storage, offices and the like. Each building and use contributes to the value of the whole

DOI: 10.1201/9781003297369-12

property in a different way. For all but the simplest of properties a plan will be produced to show the site of the property and the location of the individual buildings on the site.

A typical survey will include at least the following issues.

12.2.1 Location

- the availability of bus and rail connections for workforce and visitors;
- rail connection for freight – may be direct to the property or near a road/rail interchange;
- visibility of the property;
- road connections, with special emphasis on access to the property by heavy articulated vehicles and proximity to motorways and the main road network;
- workforce;
- private access roads – many estates have a private access road and there may be a maintenance charge payable.

12.2.2 Site details

- size, boundaries;
- vehicular access, car and lorry parking facilities, facilities for turning lorries;
- site coverage – often quoted as a percentage of site covered by buildings;
- the layout of the site with special reference to how the buildings work together to form an efficient unit;
- landscaping.

12.2.3 Buildings

- construction;
- main structure – single storey/multi-storey;
- roof type and covering – flat roof, pitched and angle of pitch, roof lights (% lights to roof size), insulation;
- walls – construction and material used;
- windows – type, size, position;
- natural light;
- floors – floor loading capacity;
- loading bays – type, size, number;
- condition of property;
- age of buildings and whether they have been refurbished, if so date of refurbishment;
- internal details including details of wall, ceiling and floor finishes; heating; use of each building/room;
- other details such as mezzanine floors, clean rooms, laboratories and other specialist facilities.

12.2.4 Services

- heating – type; extent; age; boiler type, age, capacity;
- lighting – type;
- sprinklers – number of heads, area protected;
- backup and standby service plant;
- waste disposal facilities especially for trade effluent. This may include onsite treatment works for waste prior to it being disposed of elsewhere;
- electrical supply, may be more than one supply coming into the property, capacity of supply, transformers, switch rooms;
- air conditioning – type of system;
- security systems;
- cranes and gantries;
- plant and machinery identifying rateable and non-rateable items.

12.2.5 Other issues

Enquiries will have to be made in respect of rented premises regarding whether any improvements have been made which are not reflected in the rent due to the Landlord and Tenant Acts.

In addition, the valuer should record how each of the buildings fit and work together onsite and whether there are any specific features that may affect the value of the property as a whole as opposed to the individual buildings.

12.3 Basis of valuation

Generally, there is ample rental evidence available in the locality and this will form the primary basis of valuation for these types of property. For the more specialised types of property, rental evidence may be less readily available or less reliable, and the valuer may have to resort to the use of the contractor's basis of valuation.

In most cases each building onsite will be valued individually at a given rent per m² reflecting its use, age and construction, and all the other factors that go to affect its value. In addition, consideration will also have to be given to the value that items such as car parking, lorry parks and storage land add to the value of the property.

One of the problems associated with the valuation of individual buildings is that the sum of all these values may not represent the value of the property as a whole. Individual buildings may not be located in the best position onsite and space between buildings may be cramped leading to poor vehicular circulation between buildings. Consequently, the valuer needs, as with any valuation, to stand back and look at the figures to ensure that the resultant value does indeed reflect the value of the property on the statutory terms of rateable value.

12.4 Plant and machinery

The valuation of industrial and warehouse property will often include a wide range of plant and machinery. The valuer will need to decide what plant and machinery is rateable under the Plant and Machinery Regulations (see Chapter 13).

12.5 Factors affecting valuation

12.5.1 Location

The location of the property can have a significant impact on value. While traditionally warehouses were located near ports, today the emphasis is on location near motorways or other main roads. Some of the more successful locations are in proximity to the intersection of north/south motorways with those running east/west. Many successful developments have also taken place away from motorways on other major routes which do not get as congested as some areas of the motorway network.

Industrial premises are no longer tied to the location of raw material but are now located far more widely across the country depending on their market. Prior to the mid-1960s, most industrial property was built to an industrialist's own specification and for a specific purpose. From the mid-1960s, industrial estates started to be developed and are now a common feature of the market all over the country. As with any type of property the type, nature, size and value of these estates can vary considerably.

12.5.2 Size and quantity allowance

The size range of industrial and warehouse units vary from the small nursery unit to vast warehouses and industrial premises.

In most aspects of business where large numbers of an item are purchased it is common for a discount to be given for bulk purchase or discount for quantity. In the property world the situation is not quite as simple and it is not possible to assume that just because a tenant rents a large area of space, the rent per m^2 will be lower than if the tenant rented a far smaller area. Market conditions will determine the rent that will be paid and this will be influenced by the amount of demand for that unit. If there are a number of tenants competing for a large unit, then the amount of any 'quantity allowance', if any, will be small. Conversely, if there is only one tenant in the market the amount of allowance needed to be made to induce the tenant to take the property is likely to be greater.

The key issue with such an allowance is to analyse what the market is doing, rather than to assume that there will automatically be a quantity allowance.

A number of cases have considered the issue of quantity allowance for industrial property: *Cumbrian Newspapers Group Ltd* v *Barrow (VO)* (1983) 269 EG 859, *Fife Regional Assessor* v *Distillers Co. (Bottling Services) Ltd* [1989] RA 115, *Harvey (JP) Ltd* v *Hockin (VO)* [1965] RA 573, *Hurcott Paper Mills Ltd* v *Hockin (VO)* [1965] RA 557 and *Austin Motor Co. Ltd* v *Woodward (VO)* (1966) [1968] RA 133; 15 RRC 3; on appeal [1970] RA 181; 16 RRC 1; (CA).

12.5.3 Obsolescence

Obsolescence is an important issue to consider when valuing industrial property. As already discussed, three types of obsolescence can be identified – physical, economic and technical.

Obsolescence is a factor that will need to be considered in all valuations of this type, but in those cases where the property is designed to produce a specific single product then its impact on the valuation of the property may be more significant. In such cases the valuer will also have to understand the nature and state of the industry in order to identify whether any future changes in the industry will impact on the value of the subject property.

12.5.4 Disabilities

In many cases the valuer will value each building individually, reflecting its own advantages and disadvantages. Where this approach is adopted it may then be necessary to consider whether some allowance needs to be made to the valuation to reflect particular disadvantages of the property, when considered as a whole, as opposed to a collection of individual buildings.

While not intended to be a comprehensive list, some typical reasons for allowances are as follows:

- *Bowaters PKL (UK) Ltd* v *Worssam (VO)* [1985] 1 EGLR 221 – an allowance of 8% was given for the fact that the property, which comprised two buildings and associated offices, was on either side of a service road.
- *Austin Motor Co. Ltd* v *Woodward (VO)* (1966) [1968] RA 133; 15 RRC 3; on appeal [1970] RA 181; 16 RRC 1; (CA) – a 10% allowance was given for the property being on four separate sites which were divided by both roads and a railway line.
- *Arts (G and B) Ltd* v *Cowling (VO)* [1991] RVR 192 – 10% allowance for poor access, cramped site conditions, poor layout and poor loading.
- *Clark (JE) and Co.* v *Rogers (VO)* [1994] RA 169 – an allowance of 5% for poor layout of site (only for one of the assessments under consideration).
- *Farmer Stedall Plc* v *Thomas (VO) and Liverpool City Council* [1985] 2 EGLR 221 – a 33.3% allowance for disabilities and surplus capacity of the property. It should be pointed out that the circumstances in this case were exceptional and this size of allowance should not be thought of as typical.

- *Fusetron Ltd* v *Whitehouse (VO)* [1999] RA 295 – 10% allowance for lack of off-street parking and loading and unloading facilities. In addition, a 10% allowance was made in this case for the effect of contamination of the site and its effect on the way that the hypothetical tenant could use the site.
- *Proderite Ltd* v *Clark (VO)* [1993] RA 197 – 12.5% for the layout of the buildings on the site, the restricted vehicular access to the premises and the overall size of the hereditament.
- *Royal Bathrooms (Formerly Doulton Bathrooms)* v *Maudling (VO)* [1984] 2 EGLR 203 – end allowance of 30% considered appropriate to reflect not only the physical disabilities formerly taken into account, but also a further diminution in value due to the subsequent restrictions on output and the disturbance in the balance of the buildings that formed the hereditament.
- *Hurcott Paper Mills Ltd* v *Hockin (VO)* [1965] RA 557 – the Tribunal decided upon an end allowance of 20% made up as follows:

  1   poor levels and layout, 5%
  2   remote rural position and distance from buses and houses, 7.5%
  3   lack of main drainage, 2.5%
  4   redundancy, 5%.

- *Chilton* v *Price (VO)* [1991] RVR 5 – in this case, which concerned the valuation of an isolated workshop, the Lands Tribunal made an allowance for the isolated position by adjusting the rent per m² rather than giving a specific end allowance.

This case highlights that as far as allowances are concerned there may be more than one means of allowing for the factors under consideration. In all but one case quoted, the end allowance approach has been adopted. However, it is equally valid to adjust the rate per m² to reflect all the advantages and disadvantages of the building on the actual site. From a practical point of view it is essential to know what approach is going to be adopted by each party before starting the negotiations, in order to save time and any future misunderstanding. The adoption of the end allowance approach does mean that it is easier to compare buildings on one property with those of another which can be a considerable advantage.

12.6 Example 1 – rental analysis

Analyse the rent of the following property for the purpose of a rating revaluation.

The property is currently used as a light engineering works and contains the accommodation listed in Table 12.1.

Table 12.1 Valuation of light engineering works

Ref	Use	Heated	Area m²	Remarks
1	Production	Yes	375	Portal framed building 4.0 m eaves height
2	Offices	Yes	57	Painted plastered walls, vinyl tiled floor, utilitarian finish
3	Shed	No	10	Wood construction
4	Production	No	124	1935 building, brick walls, steel trussed roof with corrugated asbestos sheeting
4a	Stores	No	35	Brick walls, flat concrete roof
5	Yard	–	270	Rough cindered surface
6	Yard	–	300	Rough cindered surface (unused)

The property was recently let on a 21-year internal repairing lease with seven-year rent reviews, at £25,000 per annum (pa).

The first stage of the analysis is to convert the rent to the terms of rateable value. The actual property is let on an internal repairing lease and thus the landlord is responsible for the external repairs and insurance.

Conversion of rent to full repairing and insuring (FRI) terms is:

Rent	£25,000
Less	
External repairs @ 5%	£1,250
Insurance @ 3.5%	£875
	£22,875

12.6.1 Adjustment of rent to FRI terms

The second stage of the analysis is the detailed analysis of the price per unit. In order to undertake this analysis the valuer will use his or her knowledge of the relativities in value of one type of accommodation to another. This knowledge will come from the type of property under consideration, its location and the market conditions at the time the rent was agreed. What is being sought is a rent analysed per m² in terms of main space (ITMS).

Table 12.2 Analysis of price per unit

Ref	Use	Heated	Area m²	Factor	Adj area	Remarks
1	Production	Yes	375	1.00	375.00	Portal framed building 4.0 m eaves height
2	Offices	Yes	57	1.20	68.40	Painted plastered walls, vinyl tiled floor, utilitarian finish
3	Shed	No	10	0.20	2.00	Wood construction

Table 12.2 (Continued)

Ref	Use	Heated	Area m²	Factor	Adj area	Remarks
4	Production	No	124	0.60	74.40	1935 building, brick walls, steel trussed roof with corrugated asbestos sheeting
4a	Stores	No	35	0.50	17.50	Brick walls, flat concrete roof
5	Yard	–	270	0.10	27.00	Rough cindered surface
6	Yard	–	300	0.10	30.00	Rough cindered surface (unused)
					594.30	
Rent on FRI terms					£22,875	
Analysis					£38.49	per M2 in terms of main space (ITMS)

Table 12.3 Analysis to unit value

Ref	Use	Heated	Area m²	£m²	£	Remarks
1	Production	Yes	375	38.49	14,434	Portal framed building 4.0 m eaves height
2	Offices	Yes	57	46.19	2,632	Painted plastered walls, vinyl tiled floor, utilitarian finish
3	Shed	No	10	7.70	77	Wood construction
4	Production	No	124	23.09	2,863	1935 building, brick walls, steel trussed roof with corrugated asbestos sheeting
4a	Stores	No	35	19.25	673	Brick walls, flat concrete roof
5	Yard	–	270	3.85	1,039	Rough cindered surface
6	Yard	–	300	3.85	1,154	Rough cindered surface (unused)
					£22,875	

12.7 Example 2 – Valuation of a specialised property

This example concerns the valuation of a specialised industrial property. The rental approach adopted in the previous example is not appropriate as such specialised properties do not normally let in the market. In such cases it is common to adopt the contractor's basis of valuation.

Table 12.4 Valuation of specialised industrial property

Ref	Description	Area m²	Cost/£m²	Cost £	Allowance %	Adj cost £
Stage 1 & 2						
1	Office block	650	£750	£487,500	10.00%	£438,750
2	Factory	5,000	£325	£1,625,000	5.00%	£1,543,750
3	Sub station	15	£1,250	£18,750	12.50%	£16,406
4	Workshop	500	£500	£250,000	0.00%	£250,000
5	Boiler house	1,500	£1,750	£2,625,000	0.00%	£2,625,000
						£4,873,906
	Site works					
	Roads	50,000	£50	£2,500,000	5.00%	£2,375,000
	Fencing	2,500	£65	£162,500	0.00%	£162,500
	Yard	1,700	£45	£76,500	0.00%	£76,500
						£2,614,000
	Total construction cost					£7,487,906
	Plant and machinery					
	Generators ×2 3,000 KVA		£50	£300,000	75.00%	£75,000
	Steelworks 25 tons		£1,500	£37,500	0.00%	£37,500
Stage 3						
	Total cost including plant and machinery			£7,600,406		
	Developed land	15	£500,000	£7,500,000	5%	£7,125,000
	Undeveloped land	2.5	£250,000	£625,000	0%	£625,000
	Land and buildings					£15,350,406
Stage 4						
	Decapitalise @ 4.4%				£675,418	
Stage 5						
	Less 5% for piecemeal development				−£33,771	
						£641,647
					Say	£641,500

12.8 Wales

Planning powers have been devolved to Welsh Ministers. The Planning Act (Wales) 2015 is currently the primary planning legislation.

13 The valuation of plant and machinery

13.1 Introduction

As discussed in earlier chapters, a hereditament will usually comprise land either with or without buildings. Personal property is normally not rateable, but certain items which are neither land nor buildings may be taken into account if they comprise either:

- chattels that are neither plant nor machinery but are permanently attached to and enjoyed with land so that the land's value is enhanced (builders huts, for example);
- plant and machinery.

13.2 Historical development

Until 1925 there was no specific statutory provision for the rating of plant and machinery, but it had been long established that certain items of plant and machinery that were intended to remain on the premises and were there to make the premises fit for the purpose for which they were used were rateable as part of the property. This view was upheld by the courts in a series of cases including *R* v *St Nicholas, Gloucester* (1783) 1 Term Rep 723, *Tyne Boiler Works Co* v *Longbenton Overseers* (1886) 18 QBD 815 and *Kirby* v *Hunslet Union* [1906] AC 43. Tools and other items, such as stock in trade, were not rateable.

In 1925 the position was formalised following the report of the Shortt Committee with the passing of the Rating and Valuation Act. Section 24 and Schedule 3 of the Act provided a list of classes of plant and machinery that were deemed to be part of the hereditament and therefore rateable, and this led to the making of the Plant and Machinery (Valuation for Rating) Order 1927. Following the Shortt Committee report, it had been envisaged that the list of rateable plant and machinery would be reviewed at frequent intervals to ensure it reflected the development of plant and machinery used in industry. However, the list of rateable plant and machinery was not reviewed again until it was considered by the Ristson Committee in 1957 which led to the Plant

DOI: 10.1201/9781003297369-13

and Machinery (Rating) Order 1960. In 1961 certain pipelines were made rateable under the 1960 Order. In 1974, following the report of the McNairn Committee, major changes were made to Class 4 of the 1960 Regulations.

Following the passing of the Local Government Finance Act 1988, the Valuation for Rating (Plant and Machinery) Regulations 1989 were made, but these effectively re-enacted the amended 1960 Regulations.

It was not until 1991 that the Wood Committee was established to review the Regulations for the purposes of the 1995 Rating Revaluation and this led to the Valuation for Rating (Plant and Machinery) Regulations 1989.

The Committee recommended that rateability of plant and machinery should be determined in accordance with the following rules and principles:

- that land and everything that forms part of it and is attached to it should be assessed;
- that process plant and machinery which can be fairly described as 'tools of the trade' should be exempt within certain limits;
- that process plant and machinery (in certain cases exceeding a stated size) which is in the nature of a building or structure, or performs the function of a building or structure, should, however, be deemed to be part of the hereditament or subject;
- that service plant and machinery, and items forming part of the infrastructure of the property, should be rated;
- that, in the case of plant and machinery which performs both a service and process function, sensible lines have to be drawn that will indicate exactly how much falls to be rated and how much does not.

The Regulations were again reviewed by the second Wood Committee for the purposes of the 2000 Rating Revaluation with a view to identifying any changes that would be required to the 1989 Regulations as a result of many of the 'prescribed industries' coming out of prescription. (Certain industries such as gas, electricity supply and generation, canals, railways and the like have their rating assessment prescribed by a statutory formula, rather than being assessed by normal valuation techniques.)

As a result of the recommendations of the Wood Committee, the Valuation for Rating (Plant and Machinery) (England) Regulations 2000 (2000/540) (2000/540) were made. The Valuation for Rating (Plant and Machinery) (England) (Amendment) Regulations 2001 (2001/846), Valuation for Rating (Plant and Machinery) (England) Regulations 2008 (2008/2332) and the Valuation for Rating (Plant and Machinery) (England) (Amendment) Regulations 2022 (2022/405) have made minor changes to the 2000 Regulations.

13.2.1 Rateability

The Regulations apply to the valuation of all property irrespective of the method of valuation used.

The approach to rateability follows a two-stage process:

- consider whether the item being considered is 'plant or machinery': if it is not plant or machinery the Regulations do not apply;
- where the item is plant or machinery, consider whether the item of plant or machinery is 'named' in the appropriate regulations (Valuation for Rating (Plant and Machinery) Regulations 2000 (SI 2000/540), as amended).

If 'plant' is not named in the Regulations then it is not regarded as part of the hereditament and is not rateable.

What is plant has been considered in a number of cases. In *Yarmouth* v *France* (1887) 4 TLR 561, which concerned a rather vicious horse and liability under the Employer's Liability Act 1880, it was held that plant means:

> *in its ordinary sense (plant) includes whatever apparatus is used by a businessman for carrying on his business – not his stock-in-trade which he buys or makes for sale but all goods and chattels fixed or moveable live or dead which he keeps for permanent employment in his business.*

Plant is 'apparatus used by a businessman for carrying on his business'. This can be distinguished from items which are 'more a part of the setting than part of the apparatus for carrying on a trade' (Donovan LJ in *Tyne Boiler Works Co* v *Longbenton Overseers* (1886) 18 QBD 815).

In *British Bakeries Ltd* v *Gudgion (VO) and Croydon London Borough Council* [1969] RA 465, the Lands Tribunal considered whether various types of partitioning in an office block were rateable. Partitioning is not mentioned in the Plant and Machinery Regulations and the ratepayers argued, as a consequence, that it was not rateable and its value should not be included in the rateable value. The Tribunal accepted that half-height partitioning in the office was plant because it was 'no more than furniture' being very easily moved and had the function of a low screen. It was indeed 'apparatus' used for carrying on a business. It was not rateable because it was not mentioned in the Regulations.

However, the Tribunal decided the full-height partitioning was not plant because '[i]ts function is I think to make offices for the staff, to provide places in which the business is carried on, not to be plant with which it is carried on'.

As part of the setting, the full-height partitioning formed part of the hereditament and was, as a consequence, rateable. In a more recent case, *Rogers (VO)* v *Evans* [1985] 2 EGLR 217, the Lands Tribunal considered that a mezzanine floor was not plant but 'an additional floor ... and until it is moved it stands firm and solid, fully performing the function of a floor and is more a part of the setting than part of the apparatus for carrying on the trade'.

It is important to note from these cases that neither partitioning nor a mezzanine floor is named in the Plant and Machinery Regulations. Had they been regarded as plant they would not have been rateable.

Where the item is considered to fall within the definition of plant, then the valuer must consider whether the item is named in the Plant and Machinery Regulations.

The term 'machinery' seems to have its ordinary, everyday meaning and where valuers are discussing plant and machinery they usually refer to them together as 'plant'. This approach is followed in the rest of the chapter.

The effect of the Regulations is that if an item of plant belongs to any of the classes in the Regulations, then it is assumed to be part of the hereditament and thus the value of that plant is reflected in the valuation of the property. If the item of plant is not mentioned in the classes, then it is to be assumed that it will have no effect on the value of the property.

The Regulations are divided into four main classes of plant and machinery.

Class 1 deals with plant and machinery for 'the generation, storage, primary transformation or main transmission of power'. Class 1 rates all items of plant and machinery listed in Table 1 of the Regulations and in paragraphs 1 and 2 of the list of accessories. It should be noted that items listed in paragraph 1 of the list of accessories are only rateable if they are used for the handling, preparing or storing of fuel required for the generation of power. In other words, the items of plant and machinery are not rateable per se but require other conditions to be satisfied.

The definition of 'power' has been considered by the courts in *ICI Ltd* v *Owen (VO) and Runcorn Urban District Council* (1954) 48 R&IT 43. 'Power' will include the generation of steam power, electrical power, pneumatic power, hydraulic power and any other form of motive power.

Note the 'excepted plant and machinery' which while used for 'the generation etc. …' is not rateable because (in broad terms but note the precise definitions) the power is mainly or exclusively for distribution for sale to consumers, or the plant and machinery is that of a combined heat and power station, or the source of energy is mainly or exclusively renewable, or the plant or machinery is used or intended to be used mainly or exclusively for storage, transformation or transmission of power for an electric vehicle charging point.

Class 2 deals with service plant, including plant and machinery for:

• heating, cooling and ventilation;
• lighting;
• drainage;
• supplying water;
• protection from hazards.

Class 2 rates all plant and machinery named in Table 2 of the Regulations and in paragraph 2 of the list of accessories.

For plant to be rateable under Class 2 it must:

- be named in Class 2;
- be used in connection with the provision of the specific services listed in Class 2;
- provide the service to the hereditament;
- mainly or exclusively be so used for that purpose;
- be not exempted by being used for manufacturing or trade processes.

In *Iceland Foods Ltd* v *Berry VO* [2018] RA 187, the Supreme Court found that the air handling system within a retail unit was plant or machinery 'used or intended to be used in connection with services mainly or exclusively as part of manufacturing operations or trade processes' within the meaning of the Regulations. Its decision was that the word 'process' was apt to cover 'the continuous freezing or refrigeration of goods to preserve them in an artificial condition', and since the services provided by the relevant plant had been held to be used 'mainly or exclusively' as part of that trade process, that plant should be left out of account for rating purposes. A trade process was simply a process (in the sense of 'anything done to the goods or materials') carried on for the purposes of a trade and there was nothing in the word 'process' which implied a transition or change.

Class 3 deals with various types of infrastructure on a property. It is divided into the following main categories of plant and machinery:

- railway and tram lines and their fixed accessories;
- lifts, elevators, hoists, escalators and travelators;
- cables for the conducting of electricity, and their accessories;
- cables, fibres, wires and conductors for the transmission of communications signals and their associated equipment;
- pipelines;
- locks, dock gates and caissons.

Class 4 rates structural process plant. Class 4 treats as rateable the items specified in Tables 3 and 4 of the Regulations, except:

- any such item which is not, and is not in the nature of, a building or structure;
- any part of any such item which does not form an integral part of such item as a building or structure or as being in the nature of a building or structure;
- so much of any refractory or other lining forming part of any plant or machinery as is customarily renewed by reason of normal use at intervals of less than 50 weeks;
- any item in Table 4, the total cubic capacity of which (measured externally and excluding foundations, settings, supports and anything which is not an integral part of the item) does not exceed 400 cubic metres and which is

readily capable of being moved from one site and re-erected in its original state on another without the substantial demolition of any surrounding structure.

To determine rateability under this class a staged analysis of the rateability needs to take place:

- Is the item being considered 'named' in either Table 3 or Table 4 of the Regulations? If the item is not 'named', then the item cannot be rated under Class 4. The Plant and Machinery Regulations use both technical terms for specific items of plant or machinery and also more generic descriptions. In deciding what is and is not included the valuer should interpret the wording in the Regulations in the way practical people experienced in the field would use the words rather than using scientific terminology.
- Is the item in the nature of a building or structure?

The Regulations do not define 'building or structure' but the matter was considered in *Cardiff County Borough Rating Authority and AC* v *Guest Keen Baldwin's Iron & Steel Co Ltd* [1949] 1 KB 385:

> *It would be undesirable to attempt, and, indeed, I think impossible to achieve, any exhaustive definition of what is meant by the words, 'a building or structure or in the nature of a building or structure'. They do, however, indicate certain main characteristics. The general range of things in view consists of things built or constructed. I think, in addition to coming within this general range, the things in question must, in relation to the hereditament, answer the description of buildings or structures, or, at all events, be in the nature of buildings or structures. That suggests built or constructed things of substantial size. I think of such size that they either have been in fact, or would normally be, built or constructed on the hereditament as opposed to being brought on to the hereditament ready made. It further suggests some degree of perman-ence in relation to the hereditament, i.e. things which, once installed on the hereditament, would normally remain in situ and only be removed by a process amounting to pulling down or taking to pieces. I do not, however, mean to suggest that size is necessarily a conclusive test in all cases, or that a thing is necessarily removed from the category of buildings or structures, or things in the nature of buildings or structures, because by some feat of engineering or navigation it is brought to the hereditament in one piece.*

- Is there any part of any such item that does not form an integral part of such item as a building or structure or as being in the nature of a building or structure? Any detachable parts of the item, such as motors, will not be rateable.
- Does the item consist of any refractory or other lining forming part of any plant or machinery as is customarily renewed by reason of normal use at

intervals of less than 50 weeks? Where such a lining is normally replaced at intervals of less than 50 weeks it will not be rateable.

- The final requirement sets a number of tests to be satisfied if the plant is not to be rateable.

1. Does the item measure less than 400 m³ measured externally and excluding foundations, settings, supports and anything which is not an integral part of the item?
2. Is the item readily capable of being moved from one site and re-erected in its original state on another site?
3. Can it be moved without the substantial demolition of any surrounding structure? (See *Cumber (VO)* v *Associated Family Bakers (South West) Ltd* [1980] JPL 38; [1979] RA 328; 252 EG 69 with regard to the consideration of the meaning of 'substantial demolition' by the Lands Tribunal.)

If the answer to all three tests is 'yes' then the item is not rateable.

13.2.2 *Microgeneration*

Regulation 2A was inserted in the Valuation for Rating (Plant and Machinery) (England) Regulations 2000 (2000/540) by the Valuation for Rating (Plant and Machinery) (England) (Amendment) Regulations 2008 (2008/2232) and applies to any plant and machinery installed on or after 1 October 2008 (1 April 2010 in Wales) which has 'microgeneration capacity'. The Government was concerned that the increase in rateable value to reflect the value of this plant would act as a disincentive to landlords and occupiers considering installing microgeneration. The Regulation requires any value attributable to a hereditament's plant and machinery's microgeneration capacity to be ignored until the next revaluation (or when the capacity ceases if earlier). So, in the simple example of the affixing of a small wind turbine or solar panel to the roof of a shop or factory, any value is ignored during the current rating list but will be included at the next general revaluation, e.g. if installed on or after 1 April 2023 its value will be excluded until 1 April 2026 or whenever the new rating lists are compiled.

The Regulation only refers to the extent that the hereditament has microgeneration capacity and does not mean that the presence of the plant and machinery is ignored if it has some other function, e.g. photovoltaic tiles forming a roof or wall covering and having the additional function of keeping the building watertight.

Microgeneration capacity includes plant and machinery for both the generation of electricity and the production of heat providing the source of energy or the technology is mentioned in section 26 of the Climate Change and Sustainable Energy Act 2006. As implied by the name, microgeneration plants are fairly small units which are usually installed to supplement the principal source of energy provided in a hereditament. In practice, most microgeneration plants will have a negligible value and would not have made any significant difference to a hereditament's rateable value.

To be a microgeneration plant, two tests are applied to each separate item requiring them to both be named and have a capacity that does not exceed specified figures:

- Source/technology – biomass, bio fuels, fuel cells, photovoltaics, water (including waves and tides), wind, solar power, geothermal sources, combined heat and power systems and other sources and technologies specified by the Secretary of State.
- Capacity – generation of electricity, 50 kilowatts; production of heat, 45 kilowatts thermal.

Excepted plant and machinery from Class 1 has also been extended by The Valuation for Rating (Plant and Machinery) (England) (Amendment) Regulations 2022 (2022/405) to exclude from rateability plant and machinery where the sources of energy or technologies relied upon are mainly or exclusively renewable or the generation etc is for electric vehicle charging points. Despite having 'England' in the SI's name, it does extend to Wales (Reg. 1(2)).

13.2.3 Valuation of plant and machinery

Often, the rent paid for a property will reflect the value of items of rateable service-type plant and machinery, for example lifts, escalators, lighting, heating and cooling equipment. Analysis and therefore valuation can be undertaken including these items in the values per m². Other types of plant will not usually be included in a rent and will be shown separately in a valuation, usually valued by taking an annualised figure of cost. While shown as separate items in the valuation, they must be regarded as part of the property and not looked at in isolation. It is the value of the hereditament as a whole that has to be ascertained, not the sum of the constituent parts which may not lead to the true rental value.

In most instances there is little rental evidence for non-service plant and machinery and valuation is undertaken using annualised notional cost. The capital cost of the item is ascertained and, after making appropriate adjustments, a percentage applied to arrive at the rental value of the item.

The valuation of plant and machinery is a highly specialised field of valuation. Two approaches to a cost valuation can be used:

- The cost of new plant can be ascertained. The cost should include not only the cost of the plant but also the full cost of installation together with any other associated work (foundations, etc.). From this amount the cost of any non-rateable elements will have to be deducted. The actual item being valued may in reality not be new so some adjustment may be needed for any technical, physical and economic obsolescence of the item. The resultant figure is then decapitalised at the statutory decapitalisation rate to arrive at the rental value of the item.

- An alternative approach is to examine second-hand costs of the item of similar plant. Where this approach is adopted, again it is the cost of the installed plant that has to be considered.

A copy of the current regulations relating to the rateability of plant and machinery is set out below.

13.2.3.1 *Plant and machinery regulations*

The Valuation for Rating (Plant and Machinery) (England) (Amendment) Regulations 2000 (2000/540) as amended by The Valuation for Rating (Plant and Machinery) (England) (Amendment) Regulations 2001 (2001/846), The Valuation for Rating (Plant and Machinery) (England) (Amendment) Regulations 2008 (2008/2332), The Valuation for Rating (Plant and Machinery) (England) (Amendment) Regulations 2022 (2022/405).

RATING AND VALUATION, ENGLAND

The Valuation for Rating (Plant and Machinery) (England) Regulations 2000 (2000/540)

The Secretary of State for the Environment, Transport and the Regions, in exercise of the powers conferred on him by section 143(2) of, and paragraph 2(8) of Schedule 6 to, the Local Government Finance Act 1988, and of all other powers enabling him in that behalf, hereby makes the following Regulations:

Citation, commencement and extent

1. These Regulations, which extend to England only, may be cited as the Valuation for Rating (Plant and Machinery) (England) Regulations 2000 (2000/540) and shall come into force on 1st April 2000.

Prescribed assumptions as to plant and machinery

2. For the purpose of determining the rateable value of a hereditament for any day on or after 1st April 2000, in applying the provisions of sub-paragraphs (1) to (7) of paragraph 2 of Schedule 6 to the Local Government Finance Act 1988–

   a   in relation to a hereditament in or on which there is plant or machinery which belongs to any of the classes set out in the Schedule to these Regulations, the prescribed assumptions are that:

    i any such plant or machinery is part of the hereditament; and

    ii the value of any other plant and machinery has no effect on the rent to be estimated as required by paragraph 2(1); and

  b in relation to any other hereditament, the prescribed assumption is that the value of any plant or machinery has no effect on the rent to be so estimated.

Prescribed assumptions as to plant and machinery: valuation for 1st October 2008 and subsequent days

2A.–(1) For the purpose of determining the rateable value of a hereditament for any day on or after 1st October 2008, in applying the provisions of sub-paragraphs (1) to (7) of paragraph 2 of Schedule 6 to the Local Government Finance Act 1988–

    a in relation to a hereditament in or on which there is plant or machinery which belongs to any of the classes set out in the Schedule to these Regulations, the prescribed assumptions are that–

        i any such plant or machinery is part of the hereditament except to the extent that it has microgeneration capacity, and

        ii the value of any other plant and machinery has no effect on the rent to be estimated as required by paragraph 2(1); and

    b in relation to any other hereditament, the prescribed assumption is that the value of any plant or machinery has no effect on the rent to be so estimated.

  (2) The exception in paragraph (1)(a)(i) applies only–

    a in relation to any item of plant or machinery that–

        i is installed on or after 1st October 2008, and

        ii on the day of installation has microgeneration capacity; and

    b in the period–

        i starting on the day that the item of plant or machinery is installed, and

        ii ending on the earlier of–

            (aa) the first day after the day that the item of plant or machinery is installed on which rating lists fall to be compiled for the purposes of sections 41(2) and 52(2) of the Local Government Finance Act 1988, and

            (bb) the day the item of plant or machinery ceases to have microgeneration capacity.

·

(3)   In this regulation "microgeneration capacity" means the capacity of plant or machinery to be used for the generation of electricity or the production of heat–

a   which, in generating electricity or (as the case may be) produ- cing heat, relies wholly or mainly on a source of energy or a technology mentioned in section 26(2) (interpretation) of the Climate Change and Sustainable Energy Act 2006; and

b   the capacity of which to generate electricity or (as the case may be) to produce heat does not exceed the capacity mentioned in section 26(3) of that Act.

Supply of written particulars

3. The valuation officer shall, on being so required in writing by the occu- pier of any hereditament, supply to him particulars in writing showing what plant and machinery, or whether any particular plant and machinery, has been assumed in pursuance of regulation 2(a) to form part of the hereditament.

Revocation and savings

4.–(1)   Subject to paragraph (2), the Valuation for Rating (Plant and Machinery) Regulations 1994 ("the 1994 Regulations") are revoked.

(2)   The 1994 Regulations shall continue to have effect for the purpose of determining the rateable value of a hereditament for any day before 1st April 2000.

Signed by authority of the Secretary of State
for the Environment, Transport and the Regions

Hilary Armstrong
Minister of State,
2nd March 2000   Department of the Environment, Transport and the Regions

SCHEDULE   Regulation 2

CLASSES OF PLANT AND MACHINERY TO BE ASSUMED TO BE PART OF THE HEREDITAMENT

CLASS 1

Plant and machinery (other than excepted plant and machinery) or, in relation to any day before 1st April 2035, excepted renewables plant and machinery or excepted EVCP plant and machinery specified in Table 1 below (together with any of the appliances and structures accessory to such plant or machinery and specified in paragraph 1 or 2 of the List of Accessories set out below) which is used or intended to be used mainly or exclusively in connection with the generation, storage, primary transformation or main transmission of power in or on the hereditament.

In this Class–

a   "transformer" means any plant which changes the pressure or frequency or form of current of electrical power to another pressure or frequency or form of current, except any such plant which forms an integral part of an item of plant or machinery in or on the hereditament for manufacturing operations or trade processes;

b   "primary transformation of power" means any transformation of electrical power by means of a transformer at any point in the main transmission of power;

c   "main transmission of power" means all transmission of power from the generating plant or point of supply in or on the hereditament up to and including–

   i   in the case of electrical power, the first distribution board;
   ii   in the case of transmission by shafting or wheels, any shaft or wheel driven directly from the prime mover;
   iii   in the case of hydraulic or pneumatic power, the point where the main supply ceases, excluding any branch service piping connected with such main supply;
   iv   in a case where, without otherwise passing beyond the limits of the main transmission of power, power is transmitted to another hereditament, the point at which the power passes from the hereditament; and

d   "excepted plant and machinery" means plant and machinery on a hereditament used or intended to be used for the generation, storage, transformation or transmission of power where either–

   i   the power is mainly or exclusively for distribution for sale to consumers; or
   ii   (aa)   the plant and machinery is that of a combined heat and power station which is fully exempt or partly exempt within the meaning of paragraph 148(2) or, as the case may be, 148(3) of Schedule 6 to the Finance Act 2000, and
       (bb)   the plant and machinery is within head (b), (c), (d) or (k) of Table 1 below, and
       (cc)   the power is at least in part electrical power.

(e)   "excepted renewables plant and machinery" means plant and machinery used or intended to be used for the generation, storage, transformation or transmission of power where the sources of energy or technologies relied upon are mainly or exclusively–

(i)   biomass;
(ii)   biofuels;
(iii)   biogas;
(iv)   fuel cells;
(v)   photovoltaics;
(vi)   water (including waves and tides);
(vii)   wind;
(viii)   solar power;
(ix)   geothermal;
(x)   heat from air, water or the ground;

(f)   "excepted EVCP plant and machinery" means plant and machinery which is–

(i)   used or intended to be used mainly or exclusively for storage, transformation or transmission of power for an electric vehicle charging point (within the meaning of section 45EA(5) of the Capital Allowances Act 2001); and
(ii)   within head (d) or (e) of Table 1 below

TABLE 1

a   Steam boilers (including their settings) and chimneys, flues and dust or grit catchers used in connection with such boilers; furnaces; mechanical stokers; injectors, jets, burners and nozzles; superheaters; feed water pumps and heaters; economisers; accumulators; deaerators; blow-off tanks; gas retorts and charging apparatus, producers and generators.
b   Steam engines; steam turbines; gas turbines; internal combustion engines; hot-air engines; barring engines.
c   Continuous and alternating current dynamos; couplings to engines and turbines; field exciter gear; three-wire or phase balancers.
d   Storage batteries with stands and insulators, regulating switches, boosters and connections forming part of any such equipment.
e   Static transformers; auto transformers; motor generators; motor converters; rotary converters; transverters; rectifiers; phase converters; frequency changers.
f   Cables and conductors; switchboards, distribution boards, control panels and all switchgear and other apparatus on any such equipment.
g   Water wheels; water turbines; rams; governor engines; penstocks; spillways; surge tanks; conduits; flumes; sluice gates.

h   Pumping engines for hydraulic power; hydraulic engines; hydraulic intensifiers; hydraulic accumulators.
i   Air compressors; compressed air engines.
j   Windmills.
k   Shafting, couplings, clutches, worm-gear, pulleys and wheels.
l   Steam or other motors which are used or intended to be used mainly or exclusively for driving any of the plant and machinery falling within this Class.
m   Aero-generators; wind turbines.
n   Solar cells; solar panels.

CLASS 2

Plant and machinery specified in Table 2 below (together with the appliances and structures accessory to such plant or machinery and specified in paragraph 2 of the List of Accessories set out below) which is used or intended to be used in connection with services to the hereditament or part of it, other than any such plant or machinery which is in or on the hereditament and is used or intended to be used in connection with services mainly or exclusively as part of manufacturing operations or trade processes.

In this Class, "services" means heating, cooling, ventilating, lighting, draining or supplying of water and protection from trespass, criminal damage, theft, fire or other hazard.

TABLE 2

(a)   GENERAL

Any of the plant and machinery specified in Table 1 and any motors which are used or intended to be used mainly or exclusively for driving any of the plant and machinery falling within paragraphs (b) to (f) of this Table.

(b)   HEATING, COOLING AND VENTILATING

  i   Water heaters.
 ii   Headers and manifolds; steam pressure reducing valves; calorifiers; radiators; heating panels; hot-air furnaces with distributing ducts and gratings.
iii   Gas pressure regulators; gas burners; gas heaters and radiators and the flues and chimneys used in connection with any such equipment.
 iv   Plug-sockets and other outlets; electric heaters.
  v   Refrigerating machines.
 vi   Water screens; water jets.
vii   Fans and blowers.
viii   Air intakes, channels, ducts, gratings, louvres and outlets.

ix Plant for filtering, washing, drying, warming, cooling, humidifying, deodorizing and perfuming, and for the chemical and bacteriological treatment of air.

x Pipes and coils when used for causing or assisting air movement.

(c) LIGHTING

i Gas pressure regulators; gas burners.

ii Plug-sockets and other outlets; electric lamps.

(d) DRAINING

Pumps and other lifting apparatus; tanks; screens; sewage treatment plant and machinery.

(e) SUPPLYING WATER

Pumps and other water-lifting apparatus; sluice-gates; tanks, filters and other plant and machinery for the storage and treatment of water.

(f) PROTECTION FROM HAZARDS

Tanks; lagoons; reservoirs; pumps, hydrants and monitors; fire alarm systems; fire and explosion protection and suppression systems; bunds; blast protection walls; berms; lightning conductors; security and alarm systems; ditches; moats; mounds; barriers; doors; gates; turnstiles; shutters; grilles; fences.

LIST OF ACCESSORIES

1. Any of the following plant and machinery which is used or intended to be used mainly or exclusively in connection with the handling, preparing or storing of fuel required for the generation or storage of power in or on the hereditament– cranes with their grabs or buckets; truck or wagon tipplers; elevating and conveying systems, including power winches, drags, elevators, hoists, conveyors, transporters, travellers, cranes, buckets forming a connected part of any such system, and any weighing machines used in connection with any such system; magnetic separators; driers; breakers; pulverisers; bunkers; gas-holders; tanks.

2. Any of the following plant and machinery which is used or intended to be used mainly or exclusively as part of or in connection with or as an accessory to any of the plant and machinery falling within Class 1 or Class 2–

    i foundations, settings, gantries, supports, platforms and stagings for plant and machinery;

   ii   steam-condensing plant, compressors, exhausters, storage cylinders and vessels, fans, pumps and ejectors; ash-handling apparatus;

   iii   travellers and cranes;

   iv   oiling systems; earthing systems; cooling systems;

   v   pipes, ducts, valves, traps, separators, filters, coolers, screens, purifying and other treatment apparatus, evaporators, tanks, exhaust boxes and silencers, washers, scrubbers, condensers, air heaters and air saturators;

   vi   shafting supports, belts, ropes and chains;

   vii   cables, conductors, wires, pipes, tubes, conduits, casings, poles, supports, insulators, joint boxes and end boxes;

   viii   instruments and apparatus attached to the plant and machinery, including computers, meters, gauges, measuring and recording instruments, automatic or programmed controls, temperature indicators, alarms and relays.

CLASS 3

The following items–

a   Railway and tramway lines and tracks (other than tracks used exclusively for the transmission of power), and relevant equipment occupied together with such lines and tracks; In this paragraph "relevant equipment" means–

   i   tracks supports and foundations;

   ii   sleepers, settings and fittings;

   iii   buffers, cross-overs and points;

   iv   power wire supports and power wire gantries;

   v   signal gantries; and

   vi   barriers, gates and crossings.

b   Lifts, elevators, hoists, escalators and travelators.

c   Cables, wires and conductors (or any system of such items)–

   i   situated in or on a hereditament used or intended to be used in connection with the transmission, distribution or supply of electricity, and

   ii   used or intended to be used in connection with such transmission, distribution or supply, other than such items or parts of such items which are, or are comprised in equipment which is used or intended to be used mainly or exclusively for switching or transforming electricity.

   iii   Poles, posts, pylons, towers, pipes, ducts, conduits, meters and any associated supports and foundations, used or intended to be used in connection with any of the items included in (c) above.

iv   Cables, fibres, wires and conductors, or any system of such items, or any part of such items or such system, used or intended to be used in connection with the transmission of communications signals, and which are comprised in the equipment of and are situated within premises;

In this paragraph–

i   "premises" means any hereditament which is used, or intended to be used, mainly or exclusively for the processing or the transmission of communications signals, excluding any part of such a hereditament within which there is equipment used mainly for the processing of communications signals;

ii   "processing of communications signals" means the conversion of one form of communications signal to another form, or the routing of communications signals by switching; and

iii   "equipment used mainly for the processing of communications signals" includes:

•   that part of any associated cable, fibre, wire or conductor which extends from the point of conversion or switching to the first distribution or termination frame or junction; and

•   that part of any associated cable, fibre, wire or conductor which extends from the last distribution or termination frame or junction to the point of conversion or switching.

iv   Poles, posts, towers, masts, mast radiators, pipes, ducts and conduits and any associated supports and foundations, used or intended to be used in connection with any of the items included within (e) above.

v   A pipe-line, that is to say, a pipe or system of pipes for the conveyance of any thing, not being–

i   a drain or sewer; or

ii   a pipe-line which forms part of the equipment of, and is wholly situated within, relevant premises; together with any relevant equipment occupied with the pipe-line; and where a pipe-line forms part of the equipment of, and is situated partly within and partly outside, relevant premises, excluding–

(aa)   in the case of a pipe-line for the conveyance of any thing to the premises, so much of the pipe-line as extends from the first control valve on the premises; and

(bb)   in the case of a pipe-line for the conveyance of any thing away from the premises, so much of the pipe-line as extends up to the last control valve on the premises;

but not excluding so much of the pipe-line as comprises the first or, as the case may be, last, control valve;

In this paragraph–

"relevant equipment" means–

i   foundations, supports, settings, chambers, manholes, pipe gantries, pipe bridges, conduits, pits and ducts;
ii   valves and flow regulators;
iii meters, pumps and air compressors (including the motors comprised in any such equipment); and
iv   apparatus for affording cathodic protection to a pipe or system of pipes;

"relevant premises" means a factory or petroleum storage depot, a mine, quarry or mineral field or a natural gas storage or processing facility or gas holder site, and for this purpose–

i   "factory" has the same meaning as in the Factories Act 1961(a);
ii   "mine" and "quarry" have the same meanings as in the Mines and Quarries Act 1954;
iii "mineral field" means an area comprising an excavation being a well or bore-hole or a well and bore-hole combined, or a system of such excavations, used for the purpose of pumping or raising brine or oil or extracting natural or landfill gas, and so much of the surface (including buildings, structures and works thereon) surrounding or adjacent to the excavation or system as is occupied, together with the excavation or system, for the purpose of the working of the excavation or system;
iv   a "natural gas storage or processing facility" includes premises used or intended to be used mainly or exclusively for the processing, storage or changing the pressure of natural gas;
v   "petroleum storage depot" means premises used primarily for the storage of petroleum or petroleum products (including chemicals derived from petroleum) or of materials used in the manufacture of petroleum products (including chemicals derived from petroleum).
vi   Lock and dock gates and caissons.

CLASS 4

The items specified in Tables 3 and 4 below, except–

a   any such item which is not, and is not in the nature of, a building or structure;
b   any part of any such item which does not form an integral part of such item as a building or structure or as being in the nature of a building or structure;

c   so much of any refractory or other lining forming part of any plant or
     machinery as is customarily renewed by reason of normal use at intervals
     of less than fifty weeks;
d   any item in Table 4 the total cubic capacity of which (measured exter-
     nally and excluding foundations, settings, supports and anything which
     is not an integral part of the item) does not exceed four hundred cubic
     metres and which is readily capable of being moved from one site and re-
     erected in its original state on another without the substantial demolition
     of any surrounding structure.

TABLE 3

Blast furnaces.
Bridges, tunnels, tunnel linings, tunnel supports and viaducts.
Bunds.
Chimneys and flues.
Coking ovens.
Cooling ponds.
Dams.
Fixed cranes.
Floating pontoons, with any bridges or gangways not of a temporary nature
     used in connection with such pontoons.
Flumes, conduits and ducts.
Foundations, settings, fixed gantries, supports, walkways, stairways, handrails,
     catwalks, stages, staithes and platforms.
Headgear for–

     mines, quarries and pits;
     wells.

Masts (including guy ropes) and towers for radar or communications signals.
Pits, beds and bays.
Radio telescopes.
Shiplifts and building berths.
Tipplers.
Transversers and turntables.
Turbines and generators.
Valve towers.
Well casings and liners.

TABLE 4

Accelerators.
Acid concentrators.
Bins and hoppers.
Boilers.

Bunkers.
Burners, converters, furnaces, kilns, stoves and ovens.
Chambers and vessels.
Condensers and scrubbers.
Coolers, chillers and quenchers.
Cupolas.
Cyclones.
Economisers, heat exchangers, recuperators, regenerators and superheaters.
Evaporators.
Filters and separators.
Gas producers, generators, purifiers, cleansers and holders.
Hydraulic accumulators.
Precipitators.
Reactors and reactor pressure vessels.
Refuse destructors and incinerators.
Reservoirs.
Retorts.
Silos.
Stills.
Tanks.
Towers and columns.
Vats.
Washeries for coal.
Wind tunnels.

13.3 Wales

The principle regulations governing the rateability of plant and machinery in Wales are the Valuation for Rating (Plant and Machinery) (Wales) Regulations 2000 (2000/1097) (W.75).

These regulations have been amended by the following regulations:

- Valuation for Rating (Plant and Machinery) (Wales) (Amendment) Regulations 2001 (2001/2357) (W.195);
- Valuation for Rating (Plant and Machinery) (Wales) (Amendment) Regulations 2010 (SI 2010/146) (W.75).

14 The valuation of other types of property

There are a wide variety of other hereditament types beyond shops, offices and industrial/warehousing properties. This chapter gives general valuation guidance about a selection of other hereditament types together with illustrative valuations.

> **The purpose of the illustrations is to demonstrate the approach rather than give actual comparable valuations. In some instances, examples for the 2017 revaluation, rather than 2023 revaluation, have been used. This is to avoid the unusual complication of COVID-19 showing abnormal falls in trade and thereby allowing a more 'normal' period of accounts to be used to give a better and clearer understanding of the approach to dealing with accounts in 'normal' circumstances.**

14.1 Advertising rights and stations

There are a large number of 'Out of Home' advertisement sites around the country and these are rateable in the usual way if they involve the occupation of land. They can also be rateable under a special provision of the Local Government Finance Act 1988 if the exhibitor only has a right to exhibit an advertisement rather than having a right to occupy land. Section 64(2) provides that a separate hereditament is created where the advertising right is let out or reserved to any person other than the occupier of the land or where there is no occupation at all of the land but the right is let out by the owner. A typical example is a hoarding on the gable wall of a shop. The advertising company does not occupy the shop, the gable wall or the adjoining land but merely has been given the right to place advertisements in return for a rent.

Advertisement hoardings vary in size and display format. They are still often described by reference to the number of sheets of paper needed in the past to make up the advertisement on the hoarding, but this is changing as technology allows for more bespoke formats to emerge. There are two broad

DOI: 10.1201/9781003297369-14

display sizes: 'small-format' and 'large format'. For large-format display types, despite a reduction in numbers since the 2017 antecedent valuation date (AVD), the most seen advertising hoarding on the roadside remains the 'landscape' 48-sheet hoarding (3 m × 6 m). There are small numbers remaining of 16-sheet 'portrait' style hoardings on the flank walls of shops and there are very long 96-sheet (3 m × 12 m) and some even longer 192-sheet (3 m × 24 m) roadside panels. However, there is no longer a need for display sizes to be formed from a multiple of a paper poster and now, with the advent of stretched banners, displays are made with fewer 'sheets' or may even be all one sheet in the case of a vinyl graphic; there are also growing numbers of electronic and digital displays. All this results in a definite move towards more variety and emerging larger size categories, such as the 'Golden Square' (6 m × 6 m), 'Mega 48' (4.3 m × 8.9 m), 'Mega 96' (4.3 m × 17.8 m) and 'Mega 6' (a 'portrait' version of the 'Mega 48'). There are also many bespoke or spectacular displays that are no longer based on the traditional 'sheet size' classification at all. At the 2023 AVD, for large-formats, the 'other' size category represented around a fifth to a quarter of all available display sizes, roughly the same as the total number of 48-sheets available.

At the smaller end of the display spectrum, the most common size of 'small format' display is the six-sheet (1.2 m × 1.8 m). in fact, the six-sheet represents over half of all available display types of any size in Great Britain. Such advertisements are seen on the ends of bus stops and in shopping centres, supermarket car parks, petrol filling stations, cinemas, etc. There are other variations of small format display, such as the four-sheet, but these are very much fewer in number than the six-sheet.

'Out of Home' advertising has changed in recent years, with several smaller companies becoming more established. There is an ongoing consolidation of sites with the advertising opportunities in the best positions being built as or converted to 'digital' displays and the poorest 'print' or 'static' display sites being removed or taken over by regional companies. Advertising sites with 'scrolling' 'back-illuminated' displays (typically showing two or three posters) are now being superseded by 'electronic' or 'digital' format displays. Such displays require minimal onsite physical need to 'change' the display and allow for maximum flexibility and impact in showing any number of displays during any particular time window.

There is rental evidence for the larger formats and this can be analysed by size. Rents will vary by position relative to the target audience, by size of poster and general location. Angle of vision, height, visibility and any obstruction to the view affects value. Sites parallel to a road will be less valuable than equivalently located sites where the hoardings are angled to the road or are situated facing a busy 'T' junction so more people see the advertisements. As with any valuation, location and position are critical to value. Sites facing major arterial routes and in highly sought locations can command very high rents, whereas similar hoardings in provincial towns will command lower rents. It is not recommended to analyse rents to £ per 'sheet' or to analyse 16-, 32-, 48- and

96-sheets to an equivalent number of, say, 48-sheets. However, analysis of rents in a particular locality may reveal a relationship based on size, all things being equal; for example, it may be demonstrated that 48-sheets will command in excess of three times a 16-sheet's rent because of impact and position, but half the rent commanded by a 96-sheet. However, whenever possible it is always best to compare 48-sheets to 48-sheets, 16-sheets to 16-sheets and 92-sheets to 92-sheets, etc. It should also be considered that the presence of internal/external illumination or multiple display formats, such as digital, is often an indication of a good position. It should be anticipated that rents for such sites will be higher and this should be factored into any analysis.

Again, there are no 'hard and fast' relationships when comparing positions which have multiple display formats or illumination; whilst it is probably safe to surmise an increase in value when compared to a 'single display', the relationship, if any, between different formats is only capable of being tested by analysis of rental evidence, comparing like with like whenever possible.

Where the site is a statutory advertising station or land used for advertising, rather than an advertising right, the value of the land should be included in the valuation. The value of any hoarding or structure put up to enable advertisements to be exhibited should also be included in the valuation. This includes a sign, hoarding, frame, post or other structure erected or to be erected on land. Very few rents will include an amount for structure; therefore, if the rental evidence is for rights only, the rateable value will need to be increased to reflect the annual value of the structure.

14.1.1 Example valuation

Advertising station situated on a busy main road in Nonsuch London Borough: the site comprises two digital 48-sheets angled so one hoarding faces each direction of traffic.

Local evidence close to AVD:

16-sheet static print panel on side of shop: £750
48-sheet digital parallel to road: £10,000
48-sheet digital facing T junction at end of the road: £21,000
48-sheet static or print display facing one direction of traffic: £4,000
96-sheet digital facing one direction of traffic, externally lit: £28,500

None of the rents includes the hoarding structures.

14.1.2 Conclusions

The 16-sheet does not assist as it is different in nature from the 48-sheets. The same is true of the 96-sheet, despite it being 'digital'. The 48-sheet parallel to the road is less attractive to advertisers as it is not as easy to see as angled sites, although it can be seen from both directions. The 48-sheet static or print display is not directly comparable and its rent is likely to indicate the particular value of that position in that it only supports a static or print display; this might require more examination if all things were equal, but in advertising individual position is very important. The T-junction site is very well positioned and will command a greater rent than the 48-sheets on the subject site.

Valuation of advertising station:

Two × digital 48-sheets @ £13,500 each =	£27,000
Structure (metal free standing supporting digital displays)	£5,000
Total	£32,000 rateable value

It is only once a value for the 48-sheet has been established that any incidental relationships between other sizes and display types can be deduced in the locality. However, caution must be exercised in making other valuations. It is always best to compare like with like if possible.

14.2 Amusement parks, theme parks and visitor attractions

These can range from major parks such as Alton Towers or Thorpe Park, to safari parks such as Knowsley and to smaller amusement parks.

All these properties are commercial in nature and are generally valued on the receipts and expenditure approach.

In the valuation of this type of property, care must be taken to correctly identify the hereditament and to see whether certain parts of the property are let out as concessions and form separate hereditaments such as catering facilities and the like.

The assessment of Alton Towers was considered by the Upper Tribunal in *Merlin Entertainments* v *Cox (VO)* [2018] UKUT 0406 (LC) in relation to the effect on the assessment of the considerable reduction in visitors following a crash on a roller-coaster ride which had seriously injured five passengers.

14.3 Bingo halls and clubs

The bingo industry has experienced tough times over recent years with revenues falling year after year. This is partly due to the continuing impact of the smoking ban (1 July 2007) and the Gambling Act 2005, which restricted the number of gaming machines, both having a large impact on attendances, and partly due to the age profile of its customer base.

In 2009 a revision to the Gambling Act 2005 (Gaming Machines in Bingo Premises) provided some key changes allowing the number and type of gaming machines to be increased. Between 2009 and 2014 the Gambling Act was amended a further three times and provided an increase to the amounts winnable from machines on the premises. This went a long way to counteracting the decreased dwell time at AWP (Amusement with Prizes) machines resulting from customers temporarily leaving the premises because of the smoking ban. It also allowed machines to be stationed in smoking shelters to further counteract the effect of the ban.

In 2014, after much lobbying from the industry, the Government reduced Bingo Duty (Gross Profit Tax) from 20% to 10% and this was expected to result in the operators using the additional funds to provide better prizes and an improved user experience by refurbishment and other changes. The industry has modernised the way in which the user experiences bingo, with tablet computers being used and entertainment on Friday and Saturday nights being provided to try and encourage a younger customer base.

In recent years several clubs have downsized or closed and the opportunity to downsize may well continue due to the increase in the online gambling sector which has reduced the need for such a substantial physical presence.

Bingo halls can be found in a variety of buildings including the following:

- converted shops within town centres;
- converted theatres or cinemas or even churches or church halls;
- purpose designed property often with associated car parking facilities.

The most usual approach to the valuation of this type of property is by references to rents; however, due to the decline of the bingo industry, open market rents are now scarce. This, coupled with nil increases at review and Retail Price Index (RPI) linked reviews not revealing open market rental value, suggests an alternative methodology is required. In analysing rents, they are usually analysed to show a percentage of the gross receipts as well as the traditional rent per m^2 where possible.

The initial valuation consideration is to determine the Fair Maintainable Trade (FMT) as at the AVD, 1 April 2021. A scheme of valuation involving two scales, depending on the type of property, derived from an analysis of the available evidence by reference to gross turnover has been adopted by the Valuation Office Agency (VOA) for the 2023 Revaluation.

Scale A applies only to bingo halls in converted cinemas, theatres or concert halls.

Scale B applies to purpose-built bingo halls, and all others not covered by Scale A.

In *Buzz Group* v *Salmon (VO)* [2020] UKUT 0116 (LC) the ratepayers reduced the size of their bingo hall because it was too large, given the decline in the game mentioned above. They argued for a lower RV because of the reduced size and argued on a £ per m^2 approach contrary to the percentage of receipts approach generally adopted; but failed to back this up with rental evidence. On the face of it, smaller premises might well suggest a lower value but there had been no decline in receipts. The Upper Tribunal noted it was likely where a bingo hall's analysis resulted in a comparatively low £ per m^2 that it was much larger than other properties with a comparable fair maintainable trade. The

Tribunal suggested such a property would be self-evidently too large for the level of trade it was able to generate, and this was precisely why the ratepayers had reduced the size. The downsizing had not reduced the value.

14.4 Bowling alleys

Since the introduction of bowling alleys in the 1960s, ten-pin bowling has gone through various phases of popularity. At the present time the industry seems to have stabilised with the numbers of hereditaments remaining static.

A significant proportion of bowling alleys are rented; consequently the rental approach is the method of valuation of choice. Rents can be analysed to a value per m² gross internal area (GIA), value per lane or percentage of gross receipts as the unit of comparison. A £ per m² is the more usual approach and accords with market practice nowadays. Care must be taken when analysing rental information as rents are often for an unfitted shell with the costs of fitting out being considerable, or may include payment for items of non-rateable plant and machinery.

14.4.1 Example valuation

The 'Bedrock Bowl' is a modern 24-lane ten-pin bowling alley situated on a leisure park in a good edge of town position. It was built in 2000 and has not been significantly modified since.

2,750 m² @ £111.5 per m² =   £306,625
(£ per m² includes a fit out addition of 15%)
Say rateable value £305,000 (approximately £12,700 per lane)

14.5 Caravan sites

As explained in Chapter 3 on 'The Hereditament', a caravan is not in itself rateable but may become rateable together with the pitch upon which it stands if enjoyed with the land in such circumstances and with sufficient permanence that annexation to the land can be inferred.

Many caravans, either on sites containing other caravans or singly, will be occupied as a sole or main residence and so be domestic and banded from paying Council Tax. Other caravans may be used for holidays, and the land and often the caravans will be valued for rating.

Sites may be run by the operator only with the site operator's caravans on them which he or she lets to holidaymakers, or the site operator may provide the site and facilities but allow others to bring their caravans on site, either permanently or as 'touring' caravans. Alternatively, there may be a mixture of site operator owned caravans and privately owned caravans.

The approach to valuation was discussed in *Garton* v *Hunter (VO)* [1969] 2 QB 37 where the Lands Tribunal considered the assessment of a leisure caravan park from a rent, receipts and expenditure, and a contractor's approach. Usually, a receipts and expenditure approach is used, but analysed for comparison purposes as a proportion of turnover.

Where an operator owns the caravans on a site and lets them for short periods to holidaymakers, it is clear that the operator is in rateable occupation and the whole site will be a single hereditament. Where individuals separately occupy caravan pitches on a long-term basis, it is likely that each will form a separate hereditament as in *Field Place Caravan Park Ltd* v *Harding (VO)* [1965] RA 521. To simplify rate collection, the Non-Domestic Rating (Caravan Sites) Regulations 1990 (SI 1990/673) provide that separately occupied pitches on a leisure site are to be treated as amalgamated in a single hereditament with the parts of the site occupied by the site operator (unless they are occupied by a charity and wholly or mainly used for charitable purposes). To allow the site operator to apportion the rate's liability, the valuation officer provides a notice under Regulation 4 listing how many caravans occupied by persons other than the site operator are included in the assessment and how much of the total rateable value is attributable to those caravans and pitches.

14.5.1 Example valuation

The 'Sandy-by-the-Sea Holiday Park' is a caravan park with site operator owned caravans, private vans, pitches for touring caravans and site shop and restaurant. There is also an amusement arcade under the control of the site operator, but run as a concession with the site operator receiving an annual fee of £80,000.

The site operator owned 'fleet' caravans typically produce an income of £6,000 pa at the AVD. Touring caravan and tent pitches produce an income of £1,500 pa.

Some of the private owners of caravans let their caravans to holidaymakers through the site operator. These tend to have a lower income per caravan because the site operator will seek to let the 'fleet' caravans first. For this caravan park it is found that they typically produce an income of £3,300 per caravan.

The site owner receives an income of £20,000 for the safe winter storage of touring caravans.

For the purposes of the valuation the number of private caravans is converted to equivalent fleet units as follows:

Number of private caravans sublet by the site operator		20
Typical income per private van	$\dfrac{£3,300}{£6,000} =$	0.55
Equivalent fleet units	$(20 \times 0.55) =$	11

Table 14.1 Valuation of Sandy-by-the-Sea caravan park

	No. of caravans Receipts	Income per caravan	Valued at	% of fleet income
Fleet hire caravans (owned by site operator)	65	£6,000	100%	£390,000
Private vans sublet by site operator (equivalent fleet vans)	11	£6,000	100%	£66,000
1–5-year-old private vans fully serviced	250	£6,000	50%	£750,000
16+ year-old vans fully serviced	80	£6,000	35%	£168,000
16+ year-old vans un-serviced	4	£6,000	25%	£6,000
Pitches for touring caravans and/or tents	40	£1,500	100%	£60,000
Winter storage (first £5,000)				£5,000
Shop gross profit		£130,000	100%	£130,000
Bar and catering gross profit		£425,000	100%	£425,000
Total estimated receipts				£2,000,000
Estimated receipts from caravans, shop, bar and catering		£2,000,000	11.0%	£220,000
Arcade concession fee		£80,000	50%	£40,000
Winter storage fee over £5,000		£15,000	25%	£3,750
				£263,750

Notes:
(1) The arcade concession is taken at a higher percentage than the income from the caravans etc. because the expenses of running the concession are not borne by the operator and the fee is therefore virtually a net profit figure.
(2) Winter storage of touring caravans can provide site operators with a significant income with potentially few overheads. For the 2017 Revaluation, valuation officers took the first £5,000 gross receipts (excluding VAT) received from the winter storage of touring caravans at the standard rental percentage for the site, but those in excess of this figure are valued at a rental percentage between 25% and 35%. The more basic the winter storage site the higher the percentage adopted.

14.6 Car parks

Many assessments will include car parking spaces as part of a hereditament. In analysing rents the valuer needs to decide whether to treat the facility as included within the analysis per m^2 of the buildings or to separately analyse the car parking per space or per m^2. Valuations for rating will have to follow the same approach in accordance with the maxim 'as you devalue so you value'.

Adjustments may need to be made in a valuation for excessive or inadequate car parking.

Car parks also exist as hereditaments in their own right and fall into four categories:

- multi-storey operated by local authority or commercial company;
- surface, in towns – as above;
- surface, other areas operated by local authority or sole proprietor;
- seasonal, rural or coastal operated by local authority, sole proprietor or environmental based charity.

14.6.1 General issues to be considered

14.6.1.1 Identification of the hereditament

The valuer needs to have particular regard to the correct identification of the rateable hereditament and the rateable occupier (*MFI Furniture Centres* v *Robinson (VO)* [1989] 29 RVR 48).

Public conveniences in or adjacent to a car park may form part of the car park hereditament or be separately assessed.

14.6.1.2 Rateable occupier

Uncertainty can arise in the case of free, regularly shared, unevenly shared and multiple use properties, such as non-market day car parking uses and annual licence spaces with physical control. In *Renore Ltd* v *Hounslow London Borough Council* [1970] RA 56, the user of a car space with a locked bollard was held to be in rateable occupation.

14.6.2 Valuation basis

14.6.2.1 Rentals

There is considerable evidence for small car parks in town centres let on short-term lease/licence to a single occupier.

There is some rental evidence on commercial multi-storey car parks.

14.6.2.2 Receipts and expenditure

Assuming normal commercial conditions apply, then, for standard surface car parks probably 25% to 35% of turnover equates to rent for a typical normal user car park, with multi-storey car parks requiring individual consideration.

More appropriately a full receipts and expenditure method will have particular regard to:

- machine minding costs/staffing;
- maintenance and repair;
- insurance.

14.6.2.3 Contractor's basis

Few free standing multi-storey car parks are now built, and most are built as a result of planning gain rather than as a commercial proposition. The contractor's basis tends to result in higher assessments for multi-storey car parks but the case for use is not as strong as for, say, public conveniences as there are commercial operators of the former.

Multi-storey car parks may not be demonstrably more popular than surface car parks. Price per space should therefore be at least related to surface car parks, with perhaps an addition for being protected from the weather.

14.6.3 General valuation issues

- percentage of occupation at peak/non-peak times;
- proportion of upper deck spaces in multi-storey;
- local authority pricing policy for traffic management reasons, rather than hereditament based reasons;
- seasonal factors;
- is the car park manned?
- is it pay and display or free?
- location;
- permit holder only spaces can result in increased income and perhaps separate hereditaments;
- limited period parking;
- limited access;
- layout;

In addition to the above, for multi-storey car parks the valuer needs to consider:

- headroom;
- lifts;
- walkways to adjoining developments which may increase occupancy rates and therefore rental value.

14.7 Cinemas

Cinemas as a class of property have seen a change in fortune over the years. Between the First and Second World Wars, people went to the 'pictures' several times a week. With the advent of television, and then video, audiences declined and cinemas were often very poorly patronised and seen as a dying concept. The large 'super cinemas' of 1,500 or more seats were usually converted to two or three cinemas or to bingo halls. The arrival of the multiplex from America (1985 in Milton Keynes) signalled a change, since when audience numbers and the numbers of cinemas have increased significantly.

Multiplexes are not the only types of cinemas in existence. The remaining converted super cinemas of the interwar years in some locations are achieving good takings; there are specialist cinemas catering for ethnic and art films, and also the large Leicester Square traditional super cinemas.

New cinema openings have continued since the arrival of the multiplexes. However, the focus for cinema building changed from building out of town multiplex cinemas, which had reached saturation point, to a new breed of cinema perhaps best described as 'small multi-screen cinemas' which fill gaps in the market not served by the larger out of town leisure complexes. This new breed of cinema can normally be found within smaller towns, where there is increasing demand for cinemas to be central rather than out of town. These cinemas predominantly form part of wider development schemes to regenerate/reinvigorate town centres as part of a town centre leisure development and will include A3/A4 units to complement the cinema occupation.

Digital technology has allowed the cinema industry to save costs on the physical distribution of film and provide an enhanced viewing experience due to the now high quality of digital recording, and allowing for variation in product, such as the expansion of the IMAX experience and the growth in 3D films. It has also enabled the concept of the 'pop-up cinema' through the use of portable screens, projectors and sound system equipment. Most often this concept utilises open car parks to create a drive-in 'movie theatre' in the summer months.

Another area which has seen growth is the further diversification of operators seeking to exploit food and drink sales. Increasingly, where physically possible, cinema premises are being altered/adapted to provide new experiences, bars and cafes, etc., in order to maximise existing and new revenue potential.

Cinemas can be split into three main categories:

1 **Traditional** – Most traditional cinemas were built in the interwar years originally with a single auditorium. Many have now been converted to alternative uses such as bingo halls or public houses and many have been redeveloped, but those still being operated as cinemas have been refurbished and converted from single to multi-screen operations, usually with between two and five screens. As these multi-screen cinemas have been constructed within the constraints of existing buildings, auditoria size and layout may not be ideal. Repair and maintenance costs for a building of this age and size are likely to be considerable, and other running costs are likely to be relatively high compared with a modern purpose-built cinema.
2 **Multiplex** – In its July 1994 Report, the Monopolies and Mergers Commission defined a multiplex cinema as 'a purpose-built cinema with at least five screens and usually offering extensive free parking'.
   The multiplex concept has succeeded because of several factors:

   • Accessibility – often out of town on good ring roads or well sited for roads in town.

- Car parking – easy, plentiful and right next to the cinema, possibly shared with other uses on a retail park: a major factor when so many old Odeons and ABCs had hardly any parking provision.
- Newness – both from the public's perception but also from ease of operation. Digital projection means no large projection rooms are needed in the modern multiplex, modern stadium style seating with bigger seats and more leg room, and larger screens with better sound quality for enhanced viewing experience. Multiplex cinemas succeed due to a combination of high numbers of admissions, carefully controlled operating expenses and flexible screening possibilities (due to the large number of screens enabling a single film to be shown on different screens and at varying times); this enables multiplex cinemas to achieve higher operating profits than more traditional style operations.

3 **Small multi-screen** – There has been an increasing trend in recent years for cinema operators to open cinemas in smaller towns as the availability of sites in larger towns and cities have become scarcer. These small multi-screen cinemas possess many characteristics of a multiplex cinema and are treated by the VOA as falling in this class if they fit the following definition:

'A small multi-screen cinema will always be located outside the M25, it can have no more than six screens and may be entirely purpose built or a modern conversion from another use. The total number of seats will never exceed 1,250.'

The valuation approach for rating for traditional cinemas has generally been by rental comparison, with rents analysed using a percentage of gross receipts. In *Rank Organisation Ltd* v *Billet (VO)* [1958] R&IT 650, an approach using gross receipts was preferred to that of comparison on a value per seat basis:

I prefer the method of arriving at the gross value by means of a percentage on estimated normal receipts, for this method of valuation enables the relationship between capacity receipts and the estimated receipts to be ascertained, and inferences to be drawn therefrom as to whether or not a cinema can be expected to be profitable.

Later, in *Associated Cinematograph Theatres Ltd* v *Holyoak (VO)* [1969] RA 305), the Tribunal stated that 'there can now be no doubt that a proper method of assessing a cinema is by reference to its gross receipts'.

For valuations for 2017 and the 2023 revaluations, valuation officers have valued cinemas using a percentage of the gross receipts at the fair maintainable level. This provides flexibility to reflect the varying likely trading performances and property characteristics of different cinemas.

In the past due to the standardised format and consistent performance of multiplex cinemas, meaningful analysis in terms of rent per seat was achievable. However, the market has evolved to such an extent with such a variety of cinema types that this is now difficult to undertake and an analysis in terms of gross receipts may now be preferable.

Valuation officers have adopted two scales for valuation, one for traditional style cinemas and the other covering both multiplexes and small multi-screen cinemas. In previous rating lists the level of gross receipts per seat determined the scale position of the subject cinema. For 2017 the box office income per seat was used as the performance indicator. It was the reference point to determine where in the percentage to rateable value range, at any particular level of FMT, the subject cinema should be valued. The change came about as many cinemas are now achieving a much more significant proportion of their receipts from ancillary 'food and beverage' sales and this revision enabled a proper reflection of the impact of this factor upon profitability.

For 2023 rather than using box office income per seat as the reference point the income per screen is used. The cinema industry has been experimenting with low density seating formats for some years and this is now becoming a mainstream reality. All three major multiplex cinema operators have been testing 'Lux' recliner seating projects at various existing localities and this has typically seen seating capacities reduce by 50% – and in some instances by 60% or more. There will also be new cinemas being developed using this low-density seating model. To reflect this and to avoid distortion between cinemas with a low-density seating format and others, the VOA has agreed with the industry that rather than using income/seat as the driver to establish the adopted % bid within a given range, box office income/screen will be adopted.

Deductions need to be made from a rent for any non-rateable items that may be included in the rent. In analysing rents, gross receipts deductions need to be made from a rent for any non-rateable items that may be included in the rent. In analysing rents, gross receipts from all sources of income should be used.

There is no reason why a full receipts and expenditure valuation cannot be undertaken, although there are difficulties, as with any property owned together with other properties, when allowing for head office expenses, etc.

When analysing rents of new multiplexes, it is important to establish whether the rent is for the basic or 'cold' shell or the partly fitted 'warm' shell. Fitting out with rateable items such as air conditioning can involve very significant costs and add substantially to rateable value.

14.7.1 *Example valuation*

The 'Multikinema' is a 14-screen multiplex cinema built in 2012 with 2,500 seats and is owner occupied.

Table 14.2 The 'Multikinema' 2017 list valuation

Receipts

Year ending	Admissions	Box office income	Box office as % of total	Total gross receipts	Spend per head	Box office per seat
2013	120,000	£10,000,000	71.43%	£14,000,000	£11.67	£4,000
2014	125,000	£11,000,000	73.33%	£15,000,000	£12.00	£4,400

Valuation

FMT	£15,000,000
Valued at	10.75% of FMT
Rateable value	£1,612,500
	SAY £1,600,000

14.8 Clubs and halls

Two main types of clubs are commonly found.

14.8.1 Members' clubs

A members' club is one where the management of the club is vested in the members or in trustees acting on their behalf. An important feature of a members' club which will have an impact on its valuation is that its purpose is to provide a benefit for its members rather than to make a profit. This type of club will include many types of social clubs, including the Royal Naval Association and the Royal British Legion clubs.

14.8.2 Proprietors' clubs

Proprietors' clubs are commercially operated clubs owned by individuals or companies.

14.8.3 Licensing

Both types of clubs will almost always have the benefit of a licence for the sale of alcohol and may also have facilities for the serving of food. Following the introduction of the Licensing Act 2003, which came into effect on 24 November 2005, licensable activities such as the supply of alcohol in a club, or the provision of regulated entertainment, require both member and proprietor clubs to obtain a club premises certificate from the licensing authorities (English district or county councils and county or borough councils in Wales).

14.8.4 Valuation – members' clubs

Generally this type of property is valued on a rental approach. It may be helpful to examine turnover when considering the possible impact of a Material Change of Circumstances (MCC), e.g. the impact of the ban on smoking in public places introduced in Wales from 2 April 2007 and England from 1 July 2007.

In *United Services and Services Rendered Club (Tooting and Balham) Ltd and the Putney Club* v *Thornley (VO)* [2002] RA 177, the Lands Tribunal held in its decision that:

> *I remind myself that I am concerned only with the rental value of the two appeal properties; … In my opinion, the available rental evidence, although limited, is sufficiently reliable to enable me to arrive at an accurate valuation of the appeal properties. It is therefore not necessary for me to consider indirect evidence, such as the rating assessments of clubs in south-west London.*

In the decision, in rejecting the use of the receipts and expenditure method of valuation, the Tribunal found:

> *In my view, those responsible for managing members' club are likely to be motivated by a desire to provide those facilities which satisfy their members' particular requirements, subject to the overall need to remain solvent.*

The rationale for this approach can be traced back to the case of *Aberaman Ex-Servicemen's Social Club and Institute Ltd* v *Aberdare Urban District Council* [1948] 1 KB 332, where it was held that profit from trading in alcoholic liquor is not a relevant consideration in assessing the value of a members' registered club:

> *… it must be remembered that rent, not profit, is the measure of value;*
> *… all possible occupiers, including the actual occupier, must be taken into account as possible tenants from year to year;*
> *… if the profits depend on the personal skill of the tenant and can be made in any other premises quite as well as in the premises in question, the expected amount of profits will not affect the rent which a tenant will give, but if the profit can be earned only on the premises to be rated and can be earned there by any ordinary tenant, then the expected profit will affect the rent which a tenant will give.*

14.8.5 Valuation – proprietors' clubs

As with members' clubs, these will usually be valued on the rentals approach. Turnover may be considered when comparing one club with another but members' clubs will usually only be valued by reference to trade if the premises operate as a public house.

The Guidance Notes on the Receipt and Expenditure Method of Valuation for Non Domestic Rating, prepared by the Rating Forum suggest:

Where open market rental evidence exists for the subject property or similar properties, and that rental evidence conforms to the statutory definition of rateable value ... or can be made to do so without adjustments of such a nature that its reliability is affected, a valuation based upon such evidence will provide the preferred method of valuation.

14.8.6 Other issues

One common problem with this type of property concerns whether Roman Catholic social clubs are exempt from rating under paragraph 11 of Schedule 5 of the Local Government Finance Act 1988 as being similar to a church hall, or whether they are to be rated in the normal way.

The main issue with regard to the issue of exemption will relate to who is the actual occupier of the property. Where it can be shown, having examined the constitution of the club, that the parish priest has overall responsibility and control over the club it is likely it will be exempt. Where it is in the occupation of the members, the club will not be exempt and will be valued as described earlier.

14.9 Education establishments

14.9.1 Local authority schools

Local authority schools are valued on the contractor's basis of valuation. In applying this approach to this type of property it has been usual for the detailed application of the method to be agreed between the VOA and representatives of local education authorities.

Typically, this will cover in detail how each stage of the valuation should be undertaken and deal with typical problems encountered in this type of property, such as the unit of assessment, changing pupil numbers, over capacity and end allowances.

In some schools there may be other facilities on the site that are jointly used by the school and the community. In these cases careful consideration will have to be given as to the unit of assessment and the identification of the rateable occupier.

14.9.2 Public and independent schools

Public schools are valued on a contractor's basis of valuation.

A number of valuation problems are common with this type of property – in valuing this type of property, regard must be had as to whether any of the property falls to be classified as 'domestic' and therefore outside the scope of non-domestic rating. Typically, the following accommodation is regarded as domestic:

- permanent boarding accommodation provided for pupils and staff regardless of whether pupils/staff are in residence throughout the term or merely during the week;
- ablution facilities used exclusively by boarders (other than in sports changing rooms);
- locker rooms used exclusively by boarders for storage of personal effects;
- common, 'prep' or study rooms and music practice rooms regardless of location within the school, provided that they are not used for teaching and that they are used exclusively by boarders;
- laundries, unless also serving premises outside the hereditament;
- kitchens and dining rooms when serving ONLY boarders and resident staff;
- staff accommodation.

In many cases parts of the school may be spread over an area and each non-contiguous property may be considered to be a separate rating assessment.

Where there is a school chapel it may be exempt – regard should be had to the Lands Tribunal decisions in *Shrewsbury School (Governors)* v *Shrewsbury Borough Council and Hudd (VO)* [1966] RA 439 and *Shrewsbury School (Governors)* v *Shrewsbury Borough Council and Plumpton (VO)* [1960] R&IT 497. In these cases the chapel was not found to be exempt as it was not available for 'public religious worship'.

By way of illustration, the valuation shown is taken from the Lands Tribunal decision in *Eton College (Provost and Fellows)* v *Lane (VO) and Eton Urban District Council* [1971] RA 186, slightly amended to show the valuation stages.

Table 14.3 Valuation taken from *Eton College* case

Stage 1		
Buildings' estimated replacement cost		£4,635,073
Stage 2		
Less age/obsolescence allowances		
(= average 63%)		£2,920,095
Effective capital value of buildings		£1,714,978
Stage 3		
Sites of buildings: 56 acres @ £7,250 per acre		£406,000
Less 63%		£255,780
		£150,220
Effective capital value of land and buildings		£1,856,834
Playing fields and amenity land (199.5 acres gross area)		£85,725
		£1,942,559
@ 3.33%		£64,687
Less end allowance 10%		£6,468
		£58,219
Less college chapels (exempt)	£500	£57,719
Say rateable value		**£57,700**

14.9.3 Universities

Universities are usually valued on the contractor's basis of valuation. As with many classes of property the detailed approach to its application is usually agreed between the VOA and representatives of the universities. The decapitalisation rate to be adopted will be the lower prescribed rate.

Some of the more important issues to be decided include:

- The unit of assessment or assessments. Many universities will be spread over a wide area and each building or group of buildings may require a separate assessment.
- The second consideration is the treatment of halls of residence and associated residential accommodation. Halls of residences and their ancillary accommodation will not be subject to non-domestic rating but rather will be subject to Council Tax. All other accommodation, including teaching, sports facilities and social facilities, will fall to be assessed for non-domestic rating. Where halls of residences are let out during out of term time for short stay self-catering accommodation, then consideration will have to be given to the period of use. If such a use is planned for more than 140 days in the coming year, it will be rateable. In this case it is usual to add a given amount per night per room to the total value of the property.

The valuation shown below is taken from the Lands Tribunal decision in *Imperial College of Science and Technology* v *Ebdon (VO) and Westminster City Council*[1986] RA 233 and reflects values appropriate at that time. Note that the decapitalisation rate used is different from the prescribed rates which have to be used today.

The above illustrates how the estimated replacement cost was calculated for all the buildings at the university. A £ per m^2 reflected the quality of each individual building.

Table 14.4 Areas and costings – valuation taken from *Imperial College* case

Stage 1 building	GIA	£/m^2	ERC (£)
170 Queens Gate	1,458.7	105	153,161
Aston/Goldsmith/Bessemer	29,453.8	135	3,976,263
Bessemer lab	2,879.5	85	244,263
Bone	3,505.8	130	455,754
Roderic Hall	7,698.9	125	962,362
Mech. Eng.	30,165.6	130	3,921,528
Physics (W1)	16,672.8	125	2,084,100
Boiler/wks	3,254.9	105	341,764
Elect Eng.	13,567.8	125	1,695,975
Civil Eng.	14,320.5	140	2,004,870
ACE Ext.	10,859.2	140	1,520,228
College block and libraries	27,087.4	110	2,979,614
Total	160,924.9		20,340,438

The different approaches that were adopted to reflect the amount of obsolescence in each building by the parties. While for some buildings there was agreement as to the percentage allowance to be given, it can be seen that for other buildings there was often quite a considerable difference in opinion between the three parties to the appeal.

Table 14.5 Respective allowances – valuation taken from *Imperial College* case

Stage 2 building	£	Ratepayer	VO	RA
170 Queens Gate	105	47.7%	40%	
Aston/Goldsmith/Webb	135	47.2%	30%	30%
L-shaped extension	135	47.2%	30%	
Bessemer link and spur	135	16.6%		
Bessemer lab	85	44.4%	35%	30%
Bone	130	54.0%	30%	25%
Roderic Hall	125	29.6%		
Mech. Eng.	130	29.3%		
Physics (Wl)	125	25.2%		
Boiler/wks	105			
Elect Eng.	125	27.1%		
Civil Eng.	140			
ACE Ext.	140	19.0%		
College block and libraries	110	12.2%		
Total allowance on buildings		25.0%	5.89%	4.84%
Total allowance on land			25.0%	5.89%
Huxley	110	12.0%		
Total allowance on buildings		23.7%	5.27%	4.28%
Total allowance on land		23.7%	5.27%	
End allowance for disabilities		7.5%		5.0%

Table 14.6 Actual costs translated to adjusted replacement costs – valuation taken from *Imperial College* case

Building	Agreed cost/ adjustment (£)	%	Adjusted cost (£)
170 Queens Gate	153,163	42.5	88,069
Aston/Goldsmith/Webb	2,560,248	35.0	1,664,161
L-shaped extension	484,623	40.0	290,774
Bessemer link and spur	931,392	10.0	838,253
Bessemer lab	244,757	35.0	159,092
Bone	455,754	45.0	250,665
Roderic Hall	962,362	10.0	866,126
Mech. Eng.	3,921,528	7.5	2,627,413
Physics (Wl)	2,084,100	5.0	1,979,895
Boiler/wks	341,764	0	341,764
Elect Eng.	1,695,975	10.0	1,526,378
Civil Eng.	2,004,870	0	2,004,870
ACE Ext.	1,520,288	2.5	1,482,281
College block and libraries	2,979,614	11.0	2,989,614
Total	£18,099,355		£20,340,438

The figures below shows how the determined allowances were translated from actual costs of construction to adjusted replacement cost, taking into account the allowances.

14.9.4 Valuation

Stage 1		
Agreed estimated replacement cost of buildings	20,340,438	
Stage 2		
Adjusted costs (−11.0%)	18,102,990	
Stage 3		
Effective capital value of land		
15.725 acres @ £400,000 less 11.0%	5,598,100	
		23,697,455
Stage 4		
Convert to annual rental value 3.5%(1)	829,411	
Stage 5		
Less for disabilities − 7.5%		2,206
		767,205
Value, say		**£767,000**

(1) Decapitalisation rate that was appropriate at the time. Now the prescribed decapitalisation rate would have to be adopted.

14.10 Golf courses

In 1988 the Royal and Ancient Golf Club commissioned a report, 'The Demand for Golf'. This concluded that a large number of new courses would be needed to satisfy demand.

Since then some 500 courses have been built, bringing the total to some 2,200 clubs. Golf was seen, rightly, as a growth sport and with the banks keen to lend money for development in the late 1980s and Japanese companies keen to invest in Britain, particularly in what they saw as the cheap or 'under charged' golf in the United Kingdom, it was not surprising that a lot of money was available for purchasing new sites, designing and building new courses and funding new clubhouses to rival the grandeur of those of some of the famous existing courses.

The crash in the early 1990s with high interest rates and a loss of confidence ended the boom in construction. Reduced spending on leisure and the high interest rates, together perhaps with insufficiently thought out business plans, saw many developers and new clubs going into receivership.

As a consequence, capital values fell and newly developed courses at the time could be bought for half or less of the original cost of land acquisition, plus money spent on development. This was a clear example of cost no longer equalling value. Looking back it is possible to see some of the mistakes that were made. Like a lot of leisure schemes, particularly sports schemes, enthusiasm for the particular activity can cloud business judgement. Location, as for all property, is obviously the important factor. A large catchment of golfers within 20 minutes' drive is needed. A good site with interesting natural features is desirable and in many cases expenditure on large and impressive clubhouses can be money wasted as the additional income from these often does not support the cost of provision.

In recent years private individuals and limited companies, often new to the golf market, have accounted for a large proportion of golf course purchases and new lettings. Operators continue to look for opportunities to add to their income-producing facilities by including a professional shop, health and fitness suite, letting accommodation, and function and conference facilities.

An increasing number of municipally owned courses are now rented and privately run following competitive tendering.

The peak of the golf market was over by the early 2000s, and the industry has suffered particularly in the wake of the 2008 financial crisis. From a one-time position of huge demand for golf there is now a general oversupply of courses in many parts of the country with several clubs facing falling memberships. Although golf remains a popular sport, the time commitment required coupled with the growing popularity of cycling have impacted on the number of golfers and across the country a number of courses have closed.

The decline of participation and membership of golf clubs is well documented and there are countless articles in the golf press detailing this. Following COVID-19 where golf and driving ranges provided one of the few activities which fell within permitted activities, there has been slight renewal of interest in the sport. Time will tell if this is a permanent change in demand for golf.

Certain locations, for example the South East, are bucking the trend with stronger demand brought about by the age demographic and the higher disposable income of the population.

14.10.1 *Types of course*

Courses vary considerably, not just in their topography, soil and layout – links by the sea, heathland, parkland – but also in their operation or occupation. First, there are the private members' clubs. These are clubs run by the members for the members. Profits made are ploughed back into improvements to the course and clubhouse and in keeping down the members' subscriptions. Income does not just come from members but also from green fees and golf society visits which members' clubs allow to varying degrees. This can produce a significant income. Municipal golf courses are operated by local authorities for the benefit of local people and green fees are generally charged rather than

having members. Municipal courses and private members' clubs are the traditional occupation, but pay and play or proprietary clubs have been very much
the approach for new courses. These may have members but are operated with
the aim of giving a profit to their owners.

14.10.2 *Valuation*

The appropriate methods of valuation for golf courses was considered in the
North Worcestershire and South Staffordshire golf courses cases:

Stourbridge Golf Club Ltd v *Clark (VO)* [1959] 52 R&IT 437; 30 DRA 380;
Blackwell Golf Club Ltd v *Linley (VO)*, *Barr (VO)* v *Enville Golf Club Ltd*
[1959] 52 R&IT 437, *South Staffordshire Golf Club Ltd* v *Barr (VO):* (1959) 52
R and IT 437. In these the Lands Tribunal found:

* Rents of comparable golf courses provided the most reliable guidance as
  to value.
* Ascertainment of a virtual rent using the Receipts and Expenditure basis
  was not appropriate for golf courses; but
* Assistance could be provided by reference to receipts, although they should
  not form the major approach to valuation.

The difficulties in the golf industry that are described in these cases are similar
to those that are current today and it is therefore still a relevant decision.

The Lands Tribunal came to the conclusion that there was sufficient rental
evidence for a rental valuation to be the primary method of valuation. However
it noted that assistance could be provided by receipts, which is endorsed by the
RICS Receipts and Expenditure Forum guidance note (section 7 shortened
method) which states this can be used as a means of comparison. A rental
comparison approach is, then, the primary method used today.

For rating, the requirement is to look at each course vacant and to let and
ask what its rental value is. It follows that the valuation approach, and indeed
valuation, should not vary depending on what type of occupier in fact runs
the course. The possible bid of a commercial operator for a famous private
members' course can and should be taken into account.
Golf courses vary and can be categorised as:

* grade A+ prestige courses with national prominence – may host international tournaments;
* grade A other championship standard courses;
* grade B very good courses to a county championship standard;
* grade C standard members' clubs and best pay and play;
* grade D pay and play courses – municipal courses and best nine hole;
* grade D – poor municipal (or what you might call a field).

The grading looks at the various elements making up the quality of the course:

- How well established – new courses generally will look new not least because of the number of saplings on many new courses. A mature course is more attractive, but the designer of a new course may have been fortunate in having existing substantial trees on the site and natural hazards already present and establishment will therefore be much quicker.
- Construction – the standards of construction of new courses have varied. The better the construction of drainage and attention to subsoil detailing for the greens the better they will be and the less repair (or worse) required. Drainage will generally be important. Due to wet winters, fairway drainage systems are now increasingly common.
- Soil – the type of subsoil, clay or sand affects the maintenance costs of the course by up to a factor of two!
- Water supply – generally greens are now automatically sprinkled. The supply of water from an artesian well or water course with an extraction licence is considerably cheaper than the metered mains. Dry summers have encouraged wider use of computerised pop-up sprinklers on fairways as well as greens and tees.
- Clubhouses – an impressive clubhouse can give a stamp to a course but equally if there is a lot of superfluous space, there will be unproductive heating and maintenance costs. The able entrepreneur (hopefully our hypothetical tenant) may be able to generate extra business from a good clubhouse for wedding receptions, conferences, sports facilities or indeed use as a public house with a full licence. Maintenance is also likely to vary between new, well-designed clubhouses and old, monumental ones. Either a club premises certificate will be required if it is a 'members' club' or, if it is in the nature of a public house then, under the Licensing Act 2003, the clubhouse will be required to obtain a premises licence for the sale by retail of alcohol, have a designated premises supervisor and have alcohol supplied by individuals holding a personal licence.
- Easements, rights of way – few older golf courses seem to be without some disability. There may be footpaths crossing the course, private access drives to houses across part of the course, wayleave agreements or, as with one course, a motorway bisecting it! These will have some impact on value.
- Ancillary buildings – ancillary buildings, such as greenkeepers' stores and pro's shops, will vary in quality and number.
- Driving ranges and other additional revenue sources – the location of the golf course will determine whether additional facilities such as a driving range are a major commercial factor when considering its rental value. Clubs may also have golf hotels, a couple of squash courts or other leisure facilities attached. These may or may not add significantly to income or members' enjoyment.

- Size – 125 acres/50 ha is generally regarded as necessary for an 18-hole course (circa 6,500 yards, with a par of 72). Larger acreages tend to have substantial areas of woodland improving appearance rather than playing area as such.
- While, as with many classes of leisure properties, freehold sales are more plentiful than rents, there is rental evidence, and, as always for rating, if there is rental evidence this is the starting point for valuation.
- Devaluation and therefore valuation are now usually approached on a value per course, rather than per hectare, basis in keeping with market practice – ranging from £175,000 for the best to £10,000–£20,000 in less favoured areas. A fair amount of discussion has taken place over the years on whether it was best to analyse rents by reference to the acreages of courses, the number of holes or a whole course value. The problem with an acreage approach is that additional area as amenity land over 125 acres adds to appearance but is likely to add a declining value per additional area. However, a value per course needs to reflect the variations in amenity land, and indeed course size within it, and is rather more rough and ready than scientific.
- Clubhouses and other buildings are taken at a £ per m² varying on quality and the grade of course. Driving ranges can similarly be valued at a £ per bay or £ per m² with other facilities such as squash courts at a value per court.

14.10.3 Example valuation

The 'Nonsuch' Golf Course is an established 18-hole course with driving range, good quality clubhouse and two squash courts.

Table 14.7 Valuation of Nonsuch Golf Course

Course	55 ha		£70,000
Driving range	3 ha	£450/ha	£1,350
Bays	200 m²	£10/m²	£2,000
Public rooms (clubhouse, committee, dining)	300 m²	£40/m²	£12,000
Secondary accom. (changing, kitchen, office)	150 m²	£30/m²	£4,500
Ancillary accom. (stores, cellar)	50 m²	£20/m²	£1,000
Greenkeeper's accom.	230 m²	£15/m²	£3,450
Store	52 m²	£5/m²	£260
Squash courts	2	£1,500	£3,000
			£97,560
Allowance for public footpaths crossing fairway 7.5% of £70,000			−£5,250
			£92,310
Say rateable value			£92,250

An alternative approach used in appraisals for purchase and suggested also for rating valuations by some surveyors is a receipts and expenditure approach. This is discussed above. There is nothing wrong in using this approach, although with rental evidence available, it is normal to use a comparison approach. A receipts and expenditure approach may be used for separately assessed golf driving ranges where the rents for these hereditaments usually disregard the, often substantial, tenant's improvements.

For members' clubs, a receipts and expenditure approach is difficult because notional accounts need to be constructed by the valuer on the assumption that the hereditament is run on commercial lines. This is introducing a further element of judgement and, again, it will be better to look for a direct comparison approach.

14.11 Hotels

The type and quality of hotels vary considerably and fall within a number of different categories, each of which has its own individual market. Any change in that market will have a substantial effect on the hotels in that sector. The various categories include:

- 5-star hotels;
- country house hotels;
- boutique hotels;
- town house hotels;
- commercial 3- and 4-star hotels;
- airport hotels;
- seaside hotels;
- older and historic 3-star hotels;
- budget hotels/lodges;
- aparthotels;
- small independently owned hotels;
- boarding houses, guest houses and bed and breakfast accommodation.

During the late 1990s the hotel industry in England and Wales enjoyed a period of strong and sustained growth in trading performance. A series of events between 2001 and 2003 brought this to a halt, and for hotels in certain locations and market sectors it prompted a rapid and dramatic reversal of fortunes. The outbreak of foot and mouth disease in early 2001 resulted in many areas of the countryside being closed and hotels in rural areas suffering a downturn in trade, and the events of 11 September 2001 resulted in a further downturn in the industry. Hotels in central London, particularly at the luxury end of the market, suffered the most. Generally, the wider hotel market then experienced a solid and improving performance trend and at the 2010 lists' AVD of 1 April 2008 the outlook for hotels looked very positive. The budget sector had dramatically expanded, with new entrants aggressively developing their brands.

But then came the recession. The years 2009 and 2010 were difficult, with 2011 and 2012 remaining relatively flat. The year 2013 saw an improvement, with 2014 very much better with provincial hotel occupancy figures returning to 2000 levels and the outlook for 2015 good with increasing occupancy and revenue together with the consequent upgrading and new building of hotels. The years 2015 to 2019 were generally a good time for hotels, although in some areas growth was restricted by a significant increase in the supply of new, mainly branded hotels. The COVID-19 pandemic in 2020 resulted In hotels being closed by law, and subject to various restrictions and further lock-downs through 2020 and 2021.

14.11.1 Valuation basis

The usual methods for valuing hotels have been agreed between representatives of the VOA and trade representatives.

The following sections contain only a brief summary of the approaches agreed.

14.11.2 4- and 5-star and major chain operated hotels

These properties are valued by taking a percentage of the FMT. The percentage to be adopted is dependent upon the proportion that accommodation receipts represent out of total turnover and the level of those receipts in terms of 'double bed units' (DBUs). The DBU is a method of comparison where the number of double and single beds in the hotel is converted to DBUs, which are then expressed as a value per DBU.

The percentage ranges for the 2023 revaluation run from about 3.5% to 10.75% in provincial locations and 6% to 12% in central London.

14.11.3 Hotels – non-chain 3 star and under

These properties are valued on a similar basis but different percentages are adopted. For these hotels the percentages for the 2023 revaluation vary from around 5.75% to 16.75% in the provinces and 7.5% to 15% in central London.

The percentages adopted have been derived primarily from the analysis of rents and the accounts from a range of hotels.

14.11.4 Leisure centres

This type of property falls into two main categories.

14.11.4.1 Private sector properties

The private sector provides a larger range of different types of property, ranging from fully equipped leisure centres to smaller gymnasium and fitness centres.

Where there is sufficient rental information available then the rental approach will, as usual, be the preferred method of valuation. In the absence of reliable rental information, the use of the receipts and expenditure method will be adopted.

14.11.5 *Public sector properties*

Public sector leisure centres typically include a swimming pool, multi-functional accommodation including gymnasium and squash courts, and other ancillary accommodation including snack bar. In some centres there may be additional external facilities such as all-weather pitches. Most centres will have some onsite car parking.

In most instances the public is admitted on a pay as you go basis depending on the facilities to be used. This category of property has been the subject of considerable debate as to the correct method to be adopted for its valuation and whether the contractor's or receipts and expenditure basis should be adopted. In *Eastbourne Borough Council and Wealden Borough Council* v *Allen (VO)* [2001] RA 273, the contractor's basis was preferred to that of the receipts and expenditure basis.

Many council-owned centres are now contracted out to private operators.

14.12 Marinas

Marinas are natural or artificial harbours for privately owned pleasure boats, both motor and sail. They may be next to or within a river, estuary, natural harbour, lake, canal or even gravel pit. Usually they consist of an enclosed area of water, with a number of jetties and landing stages for boats to berth either alongside or at right angles. The modern style is to have pontoons or jetties with fingers coming off them with the boats being moored alongside these fingers. The marina provides a secure place for the boats, both protecting them from the elements and, perhaps increasingly, providing security for what are or can be very expensive chattels. Various facilities are provided, although a shower and lavatory block would be the basic requirement.

There is some limited rental evidence in existence, but given its paucity the receipts and expenditure approach seems the most appropriate method of valuation, as marinas are generally run for profit, although the enthusiast may feature whose motive is not solely profit.

Valuation for rating is usually undertaken by taking a percentage of maximum berthing income assuming full occupancy. The percentage varies depending on the reasonably to be expected occupancy level at the AVD and is derived from an examination of the profitability of a selection of marinas.

Berthing charges are levied according to the length of the boat and are at so much per metre run. A 12 m or 40 ft boat is the average length and it is usual to base valuations on this length for coastal marinas in arriving at the berthing income. A scale of charges is usually provided with break points for the length of the boat, the length of time being taken and the method of payment. The charges used are, of course, those for the AVD year and which would be known or anticipated at the AVD. Charges do vary considerably, not just because of the facilities provided but also because of their location around the country and the ease of access into the marina.

Some coastlines and inland cruising areas are very much more popular than others. Certain fixed costs to a tenant will be fairly similar for all marinas, although such items as lock maintenance, breakwaters, repair and dredging (of access channels as well as the marina basin) may vary significantly. As a result the net income per berth or ratio of berthing charge to rent income will vary and, it could be argued, also needs to be reflected in the percentage used. For instance, the south coast and particularly the Solent area is better regarded than, for example, the Suffolk coastline.

An annual agreement will usually include car parking and the use of showers and lavatories.

However, there are 'extras' such as launching charges (crainage) and other sources of income which need to be considered and are not part of the value arising from potential berthing income. These include fuel sales, use of laundry, pump out and chemical toilet disposal facilities, and additional land for winter boat storage for people not using the berths. The marina may also run other facilities on the site, such as a chandlery, workshop for repair and maintenance of boats, yacht sales (brokerage), café and more recently dry stacking. It is possible to add these by taking a percentage of the likely AVD income along-side the berthing income. These additional facilities may, of course, be let out to other occupiers and if so should not be included in the marina assessment as they will form separate hereditaments. Hopefully the evidence of rents will then provide direct assistance in valuing these.

Marinas do vary considerably in design. They may be dug out of what was dry (or indeed muddy!) land, they may be built out into an estuary or they may simply be moorings from pontoons in a river. This will affect maintenance costs. Some require frequent dredging to maintain the channels and indeed to keep the marinas sufficiently deep for boats to moor.

As always location is important – not just near to centres with an affluent population, but near or in attractive sailing or cruising areas. Some coastal marinas have limited access in or out of the marina depending on the state of the tide and some have a sufficient rise and fall of the tide to require lock gates into the marina.

The quality of the facilities and the 'style' of the marina will influence value. Are the facilities of good quality, are the pontoons good, what repair facilities are available, what is the quality of club rooms and bars on site?

14.12.1 *Example valuation*

The 'Peter Duck' Marina is an established marina in a tidal river, on the South Coast. Its capacity is 400 berths, but it has a reasonably expected (and actual) occupation at the AVD of 90%. It has a berthing length of 4,850 linear m.

Table 14.8 Valuation of Peter Duck Marina

1. Berthing income	£575,000
2. Non-berthing income	
On-site laundry	£5,000
Workshop	£20,000
Chandlery store	£15,000
Small restaurant	£15,000
Dry stacking (i.e. storage)	£25,000
Fuel	£30,000

Due to the river being tidal, there is an annual cost of £50,000 to the marina operator for dredging.

The annual tariff for a 10 m boat (excluding VAT) as at the AVD (i.e. 1 April 2021) was £125/linear m.

The actual occupancy rate was 95%.

Table 14.9 Valuation for a 2023 rating list

Berthing income
4,850 Linear metres @ £125/LM

£606,250 @ 13%	£78,812
Less: a calculated amount to reflect that the outgoings (i.e. in this case for annual cost of dredging which is greater than the normal range of up to 5%) reduce the RV by 3.25	£2,561
Adjusted RV for berthing income	£76,251

Non-berthing income
For non-berthing income, each stream of income is valued on a 'sliced' basis.

Income from storage of £25,000 @ 13	£3,250
Income from the on-site restaurant of £15,000 @ 10.5	£1,575
Income from workshop of £20,000 @ 9	£1,800
Income from fuel sales, chandlery and laundry of £50,000 @ 3.25%	£1,625
Total income £716,250	
Total RV	**£84,501**
SAY	**£84,500**

14.13 Mineral hereditaments

A wide range of properties involve mineral extraction:

- mines (coal and other);
- quarries;
- gravel pits and chalk pits;
- oil/gas deposits;
- slag-heaps (where minerals are extracted for use);

- landfill sites from which methane is extracted for beneficial use;
- brickworks;
- landfill (waste tips) is usually classed with mineral properties because both are annually revised;
- peat.

Mineral valuation is a specialised area of valuation practice. There are a number of significant differences from 'normal' classes of hereditaments including the need to revise rateable values annually and for an 'adjusted rateable value' to be shown in a rating list.

In granting a lease to extract minerals, a landlord usually gears the rent to the output of the property. This approach suits the tenant because, if due to economic circumstances the tenant does not wish to extract so much in a year, the rent is reduced accordingly. These rents or royalty payments tend to have a fixed or non-variable part, as well as a variable element and usually include the following:

- Dead rent or certain rent – a minimum fixed rent which has to be paid irrespective of the year's output.
- Royalty – a payment of so much per tonne or m³ of the mineral extracted and sold. Usually this is not paid in addition to the dead rent but replaces it when the royalty payment level exceeds the dead rent.
- Surface rent – a rent of £X per acre/ha of land handed over to the mineral tenant or rendered unusable by the landlord due to the mineral tenants' works. This will include land used for storage areas, sites of buildings and lorry parking, and land rendered unusable for normal agricultural use due to the proximity of the mining operations.

While mineral valuation is a specialised area, the object in a valuation for rating purposes is the same as for any other type of hereditament: to ascertain the rent which might reasonably be expected for the hereditament on the statutory basis.

14.13.1 *Valuation*

There is some difficulty in applying a normal rental basis to this type of hereditament as rents in the market usually vary directly with output. Rents therefore need to be analysed by reference to output, and in practice valuations for rating are made on the basis of applying an estimated royalty per unit of output to the actual output of mineral.

To this mineral figure is added the rental value of the land in the hereditament and the value of any buildings and rateable plant and machinery found by using the contractor's method of valuation.

The following section looks at the valuation approach in detail.

14.13.1.1 Mineral element

Output – the quantity of minerals extracted is a stated matter in Schedule 6, paragraph 2(7), of the Local Government Finance Act 1988. It follows from this that the output figures to be adopted in making a rating valuation are those current at the material day for the valuation and not the AVD.

A change in the quantity of minerals extracted from or refuse deposited on the hereditament is defined as a MCC in paragraph 2(7) of Schedule 6 to the Local Government Finance Act 1988. It is therefore possible and, indeed, has been the practice for many years, to annually revise valuations to reflect the degree of working of mineral hereditaments.

While rating valuations are properly a forward estimate of rental value for the year after the valuation date, it has become common practice to base valuations on the preceding year's output (usually 1 January to 31 December). This does provide a fair and practical approach, as it is usually likely that conditions in one year give a good guide to the next. (However, last year's output is evidence which can be used only to estimate what may be anticipated for the next year and adjustments may need to be made to last year's output to reflect changes expected in the coming year.)

'Grossing up' output to an annual equivalent – mines and quarries do not last forever and some do not even exist for a year. Where a hereditament is not likely to exist for the whole of the forthcoming year, the approach has to be altered. In *Gilbard (VO)* v *Amey Roadstone Corporation Ltd* [1974] RA 498, the Court of Appeal held that the notional rent for a quarry which would be exhausted long before the end of the rating year must be based on the hypothesis of a full year's working. This approach avoids the anomalies that would otherwise arise through the application of the procedural rule that a proposal is operative from the beginning of the rate year. It was held that such a modification would not work unfairly on the ratepayer because the ratepayer would only pay rates during the short period in which he or she would occupy the working.

Mineral hereditaments which are occupied for less than one year – the rateability of short-life mineral hereditaments was considered by Lord Denning in the Court of Appeal case *Dick Hampton (Earth Moving) Ltd* v *Lewis (VO)* and *United Gravel Co. Ltd* v *Sellick (VO)* [1975] RA 269.

Lord Denning decided:

It cannot depend how many machines the contractors have available; or whether they do it in 11 months or 13 months. I cannot accept the supposed 'working rule' of 12 months. No matter whether the extraction takes only 6 months or 9 months, they are in the rateable occupation of the contractors.

Royalty – the amount of the royalty payment per unit of mineral varies not only between different types of mineral but also between:

• Grade or quality of mineral, e.g. sand and gravel may be contaminated by clay which entails additional cost in processing the aggregate.

- Ease of working – problems include depth of overburden, faults and potential for flooding. It is therefore important to compare pits or mines that have similar working conditions.
- Proximity to the market for the particular type and grade of mineral.

It can be argued that the royalty paid for extracting minerals is not really a rent but a capital payment for the purchase of an asset. The minerals once worked have gone for all time. The courts have not accepted this as indicating that royalty payments do not constitute evidence of rental value for rating purposes. However, it seems reasonable that royalties, at least in part, do represent a capital payment and Parliament has recognised this in regulations, currently the Non-Domestic Rating (Miscellaneous Provisions) Regulations 1989 (SI 1989/1060).

The Regulations apply to any hereditament:

- which consists of or includes a mine or quarry; or
- the whole or part of which is occupied together with a mine or quarry in connection with the storage or removal of its minerals or its refuse.

It should be noted that peat and material extracted from slag heaps are not classed as minerals under the Mines and Quarries Act 1954 and do not therefore receive the benefit of the 50% deduction described earlier.

It provides that in arriving at the rateable value, 50% of the value attributable to the occupation of the land for the purpose of the 'specified operations' of winning and working, grading, washing, grinding and crushing of minerals shall be treated as being attributable to the capital value of the minerals and excluded from the rateable value.

Land is defined as excluding buildings, structures, road, shafts, adits or other works.

Thus, in practical terms, the rental value attributable to the royalty element and land used for the specified operations has to be halved to give rateable value. Only the rateable value now appears in the rating list and only the land used for specified operations has the 50% deduction.

14.13.2 Land

The site for specified operations and land for other purposes, e.g. office site, lorry park, site of concrete batching plant, is usually taken at a rental figure per acre/ha. Care should be taken to ensure it is valued in its enhanced state with the benefit of water supply and drainage. A figure is not placed on the actual land being worked as this is considered to be covered by the royalty payments element.

14.13.3 Buildings, plant and machinery

A contractor's basis approach is used for buildings, rateable plant and machinery and site improvements.

Typical buildings include offices, workshops, stores, switch houses and mess rooms.

Typical plant and machinery includes foundations, settings, fixed gantries, supports, walkways, stairways, handrails, catwalks, stages, staithes and platforms, headgear for mines and quarries, bins and hoppers, bunkers, pits for weighbridges, electrical equipment and pumps for drainage.

Where the plant at a mineral-producing hereditament has a capacity greater than the output, the figure ascribed to the plant and machinery is sometimes reduced, but not in direct proportion to the ratio of the actual to potential optimum output of the plant. A superfluity allowance is only appropriate where the output and that superfluity would not attract a bid from a hypothetical tenant who would have regard to the need to deal with peaks in production and the benefits of shorter working hours versus overtime payments, etc.

14.13.4 Landfill

In the past a typical associated operation, particularly with sand and gravel pits, was the use of worked out areas for landfill. With much greater regulatory scrutiny surrounding the environmental control and operation of landfill sites this natural extension is now becoming more problematical. Nevertheless, the right to tip is valuable (particularly near towns) and the planning consent still may require the land to be restored to agricultural use; the pit therefore needs to be filled in some way.

The royalty figure depends on the type of waste deposited, as the site licence will limit the types of waste to be deposited.

14.13.5 Example valuation

The Flintstones Quarrying Co. occupies a sand and gravel pit in the south of England. The site's planning permission requires the land to be restored after working is complete, and tipping is already being undertaken. A concrete batching plant is also situated on the site.

Last year's output of sand and gravel was 100,000 tonnes and 50,000 tonnes of household rubbish and 20,000 m^3 of clean fill were deposited.

Of the site, 1.0 ha is used for washing and separating the aggregate into sand and different sizes of gravel. The concrete batching plant occupies 0.25 ha and the site office, mess room and car park occupy 0.5 ha.

The buildings and rateable plant and machinery have a 1 April 2021 value of £330,000, the site of the concrete batching plant has a value of £15,000 pa per hectare and the other land has a spot value of £5,000 pa.

Only the adjusted rateable value figure appears in the rating list.

Table 14.10 Valuation of quarry

Mineral element	Unadjusted rateable value	Rateable value
Sand and gravel output last calendar year 100,000 tonnes @ £2.30 per tonne	£230,000	£115,000
Land subject to the Non-Domestic Rating (Miscellaneous Provisions) Regulations 1989 1 ha @ say £15,000 per ha pa	£15,000	£7,500
Total subject to 1989 Regulations reduction	£245,000	£122,500
Land (not subject to reduction) Site for concrete batching plant 0.25 ha say	£5,000	£5,000
Buildings, plant and machinery (including concrete batching plant) Value at 1/4/21		£330,000
Decapitalise @ 4.4% (statutory decapitalisation rate)	£14,520	£14,520
Landfill Input last calendar year Domestic, commercial and inert landfill 60,000 tonnes @ £1.40 per tonne	£84,000	£84,000
Clean fill 20,000 m³ @ 55p per m³	£11,000	£11,000
Totals	£359,520	£237,020
Unadjusted rateable value		£359,520
Adjusted rateable value		**£237,020**

14.13.6 Wales

The Non-Domestic Rating (Miscellaneous Provisions) (No. 2) (Amendment) (Wales) Regulations 2021 (2021/1141) W.85 relates to the valuation of mineral hereditaments in Wales.

14.14 Petrol filling stations

The petrol filling station market changed very significantly during the 1990s with the development of hypermarket filling stations operated by the large food retailers. Some 45% of retail fuel sales are now at hypermarket or super-store sites. The price competition between the traditional sites and the food store sites has resulted in the closure of a large number of poorer petrol filling stations over the years. Between 1990 and 2022 around 11,000 petrol filling stations closed. There are now in the region of 8,400 sites operating in the United Kingdom; the reduction has happened despite the rapid increase in car ownership during the 1990s through to the present day.

The value of a petrol filling station is very much determined by its potential for trade and therefore rents are analysed having regard to the maintainable throughput of a station. This is the throughput that could be achieved by a competent operator pursuing pricing policies that seek to maximise profitability.

For the 2017 and 2023 revaluations, maintainable throughput was therefore based on the actual throughput achieved by the forecourt operator and the pricing policy adopted by the operator. This may be high or low in relation to competitor sites. Value is dependent on the two factors of throughput and price, the latter reflecting profitability. Most of the price paid by motorists for fuel is duty and VAT and the operator typically receives only a gross margin (the difference between the pump price and the wholesale price together with the duty/VAT) of a few pence per litre.

In 2015 the net margin in the United Kingdom averaged around 4p per litre. Therefore, price cutting, even by 1p per litre, has a very significant effect on an operator's profit.

Throughput will be influenced by passing traffic volume, ease of access, visibility, size of forecourt and the nearness of competing stations.

The analysis of rents shows a progressively higher value per thousand litres of maintainable throughput. This is due to the lower wholesale prices charged to petrol filling station operators for greater volumes of fuel and the basic running costs of a petrol filling station being, to a large extent, fixed. Typically, a maintainable throughput, achieved at average pricing, of only 2.5 million litres would show a rental value for the 2017 Revaluation of £4.42 per thousand litres, a throughput of 5 million litres would show a rental value of around £6.18 per thousand litres and a throughput of 10 million litres would show a rental value of £8.72 per thousand litres.

The traditional petrol filling station kiosk has been replaced by varying sizes of forecourt shops, some forming small supermarkets. These can add significantly to the value of the site and in many cases may be the more valuable element of the hereditament. This is due to the continued downward pressure on fuel margins and the growth of the convenience retail market in the early 2000s. Forecourt shops are usually valued by reference to turnover, based on a sliding scale of 2–5%, adopting a higher percentage the higher the turnover, with a discount applied to the turnover percentage for larger shops.

Car washes can also add significantly to the profitability of a site. However, the car wash plant is not rateable, as it is not mentioned in the Plant and Machinery Regulations. The valuation will be an assessment of the rental value of the car wash site, together with any building, settings or rateable supports.

14.14.1 Example 2017 Revaluation valuation

The Five Ways Service Station is a well-positioned petrol filling station on a main arterial road. It is open 18 hours per day and the pump prices are in line with national average prices. The throughput for the three years prior to the 2017 AVD was:

2012–2013	7,800,000 litres
2013–2014	8,250,000 litres
2014–2015	8,100,000 litres

There is a well-used forecourt shop and car wash.

Forecourt

Maintainable throughput		8,000,000 litres
@ £/1,000 litres		£8.18
Forecourt value		£65,440
Forecourt shop		
In terms of sales area 70 m²		
Turnover in three years to AVD consistently around £1,000,000		
Adopted turnover		£1,000,000
% age bid		5.25%
Shop value		£52,500
Car wash		
Turnover in three years to AVD consistently around £60,000		
Wash value	20%	£12,000
		£129,940
Say rateable value		**£130,000**

14.15 Public houses

There have been very substantial changes in the way traditional public houses are let and run since the 1980s. The large, brewer owned estates run largely through tied tenants have been broken up, although there are now large estates of public houses owned by companies which are not brewers. Tied tenancies still exist to some of the smaller, local brewers, but most houses are now managed by employees of the owning company, let on full repairing and insuring (FRI) leases of up to 20 years or are owner occupied.

There is plenty of rental evidence and traditional public houses are valued on a rentals approach. Rents are not analysed on a m² basis because the size of the public house is not what a prospective tenant would primarily be interested in when taking a tenancy. A large public house in an inferior location may generate a much lower turnover than a small one in a premium spot. The hypothetical tenant would want to know the likely trade that could be achieved in running the house. Rents are therefore analysed by reference to the FMT likely to be achieved by a reasonably efficient publican at the AVD. The analysis produced is a percentage of gross receipts which will show a variation depending largely on the level of receipts. Usually, the analysis is made separately for drinks and the food and letting trade known as the dry trade. The percentages for the different income streams may vary due to their differing costs and profitability.

Information needs to be collected about the public house to be valued. Gross Receipts All Sources (GRAS) excluding VAT should be obtained ideally for the previous three years to show how well the house has been trading and to identify any trends. Receipts may be affected by a range of factors which mean they are not the reasonably to be expected FMT. The actual licensee may be exceptional, generating trade beyond the norm that would be expected from a normal licensee. Changes in licensee or ownership or frequent changes of manager may have affected sales or mean the history of receipts is not available. The state of the house, whether it has recently been refurbished, price structure and brand of beer sold all affect sales. Competition from other public houses and any change to them, including extension or refurbishment, can affect trade and need to be taken into account.

The FMT from the wet trade should be taken to include receipts net of VAT from the sale of intoxicating liquor, soft drinks and incidental bar sales such as crisps, sweets and nuts. The takings from fruit machines and other AWP machines has become less significant in recent years and for 2017, where significant, was reflected in the bid percentage adopted but for the 2023 revaluation is disregarded.

The dry trade's FMT is the likely annual FMT excluding VAT from catering excluding incidental bar sales. This can be a substantial element of today's public house trade with many trading almost as restaurants. Some traditional public houses also have letting bedrooms and these may be treated as part of the dry trade potential or valued separately.

It is important to note, as with any receipts and expenditure valuation, or valuation involving taking a percentage of likely receipts, that the valuation is of the property and not the present occupier. Actual receipts should be used as a guide to the likely gross takings the hypothetical tenant could achieve. It may be that the actual tenant is achieving exceptional trade due to the personal characteristics of the licensee, who may achieve the additional trade through perhaps being a famous ex-footballer or TV chef or having an exceptional personality or business acumen. The likely trade needed for the valuation is what a reasonably efficient licensee would achieve, not what a particular individual can achieve. On the other hand, if the licensee is morose and unfriendly it is reasonable to assume that a different licensee could achieve a better trade.

14.15.1 Example valuations

14.15.1.1 'The Paynter's Arms' 2017 Revaluation valuation

'The Paynter's Arms' public house is a traditional village public house that has expanded its catering operations in recent years to attract trade from a nearby town.

Table 14.11 Recent trade figures from The Paynter's Arms

Year	Wet £	Dry £	Total £
2012–2013	175,000	170,500	345,500
2013–2014	180,000	192,000	372,000
2014–2015	188,000	216,000	404,000
Estimated FMT at 1 April 2015 AVD			
Liquor	£185,000	@ 8.28%	£15,318
Catering	£215,000	@ 7.07%	£15,201
			£30,519
Say rateable value			£30,000

Table 14.12 Recent trade figures from Westbrooks

Year	Liquor	Catering	Accommodation	Total
2012–2013	879,000	1,240,000	983,000	3,102,000
2013–2014	934,000	1,420,000	1,118,000	3,472,000
2014–2015	966,000	1,530,000	1,238,000	3,734,000
Estimated FMT at 1 April 2015 AVD				
Liquor		£950,000	@ 11.63%	£110,485
Catering		£1,500,000	@ 8.39%	£125,850
Accommodation		£1,200,000	@ 12.83%	£154,000
				£390,335
Say rateable value				£390,000

14.15.1.2 Westbrooks 2017 Revaluation valuation

'Westbrooks' family pub and restaurant is a 1930s roadhouse extended and refurbished in 2011 to include an expanded restaurant and a travel lodge style 80-bedroom block.

14.16 Riding schools and livery stables

While there is a distinction between riding schools and livery stables, in practice the use of many properties overlap.

14.16.1 Livery stables

A livery stable offers facilities for owners to stable their horses, often offering the following options:

- full livery – the horse is stabled, fed, watered, groomed, mucked out and exercised by the proprietor of the livery yard;

- part livery – the horse is stabled but the owner might undertake to exercise, groom or muck out and feed, in any combination;
- stabling only (often referred to as DIY livery) – the horse is stabled and the owner is responsible for everything else.

In addition, the owner may offer a range of additional services which could include hiring out of horses and sales of feed, bedding and ancillary equipment.

14.16.2 Riding schools

A riding school, which must be licensed, operates on the basis of hiring out horses or ponies on a pay per hour basis. Typically:

- clients go out together on a ride accompanied by a member of staff;
- more experienced riders go out in groups or alone;
- clients receive tuition alone or in groups from a member of staff.

14.16.3 Exemptions – agricultural buildings

The buildings used for the stabling of horses and ancillary buildings are not exempt as 'agricultural buildings' following the decision in *Hemens (VO)* v *Whitsbury Farm and Stud Ltd* [1988] AC 601, 13 EG 43 where the House of Lords found that horses and ponies, other than those used for farming the land or reared for food, are not 'livestock' within the definition in section 1(3) of the Rating Act 1971, which has now been re-enacted in paragraph 8(5), Schedule 5, of the Local Government Finance Act 1988.

14.16.4 Exemptions – agricultural land

The exemption of agricultural land is covered in Chapter 5. The definition of agricultural land in Schedule 5, paragraph 2, of the Local Government and Finance Act 1988 includes land used as arable, meadow or pasture ground only. Generally it is considered that land used for the grazing of horses will be exempt under this definition. Minor use of the land for the occasional use of jumps will usually not affect the exemption but this is a matter of fact and degree.

14.16.5 Valuation

The normal valuation approach is on a rental basis with a unit price being attributed to a standard loose box (13.5 m^2). The value of ancillary buildings are then related to the standard value of a loose box.

14.16.6 Case studies

The case of *Pritchard* v *Gregory (VO)* [1999] RVR 23 concerned the valuation of five stables and a tack store for a local 1995 rating list. The valuation given is for illustrative purposes only.

Table 14.13 Valuation of five stables and a tack store

	m^2	£	£
Stables	5.00	120.00	600.00
Tack store	6.82	9.17	62.50
Open fronted hay store (reflected)			662.50
Rateable value			£660.00

In the case of *Ellis* v *Broadway (VO)* [1998] RA 1, the property comprised 24 loose boxes in three groups, an indoor riding school and ancillary buildings. One group of ten loose boxes was timber built with corrugated, asbestos clad, pitched roofs. A second group of six loose boxes was similarly constructed. A third group of eight loose boxes was timber built with pitched, felt clad roofs. Each loose box was about 3.5 m², although there were minor variations in width and depth.

The indoor riding school was of a concrete portal frame construction with a pitched, felt clad roof. The sides of this building were partially clad with corrugated asbestos. It contained a small viewing area. Ancillary buildings included a timber and felt office/tack room, a lecture room constructed of concrete block single skin walls and a pitched, corrugated, asbestos clad roof, and a timber-framed and partially timber clad hay store. There were concrete yards and walkways to all the buildings with the exception of the hay store.

Apart from the actual valuation of the property, a further issue was its state of repair and how this should be allowed for in the rating assessment. It should be noted that this case was decided prior to the passing of the Rating (Valuation) Act 1999 and it is unlikely that a reduction for the state of repair would be allowed under the current definition of rateable value.

Table 14.14 Valuation of riding school

		£	£
Loose boxes	24 @175.00	4,200	
Feed store/office/store and lecture room	30.9 m²	17.50	541
Hay store	say		50
Indoor riding school	1,307.0 m²	3.85	5,032
			9,823
Less			
Allowance for access and disrepair		22.5%	
Say rateable value			£7,600

14.17 Snooker halls

Snooker halls can come in a range of different types of property. They are usually situated in poorer and converted property, and often in less desirable areas. It is common for this type of property to be licensed and often food is available. In many instances considerable income is made from the sale of food and drink.

The properties are usually valued on a rentals approach with rents being devalued to a rent per m^2.

The case of *Hearne* v *Bromley (VO)* [1984] RVR 81 concerned the valuation of a new snooker club in a converted office building and may be a useful reference as to the approach taken.

14.18 Sports grounds

Sports grounds can take many different forms, ranging from first class amateur rugby, football and cricket grounds to those provided by companies for their employees.

One question to be considered prior to looking at the valuation is whether the property is exempt from rates under paragraph 15 of Schedule 5 of the Local Government Finance Act 1988. It is quite common for sports grounds to be located in or adjacent to a park and, subject to them satisfying the requirements of paragraph 15, they may be exempt as being or forming part of a park. Where part of the park is reserved for a particular club then it may well be that the club will be in rateable occupation and the reserved area will form a separate and rateable hereditament.

14.18.1 Method of valuation

The rentals approach will always be the preferred method of valuation, but at times it may be difficult to verify whether a passing rent is truly at market value or a lower figure has been accepted because of the status of the user. For example, a local authority may rent out a ground at less than its market value to a local good cause.

The receipts and expenditure approach is often not available as there is no intention of making a profit. In the absence of rents, a contractor's basis may have to be adopted. In such cases the ability of the tenant to pay the resulting rent will need to be very carefully examined.

14.19 Swimming pools

Generally most swimming pools are provided by local authorities and can take a number of common forms:

• traditional stand-alone heated indoor pools;
• open air pools, often in seaside areas or adjacent to public parks;

- leisure centres comprising a swimming pool together with a range of other facilities. This type of property is dealt with under the heading 'Leisure centres' in section 14.10.

14.19.1 Traditional stand-alone pools

Many of the pools are now of a considerable age, although there was a spate of development of such stand-alone facilities in the 1970s around the time of local government reorganisation.

The traditional stand-alone pools tend to be provided by the local authority and will almost certainly be run at a deficit. They are provided by the authority as a service to the community and are often used by school children, swimming clubs and other similar organisations, as well as the general public.

Given that these properties are loss making and it is not usually the intention of the local authority to make a profit, they are valued on a contractor's basis of valuation. When arriving at the effective replacement cost at stages 1 and 2, it will reflect the fact that new properties of this type will be better designed and will often provide a better range of services to the public.

The valuation will also reflect the presence of rateable plant and machinery that is found on this type of property.

14.19.2 Open air pools, often in seaside areas or adjacent to public parks

The approach to the valuation of this type of property will be similar to the traditional stand-alone pool. One additional factor that will often need to be considered is the unit of assessment. The valuer will often have to examine whether the pool comprises a single hereditament with an adjoining park or foreshore or whether there are two or more hereditaments.

14.18.3 Private swim schools

In recent years there has been a trend for private swim schools to be established. Some operating from industrial buildings and others operating from domestic premises, whether within the house or an outbuilding. The approach to valuation is rentals, or in its absence a receipts-based approach. For those operating from domestic properties there Is also the domestic/non-domestic borderline to consider.

14.20 Telecommunication masts

Traditionally, masts were erected for transmitting radio and TV signals, but from the mid-1990s there has been a rapid increase in the number of masts due to the rollout of mobile phone networks. In many cases operators have shared sites rather than develop new sites themselves. Sites have also been constructed on the rooftops of tall buildings to cover urban areas.

From 2008, broadcast TV sites underwent a transformation from analogue to digital broadcasting. A desire to save costs has led mobile phone operators to develop Radio Access Network sharing (RAN sharing) where two operators share the transmission equipment and the aerials. RAN sharing is in its early days but as many as 2,000 existing sites could be de-commissioned and many site shares removed from existing sites as the networks consolidate.

The valuation method is the same irrespective of the use of the mast. However, use has some influence on value, as does location, height and demand for site sharing.

The valuation method used is primarily a rental method for the land with a contractor's basis addition for rateable plant and machinery and site improvements. There are three basic elements to consider when valuing a mast site:

- base site rental value – rental value of undeveloped site
- additional site rental value for site sharers – known as a 'payaway'
- rateable plant and machinery and site improvements – decapitalised cost at the statutory decapitalisation rate.

Most modern site rental agreements include a base site rent, which reflects height and location, etc., with additional rent paid to the landlord for site sharers subsequently attracted to the site. The site sharer will pay a fee to the mast operator and a proportion of this is usually passed onto the landlord as additional rent. This additional rent is known as a payaway and can considerably increase the overall rental value of the site. It is expected that RAN sharing will remove one payaway from the rental value even though two operators share the same equipment.

The increase in site sharing has led to difficulties in identifying how many hereditaments exist at a mast site, when sometimes operators had their own exclusive equipment cabins at the site but shared the mast, and in other cases they shared the mast and the existing equipment building or site cabin. These difficulties led to the Non-Domestic Rating (Telecommunications Apparatus) (England) Regulations 2000 (SI 2000/2421) and the Wales Regulations 2000 (SI 2000/(W)). With effect from 1 October 2000 in England and from 1 April 2001 in Wales, all site sharers (excluding central lists site sharers) on a telecommunications mast were treated as being included in the 'host' mast assessment. The host was therefore made responsible for the whole aggregated assessment. The 'host' has to be a telecommunications operator or in the business of hosting sharers on a mast, for example a hosting operator who takes a lease of the whole of a roof top site with the intention of attracting site sharers. The landlord of a larger hereditament, for example a water tower or a hospital, would not be considered as a host under the Regulations as the primary hereditament is not a telecommunications hereditament.

Many mast sites for mobile phones have been developed on the public highway under the Telecom Code Powers (Schedule 2 of the Telecommunications Act

1984. These powers allow masts to be erected on highway land with no rent or charge being payable for the land. In these cases a rental value is taken from similar sites on 'private' land where rents are paid (*Orange PCS Ltd* v *Bradford (VO)* [2004] EWCA Civ 155).

The site rent does not usually include the developed infrastructure, and an addition has to be made to reflect this to arrive at the full annual rental value. This is done by decapitalising (at the statutory decapitalisation rate) the cost of construction of the rateable elements of plant and machinery and relevant site improvements. Such items would include the mast, buildings and cabins, telecommunications cables, generators, power supply, fencing and hard standing. The transmission and electronic equipment and the aerials themselves are not rateable items of plant and machinery and are therefore not included in the valuation.

The 2017 reforms to the Electronic Communications Code were intended to speed up investment in the UK's mobile infrastructure, and the rollout of 5 G. The key change was a new valuation mechanism, which aimed to significantly reduce rents paid to providers of sites for telecoms infrastructure. Unsurprisingly this was not well received by landlords and has resulted in delays to rollout, as landowners were reluctant to accept rents at these levels, resulting in protracted negotiations and litigation. Payments of premiums in addition to the lower rents has been one of the results.

14.20.1 Example valuation

A 15 m lattice mast on a green field site adjacent to a main road was developed by mobile operator 1. Operators 2 and 3 share the site.

Rental value	
Base site rent (includes one operator)	£4,500
Site sharer addition for two sharers (two @ £2,250)	£4,500
Capital cost	
Cost of rateable plant and machinery and site improvements:	
15 m lattice mast including foundations	£16,500
Equipment Cabin 1	£10,000
Equipment Cabin 2 (shared)	£10,000
Telecoms cables	£2,000
Electrical connection	£2,000
Hard standing	£1,500
Fencing and gates	£2,000
Site preparation	£1,000
Total capital cost:	£45,000
Decapitalise @ 4.4% (statutory rate)	£1980
Rateable value	£10.980

If operators 2 and 3 enter into a RAN sharing agreement and operator 2 removes his aerials and equipment from the site, this is a MCC and the value will be reduced by one payaway of £2,250 resulting in a revised rateable value of £8,730, say £8,700.

14.21 Theatres

There are a wide range of types of theatre in the country. They range from small amateur dramatics theatres to smaller town theatres and seaside theatres to large provincial and West End theatres. Running a successful theatre is costly, and many are subsidised either by being civic theatres or by receiving grants.

Many are owner occupied, although there is some rental evidence, particularly in London. Care needs to be taken to ensure any rent is truly at arm's length and to exclude from the rent any element attributable to items which, although included in a letting, do not form part of the rateable hereditament, such as the seating and equipment. Rents are usually analysed either on a per seat basis or by reference to gross receipts. For the larger theatres valuations are usually in the range of 2–6% of gross receipts.

Many of the larger provincial theatres have now added a café/restaurant facility to boost catering receipts beyond confectionery and ice cream.

14.21.1 Example valuation for 2017 Revaluation

The 'Theatre Royal' is a well-positioned provincial theatre with 1,310 seats. Trade figures leading up to the 2017 Revaluation are given below.

Table 14.15 Valuation of Theatre Royal for 2017 Revaluation

	Box office	*Confectionery/ices*	*Total*
2012–2013	£4,200,000	£165,000	£4,365,000
2013–2014	£3,850,000	£150,000	£4,000,000
2014–2015	£4,150,000	£160,000	£4,310,000

Valuation
Likely gross receipts £4,300,000 @ 3.5% = £150,500 rateable value.
Equates to £115 per seat.

14.22 Village halls, community halls

This type of property will include village halls, community halls and centres, as well as the types of property used by local groups such as the Scouts and Guides.

Generally, village halls are administered by a local management committee which may be acting on behalf of a charitable trust which originally provided the hall. Where the property is occupied by a charity, then the amount of rates payable may be reduced by charitable relief.

In looking at a community hall or centre, the valuer must have particular regard to who the rateable occupier is. Depending on the actual circumstances, it may be that the local authority is found to be in 'paramount occupation' of the property, and therefore the rateable occupier.

Some of the factors to take into account in arriving at the valuation include:

- the location, especially if the property is in a rural area;
- the nature of the hall;
- the construction, size, maintenance liability and facilities provided;
- the population served;
- possible alternative uses having regard to *rebus sic stantibus*;
- other possible tenants, e.g. the parish council or an independent club;
- the ability to pay of likely tenants;
- rental evidence.

Where possible, a rental approach to the valuation will be the preferred approach, but this may not be available for all properties where there is insufficient rental evidence on which to base the valuation. In these cases the receipts and expenditure approach could be considered. If the hall had only recently been built one could attempt a contractor's approach but it should be remembered that cost does not necessarily equal value.

Whatever approach is adopted it will be necessary to consider the hypothetical tenant's ability to pay (see *Addington Community Association* v *Croydon Borough Council and Gudgion (VO)* [1967] RA 96).

15 The 2023 rating lists

Preparation, alteration and appeals

15.1 Introduction – revaluation and rating lists

Every five years the Local Government Finance Act 1988 requires the valuation officer for each billing authority to compile and then maintain a local non-domestic rating list for the billing authority's area: section 41(1) and (2) of the Local Government Finance Act 1988.

Valuation officers are valuers appointed by the Commissioners of HM Revenue & Customs. They are employed by the Valuation Office Agency (VOA) of HM Revenue & Customs and have been responsible for preparing lists of rateable values for local authority areas since 1950. Valuation officers are independent of both central government and local councils and their task is to act in an impartial manner to maintain fairness in the rating system.

> *The position of the valuation officer in rating is that he is a neutral official charged with the statutory and recurring duty of bringing into existence a valuation list and maintaining its contents in correct and legal form.*
> (*Society of Medical Officers of Health* v *Hope (VO)* [1960] 1 All ER 317)

In a more recent case, *Hussain* v *Turner (VO)* [2017] UKUT 0417, where the valuation officer had first divided an assessment and then decided this was incorrect and remerged, the ratepayer criticised the VO for changing her mind but the Upper Tribunal did not place weight upon this, commenting an argument that the valuation officer has acted inconsistently seldom succeeds, even where a change of tack disadvantages the ratepayer, as the VO was under a statutory duty to maintain an accurate rating list, including by correcting earlier errors.

In *ABC Production Services* v *Hodson (VO)* [2015] UKUT 0015 (LC) the appellant suggested the valuation officer was neither impartial nor independent and his opinion evidence should therefore be ignored. The Upper Tribunal regarded this suggestion as 'untenable' noting the ratepayer, who proposed to give his own opinion evidence had, unlike the valuation officer, a direct financial interest in the outcome of the appeal

DOI: 10.1201/9781003297369-15

The Local Government Finance Act 1988 requirement for new rating lists to be compiled and all hereditaments to be revalued every five years was a return to the aim of quinquennial revaluations which had been removed by the Rates Act 1984. In England and Wales this objective had not been achieved in the past, with revaluations being postponed many times. Since 1950, revaluations have taken place in 1956, 1963, 1973, 1990, 1995, 2000, 2005, 2010, 2017 and, most recently, 2023. The 2015 Revaluation was postponed until 2017 marking the first break in a quinquennial revaluation for 25 years, somewhat repeated with the six-year gap to 2023 due to the COVID-19 pandemic. The Government's intention is to move to a three-yearly cycle.

The old lists before 1990 were substantial A3-sized folders kept by the local authorities and amended, in various coloured inks, by the direction of valuation officers. The modern rating lists are held by the appropriate valuation officer in the form of a computerised record. Copies are held by the billing authorities and are available, for a charge, to members of the public. The information in them may also easily be viewed on the VOA's website, www.gov.uk/correct-your-business-rates, together with other information about the assessment and valuation, including a summary of how most individual rateable values were assessed. Certified extracts of rating lists may also be obtained from the appropriate valuation officer.

Having compiled a new rating list a valuation officer is required 'as soon as is reasonably practicable' to send a copy to the billing authority. The billing authority will already have received a copy of a draft list as the Local Government Finance Act 1988 also requires the valuation officer, not later than 31 December in the year preceding the revaluation, to send a copy of the list he or she then 'proposes (on the information then before him) to compile' to the billing authority.

Both the valuation officer and the billing authority are required to have copies of a rating list in force available for public inspection at a reasonable time and place without charge. A reasonable charge may, however, be made if a photographic copy or a written extract is required: Schedule 9, paragraph 8, Local Government Finance Act 1988.

15.2 Contents of local rating lists

So as to achieve uniformity, originally with the Community Charge, and now with the Council Tax, the national non-domestic rate has been established by the Local Government Finance Act 1988 as a daily liability to rates. This is made clear by the requirement in section 42 that local rating lists must show 'for each day in each chargeable financial year' entries for each rateable hereditament. Provision has been made for lists to show the date from when any new or altered entry is effective. Changes to hereditaments and rateable values during a year can be monitored by billing authorities over the year and rates bills calculated on the daily basis by an examination of the effective dates of any alteration.

The hereditaments to be included in a local list are those situated in a local authority's area that are:

- either wholly non-domestic or composite hereditaments;
- not exempt;
- not to be included in a central rating list;
- not cross boundary hereditaments required to appear in a different rating list due to the Non-Domestic Rating (Miscellaneous Provisions) Regulations 1989 (SI 1989/1060).

The rating lists are required to show information about each hereditament included in a local rating list. This information is:

- assessment number;
- address;
- description;
- rateable value;
- whether it is composite;
- whether any part is exempt;
- the effective date of any alteration to the original 1 April 2023 compiled list.

Regulation 2 of the Non-Domestic Rating (Miscellaneous Provisions) (Wales) Regulations 2017 (2017/327) W.82 makes specific provision as to the contents of local rating lists in Wales.

15.3 Central lists

Certain hereditaments have to appear in central rating lists and not in local lists. These are properties occupied by some of the large public utilities such as the gas companies, the Canal and River Trust, the power generation companies, the electricity supply companies, Network Rail, London Underground and British Telecom.

In 1990 with the introduction of the single national multipliers and rate pools, rating became a national and not a local tax. It was then considered simpler to place properties occupied by public utilities in central lists rather than giving them individual apportioned assessments in local lists as had been the practice in the past.

It is the duty of the Central Valuation Officer to compile separate lists for England and Wales. Copies of the respective central lists are available for inspection at the Department for Levelling Up, Housing and Communities (DLUHC) and the Welsh Parliament or Senedd Cymru.

The provisions for compiling, altering and appealing against the central lists and entries are similar to those for local lists.

15.4 Regulations

The reorganisation of the appeals system for rating and Council Tax in 2009 did away with the existing 56 English local valuation tribunals and replaced them with a single Valuation Tribunal for England (VTE). A similar change occurred in Wales from 1 July 2010 with the replacement of the four Welsh valuation tribunals with a single Valuation Tribunal for Wales (VTW). This change should have simplified the system, but unfortunately the new regulatory regime has been made more complicated involving three rather than two sets of appeal regulations. There have been considerable amendments to these regulations which do make following them quite challenging.

The Regulations are:

- the Non-Domestic Rating (Alteration of Lists and Appeals) (England) Regulations 2009 (2009/2268);
- the Valuation Tribunal for England (Council Tax and Rating Appeals) (Procedure) Regulations 2009 (2009/2269);
- the Council Tax (Alteration of Lists and Appeals) (England) Regulations 2009 (2009/2270);
- the Non-Domestic Rating (Alteration of Lists and Appeals) (Wales) Regulations 2005 (SI 2005/758).

15.5 Changes to the appeals system in 2017

In a consultation document in August 2016, the Government noted there was widespread agreement that the current appeals system was broken and in need of reform. Earlier in October 2015, the Government had consulted in England on the idea for a new three-stage approach to appeals, 'Check, Challenge, Appeal', and the August 2016 paper expanded on the original concept. The Regulations have since been amended to implement 'Check, Challenge, Appeal'.

The amending regulations for England are:

- The Non-Domestic Rating (Alteration of Lists and Appeals) (England) (Amendment) Regulations 2017 (SI 2017 /155);
- The Valuation Tribunal for England (Council Tax and Rating Appeals) (Procedure) (Amendment) Regulations 2017 (SI 2017/156);
- The Valuation Tribunal for England (Council Tax and Rating Appeals) (Procedure) Regulations 2009 (2009/2269) (SI 2009 / 2269) ('Procedure Regulations') as amended by (SI 2017/156), 2018 (SI 2018/911);
- Non-Domestic Rating (Alteration of Lists) and Business Rate Supplements (Transfer to Revenue Accounts) (Amendments etc.) (England) Regulations 2018 (SI 2018/1193) Part 4;

- Valuation Tribunal for England (Council Tax and Rating Appeals) Procedure) (Amendment) Regulations 2021 (SI 2021/579) (amends regs for virtual hearings);
- The Valuation for Rating (Coronavirus)(England) Regulations 2021 (SI 2021/398);
- The Non-Domestic Rating (Alteration of Lists and Appeals) (England) (Amendment) Regulations 2022 (SI 2022 /1198).

The approach, from 2017, is designed to manage the flow of cases through the system, in a structured and transparent way, allowing ratepayers to make an informed decision about how to proceed at each stage. The reforms aim to provide a system which is easier to navigate, particularly for small businesses or unrepresented ratepayers, with the emphasis on early engagement by the parties to reach a swift resolution of cases.

The three stages in basic detail are:

- A Check stage, in which facts concerning the property are agreed between the valuation officer and the ratepayer. This is intended to be both swift and to lead to an agreed position for the great majority of cases. Ratepayers can repeat the Check stage as often as they wish.
- A Challenge stage, during which the great majority of cases are resolved between the valuation officer and the ratepayer.
- an Appeal stage, submitted directly to the Valuation Tribunal for England, which focuses on issues which remain outstanding and which are material, on the basis of arguments and evidence which have already been established.

A package of incentives and requirements are designed to encourage early identification of facts and issues to establish accurately the key points of dispute from the outset, so that valuation officers (and ultimately the VTE) can deliver the best possible service for businesses and deploy resources efficiently. These include requirements on the parties to share both evidence and arguments early in the process; fees for the appeal stage; restrictions on evidence that can be presented at appeal; and penalties for knowingly, recklessly or carelessly providing false information.

There are time limits at each stage in order to keep cases moving forward, with 'trigger points' at the Check and Challenge stages, so ratepayers can feel confident that they can progress their cases even if no decision has yet been made by the valuation officer.

Part of the change was to put a greater emphasis on online input, registration and progressing of cases.

15.6 Alterations to local rating lists by valuation officers

Having compiled a rating list, a valuation officer is under a duty to maintain the list as accurately as possible. The valuation officer may alter the list to maintain its correctness, e.g. to insert a new hereditament, or to alter the value of an existing hereditament to reflect an extension or to merge two assessments where the two hereditaments now have a single occupier.

In *Re the appeal of Pearce (VO)* [2014] UKUT 0291 (LC), the Upper Tribunal regarded the valuation officer's duty as restricted to the extent that the valuation officer could not alter a rateable value determined by a valuation tribunal or higher tribunal because this would undermine the whole hierarchical framework of adjudication and appeal. The valuation officer had to respect the tribunal's decision and not replace it with the valuation officer's view of the 'correct' rateable value, however erroneous he or she might consider the tribunal's decision. The only recourse was to challenge the decision on appeal or possibly seek a review by the tribunal of its decision.

However, the valuation officer is only restricted to the decision on the particular rateable value. Should a change to the property or locality occur necessitating an alteration to the list, the valuation officer is able to approach the valuation afresh under the duty to maintain an accurate list.

Should the ratepayer be unhappy at a decision of the valuation officer then there is opportunity to challenge the rating list entry. An 'interested person' (i.e. broadly the occupier or someone with a superior interest in the property) may, in effect, object to the original compiled list entry or to the valuation officer altering the list first by the Check stage to ensure the factual basis behind the valuation officer's valuation is sound and then by a formal proposal – the Challenge stage – and, if that is not sufficient, the Appeal stage.

When the valuation officer alters the list, he or she is not required to give any prior notice of the change, but within four weeks of the alteration he or she must notify the ratepayer, the billing authority and anyone who has already made a proposal for the hereditament that has been referred to the valuation tribunal as an appeal of the effect of the alteration. The valuation officer is not required to do this if the change is merely to correct a clerical error, e.g. a spelling mistake, a change in the hereditament's address or a change in the billing authority's boundary.

The valuation officer's notice to the billing authority takes the form of a 'Schedule of Alterations', whereas each altered entry is notified to the particular ratepayer by an individual 'Valuation Officer Notice of Alteration' (VON). The VON will show for a new entry the following:

- billing authority;
- billing authority's reference number (this will be a new number unless the altered entry is for a hereditament already in the list);
- description;
- address;
- rateable value;
- effective date of the new or altered entry or the effective date of deletion where a new entry does not take the place of an old entry or entries, e.g. where a property has been demolished.

15.7 Proposals

Until recently, the method by which ratepayers and others have sought an amendment to a rating list has been by way of proposing a change to a list. This has taken the form of a written proposal for alteration of a list. In the past valuation officers also had to make proposals to alter lists but under the post-1990 system, they are able to simply alter their rating lists as they consider necessary.

The 2017 Check, Challenge, Appeal reform has not swept away proposals and the ability to propose a change to a rating list; indeed 'challenges' are still 'proposals' and the grounds of proposal have not changed, but a check stage has been introduced as a pre-requisite to making a formal challenge by way of proposal.

15.8 The Check stage

Rates are often a very substantial outgoing for ratepayers, and it is not surprising that they may wish to query their rateable values. This will not just be to check the valuation, perhaps to ensure it is in line with neighbouring properties, but also be to ensure the valuation officer has correctly identified the hereditament and rateable plant and machinery. The new three-stage appeals process envisages first checking the factual position as an initial stage, the Check stage, where the basic facts are checked and agreed, as at the date of confirmation of the Check, before any further action is taken on considering the actual valuation. It may be that the ratepayer is content at this stage.

The stage is not merely an optional check but a mandatory requirement before the Challenge period can be triggered and the ratepayer can move onto the second formal stage – the Challenge or proposal stage. It is initiated by making a request to the valuation officer for information the valuation officer holds about the hereditament (Regulation 4A and 4B). The valuation officer is obliged to provide information that reasonably relates to any of the possible grounds for a challenge to the assessment providing the valuation officer considers it reasonable to provide the information. If the valuation officer is aware there is missing factual information from the valuation officer's records, the valuation officer can ask for this to be provided. This might include rental information or details of the area of a recently constructed extension.

It is clear the revised process is, in part, not simply about the ratepayer ensuring the assessment are reasonable and not excessive but in enabling the valuation officer to gain up to date factual information about the hereditament to ensure the assessment properly reflects its current state, *rebus sic stantibus*.

Upon receipt of the information from the valuation officer, the Interested Person (IP) is required to review the information provided and confirm in writing, via the VOA portal, to the valuation officer what parts of the information are accurate and what is inaccurate (Regulation 4C). There is a duty on the IP to provide the information missing from the valuation officer's records which the valuation officer has requested to be supplied. Where information received from the valuation officer is inaccurate the person needs to provide the accurate information in its stead. This confirmation is particularly important as it provides the 'material day' for many proposals. See Chapter 6, section 6.5.2.

The valuation officer acknowledges receipt of the information (Regulation 4D), decides if the supplied information is accurate or inaccurate, amends the valuation officer's records if necessary and makes any amendment to the rating list needed to correct an inaccuracy (Regulation 4E).

The valuation officer then serves a Notice of Decision, in writing, that the Check stage has been completed (Regulation 4F) stating the date of the notice, the name of the person initiating the stage, the address of the hereditament, details of any alteration made to the rating list as a result of the check, advice on what changes the valuation officer has made to the valuation officer's records and a statement of the person's right to make a proposal and proceed to the Challenge stage.

The legislation provides a safeguard for the ratepayer. Should the valuation officer not have sent the notification within 12 months from the date the valuation officer received the confirmation of the check then the Check stage is deemed to be completed and the window in which to submit a challenge is triggered, and the IP can move on to the Challenge stage. The valuation officer and the IP can agree to extend this time limit.

The regulations require that all documentation at check and challenge has to be submitted via the VOA portal unless an alternative method has been prior agreed. The Challenge stage

There are many reasons why ratepayers and others may wish to see alterations to entries in rating lists, but the ability to make valid proposals is restricted by the Non-Domestic Rating (Alteration of Lists and Appeals) (England) Regulations 2009 (2009/2268) (2009/2268), as to both the categories of persons who may make proposals and the circumstances in which they may be made.

From 1 April 2017 an additional requirement has been added to the Regulations that a proposal may not be made until the Check stage has been completed. Challenge is a separate and distinct process by which proposals are submitted and considered.

15.9 Grounds of proposal

Regulation 4(1) lists grounds when proposals may be made because the proposer considers:

- The original entry in the compiled rating list was wrong, i.e. original entries as at 1 April 2017 for a 2017 list and 1 April 2023 for a new 2023 list.
- The rateable value in the list is wrong due to a Material Change of Circumstances (MCC) having occurred. Regulation 3 defines 'Material Change of Circumstances' as meaning a change in any of the matters in

paragraph 2(7) of Schedule 6 of the Local Government Finance Act 1988, e.g. a change in matters affecting the physical state of the locality, or a change in the mode or category of occupation of the hereditament, or a change in the use or occupation of other premises situated in the locality of the hereditament. The MCC must have occurred on or after the compiled list date and not before. A change before compilation should be reflected in the compiled list figure and therefore are not a material change for that list.

- The rateable value in the list is wrong because of an amendment to the classes of rateable plant and machinery in the Plant and Machinery Regulations that came into force after the rating list was compiled. This reason can be used only if a change is made to the Plant and Machinery Regulations.
- The rateable value resulting from an alteration to the list made by the valuation officer is wrong.

  - A proposal cannot be made on this ground if the valuation officer is only altering the list to give effect to:

  (a) the outcome of a previous proposal or
  (b) a decision of a valuation tribunal or the Upper Tribunal or court

- The rateable value or any other part of a hereditament's entry in the rating list is, or has been, wrong as shown by 'reason of a decision in relation to another hereditament' made by the VTE, a local valuation tribunal, the Upper Tribunal or a court determining an appeal from any of these tribunals.
- The Effective Day shown for an alteration to a list entry is wrong.
- A new hereditament should be inserted into the list.
- A hereditament should be deleted from the list.
- The list should show part of a hereditament as domestic (i.e. a composite hereditament) or exempt.
- The list should not show part of a hereditament as domestic or exempt.
- A hereditament or hereditaments already shown in the list should be shown as one or more different hereditaments, i.e. a division or merger.
- A hereditament's address in the list is wrong.
- A hereditament's description in the list is wrong.
- Part of the entry in the list for a hereditament has been omitted, e.g. if it is part exempt the Regulations require this fact to be stated in the rating list.

There is a restriction preventing the making of a proposal on a property where the VTE or Upper Tribunal has already considered a proposal for the same rating list on the same facts and for the same hereditament. For example, where the VTE has dismissed an appeal without ordering any alteration at all or, indeed, where the valuation officer has altered the list by direction of the VTE in making a decision.

Proposals can be made on more than one ground providing the material day and effective dates (see section 15.15) are the same for each ground.

15.10 Who can make proposals?

There is no general right to make proposals. They can be made only by people classed as 'Interested Persons'. Ratepayers cannot make proposals on other persons' properties to increase or decrease their assessments. Up until 31 March 2017 billing authorities had limited rights to make proposals, but this has now been ended. Billing authorities can, of course, make proposals where they are Interested Persons in their own right by being owners or occupiers of their town halls, etc.

Interested persons can make proposals in any of the circumstances set out in the Regulations if they have reason to believe the ground exists. In the case of challenges to the compiled list entry or an alteration to the list by the valuation officer or a MCC, they are limited to making a single proposal for any one ground and event. So, for example, an interested person can make only one proposal challenging the original compiled list in the 1 April 2023 assessment, or can make only one proposal relating to a particular physical change in the locality. A new interested person, such as a new ratepayer, has the right to make fresh proposals because he or she has not made a proposal before.

'Interested person' means the occupier of a hereditament and anybody holding a superior interest, e.g. the freeholder or head lessee, who may in due course become entitled to possession because of the expiry of under leases and tenancies. The definition also includes any person with a qualifying interest (such as a subsidiary or parent company of the occupier or body holding the superior interest) and the Crown Estate Commissioners for hereditaments forming part of the Crown Estate.

15.10.1 *Former interested persons*

People who were, in the past, interested persons for a hereditament can make proposals challenging the rateable value or effective date of alterations to the rating list made by the valuation officer if they were interested persons at any time during which the valuation officer's alteration to the list had effect.

The rights of these former interested persons is a little more complicated than the foregoing suggests Their rights are determined by when their interest in the hereditament actually ceased. If they still had an interest in the property at the date they made a request for information at the Check stage but have since left, they can make a proposal on any of the 15 grounds under Regulation 4(1) provided it relates to the time in which they were an interested person. If, on the other hand, the former interested persons left before the date of a check request for information then the grounds on which they can submit a proposal are restricted to either a change in the Plant and Machinery Regulations, a list alteration or the effective date of a list alteration, again the grounds only apply if they relate to the time in which they were an interested person.

For example, Jane Brown occupied 5 Nonsuch Street from 1 December 2021 to 1 December 2023. On 3 March 2024 the valuation officer altered the entry for 5 Nonsuch Street in the rating list with effect from 1 April 2023. The occupier, as the current interested person, can make a proposal, but so also can Jane Brown as she was an interested person during part of the time the alteration had effect, i.e. for her, 1 April 2023 to 1 December 2023.

Conversely, Doug Jones occupied 7 Nonsuch Street from 1 September 2023 to 30 August 2024. As a former IP he cannot make a compiled list proposal as it is not a ground on which he is entitled to make a proposal.

In other circumstances persons who were interested persons in the past cannot make proposals. In *Mainstream Ventures Ltd* v *Woolway (VO)* [2000] RA 395, a proposal was held to be invalid because the proposers were no longer occupiers at the date of the proposal and were therefore not interested persons.

15.11 Contents of a proposal

Having established that someone has the right to make a proposal, the next step is to ensure the proposal itself is properly made. Regulation 6 provides for certain requirements to be satisfied and which are mandated under Regulation 8. These requirements have developed over the years and are quite complicated. If a proposal misses out one of the requirements then the proposal will not have been made. It is therefore important to satisfy all of the requirements.

Until 2017, whilst proposals could be made online the more usual approach was to complete a paper form. The Regulations did not provide a set statutory form to be used, as a proposal and a letter, providing it satisfied the requirements, could constitute a valid proposal. A pre-printed form was, nonetheless, made available to guide people through making a proposal and was the method normally used.

The intention from 2017 was to require people to make proposals online and Regulation 6(2) requires proposals to be made by serving them on the valuation officer using the electronic portal provided by the VOA. 'An exception allows proposals to be made in another manner if agreed with the VO.'

To be accepted as complete a proposal must:

- give the name, address and contact details of the person making the proposal;
- state the grounds for making the proposal including particulars or details on which each of these grounds is based;
- provide evidence supporting the grounds;
- explain how the evidence supports the grounds;
- identify the alteration sought;
- identify the date from which the proposed changes should be effective;
- identify the date the proposal is served on the valuation officer.

The proposal needs to state the rent payable each year at the date of the proposal, when that amount first became payable and details of any rent-free periods, unless the proposal is only:

- proposing a deletion;
- proposing an amendment to the address;
- proposing a change to the description of the hereditament.

If the proposal relies on a decision of the VTE or the Upper Tribunal on another hereditament then the proposal also needs to state the reasons for believing the decision is relevant and for believing, by reason of the decision, that the rating list entry is inaccurate. Details identifying the decision also have to be given (identifying the hereditament, name of the tribunal and date of decision).

Proposals may deal with more than one hereditament where a division or merger of assessments is proposed, or where each of the hereditaments is within the same building or curtilage and the proposer makes the proposal in the same capacity for each of the hereditaments, e.g. as freehold owner.

Regulation 6A sets further restrictions on material change of circumstance proposals. Only one proposal may be made in respect of each MCC (Reg 6A(3)) and if a check has been submitted which refers to an MCC and no proposal has been submitted within the relevant time frame then no further proposals may be submitted in respect of that MCC (Reg 6A(5)).

15.12 Time limits for proposals

For 2017 onwards, if an interested person wishes to make a proposal it must be submitted within 4 months of completion of check as defined in Regulation 6(1) and the check and proposal have to be made by the same person. There is an exception regarding external material changes of circumstance (matters under Schedule 6 para 2(7)(d) & (e)), whereby the time limit for submission is the later of 4 months from the date of completion of check or 16 months from the submission of a challenge whichever is later (Regulation 6A(2)).

The ability to make proposals has always been time limited. Understandably the right has usually been curtailed once new lists have been compiled and in force. The current rules prevent proposals on assessments in a list being made unless a confirmation under Reg. 4C (see 15.8) has been made to the VO before a new list has been compiled (see Reg. 5A). A proposal cannot be made, except in limited circumstances, unless the Check process has been begun and confirmation made before the list ends. For 2017 lists this was 31 March 2023.

The exceptions are for proposals challenging alterations to the list by the valuation officer, including effective dates, when proposals may also be made up to the end of six months beginning with the date of alteration; and, if a proposal is made on the basis of a tribunal decision then the confirmation needs to be made no later than six months after the list has ended.

15.13 Incomplete proposals

Prior to 2017, proposals would be classed as invalid if not satisfying the requirements of the Regulations. A perhaps easier approach is now adopted under Regulation 8 which requires valuation officers to 'refuse' those that fail to satisfy the requirements.

To refuse an incomplete proposal, the valuation officer serves a 'notice of refusal' specifying the missing information and the date the notice is served. The proposer can then submit a fresh proposal within the balance of the four-month period for submission. In calculating this four-month period, the time between the date the incomplete proposal was served on the valuation officer and the notice of refusal served is ignored.

Where an incomplete proposal relates to an external MCC and it was served on the valuation officer after four months from the completion of check, the resubmission of a proposal is not possible.

15.14 Procedure subsequent to the making of proposals

Having received a complete and valid proposal, the valuation officer may accept that its contentions are correct and alter the list treating the proposal as 'well founded'. Alternatively, negotiations may ensue with the proposal being withdrawn as its contentions were not well founded, agreement being reached or agreement not being achieved and a decision notice issued. The ratepayer then has the opportunity, once the decision notice is issued, to appeal to the VTE.

The Regulations make provision for each of these circumstances and also for the valuation officer to carry out some initial procedural steps.

15.14.1 Initial procedure

The valuation officer is required within six weeks of receipt to serve a copy of the proposal, together with a statement of the provisions for appeal, etc., on the ratepayer(s) for the hereditament(s) to which the proposal relates, unless the ratepayer made the proposal ('ratepayer' means the occupier or, if the hereditament is unoccupied, the owner, i.e. the person entitled to possession of the hereditament). There is no statutory requirement for the valuation officer to serve a copy of the proposal on an owner.

15.14.2 The billing authority

The valuation officer also needs to provide the billing authority with specified information within six weeks of receipt of the proposal, namely:

- the identity of the hereditament;
- the date the proposal was made;
- the rateable value in the list;
- the rateable value and effective date proposed in the proposal;
- whether or not the proposal has been determined.

This is to assist them in forecasting the impact that proposals might have on rates payable in the future so they are aware of the degree of challenge to rateable values. The information required seems not to provide the authorities with the actual nature and reasons for the challenge. The billing authority is not entitled to an automatic copy of a proposal, though like other persons it may inspect the proposal (Schedule 9, paragraph 9, Local Government Finance Act 1988).

Because of their special knowledge of the local area, the billing authority may be able to provide additional evidence relating to a proposal. The Regulations provide for billing authorities to do this if they wish. A copy needs to be provided to the proposer who may furnish further evidence in response. This is particularly important regarding cases involving completion notices.

The billing authority is also able to receive a copy of a decision notice sent from the valuation officer to the proposer advising that the valuation officer is not going to treat the proposal as well founded or agree to a change to the list entry or entries. The billing authority, under Regulation 13, can do this by notice individually for each proposal, by reference to particular hereditaments or by notifying a whole class of proposal or hereditament to the valuation officer.

15.14.3 *Valuation officer's response*

If the valuation officer considers it reasonable to do so the valuation officer must provide the proposer with any information the valuation officer holds that relates to the particular grounds set out in the proposal. The wording of the Regulation is interesting as it imposes an imperative, 'must', but only if the valuation officer considers it 'reasonable to do so'.

The valuation officer must also provide any further information that comes into the valuation officer's possession and is relevant to the grounds, again providing it is reasonable to do so.

The proposer then has three opportunities to provide further evidence if the proposal has not yet been determined:

- In response to the valuation officer's original or additional information, the proposer may provide further evidence to support the grounds.
- Before the proposal is determined, provide further evidence relating to the grounds providing the information neither was known to the proposer nor could reasonably have been acquired by the proposer before the proposal was made.
- If the proposer and VO agree in writing that further evidence may be provided in circumstances not covered by the above.

All evidence supplied by the proposer during the Challenge stage becomes part of the proposal (Regulation 9(12)).

Penalties may be imposed by the valuation officer for knowingly, recklessly or carelessly providing the valuation officer with information in connection with a proposal which is false in a material particular. A penalty may arise in relation to Check or Challenge once a proposal is submitted. Regulation 9A provides penalties for 'smaller proposers' of £200, and for other proposers of £500. There is provision for appeal.

15.14.4 Well-founded proposals

Where the valuation officer agrees with the proposed alteration, the valuation officer advises the person making the proposal (and the ratepayer if not the proposer) that the valuation officer considers it well founded and intends to alter the list accordingly. The valuation officer must alter the list as soon as reasonably practicable after the well founded notice (Regulation 10). Once a proposal is well founded the Challenge process ends.

15.14.5 Withdrawal of proposals

A person who made a proposal can generally simply withdraw it by serving notice on the valuation officer. However, where the proposer was originally the ratepayer and made the proposal in the capacity of ratepayer but is no longer, the written consent of the current ratepayer at the date of withdrawal is also required.

Where the proposer withdraws the proposal, another interested person (such as the owner) may be able to 'take over' the proposal if the person notifies the valuation officer within six weeks. This can be done only if the interested person, who would have had the right to make the proposal, had opted in by serving notice on the valuation officer within two months of the proposal being made stating the person wished to be a party to any proceedings. In this scenario the opted in interested person becomes the 'substitute proposer' and in the event the proposal is not settled may appeal any subsequent Decision Notice to the valuation tribunal.

15.14.6 Agreement

Proposals are usually settled by the valuation officer serving a decision notice and this being effectively accepted by the proposer but may also be settled by withdrawal/agreement rather than proceeding to appeal and being decided before the VTE. When a written agreement is reached, the valuation officer is required to alter the list within two weeks of the agreement being made. The proposal is then treated as having been withdrawn: Regulation 12. For an agreement to be complete the following must agree in writing:

- the valuation officer;
- the proposer;

- the occupier (at the date of the proposal) of any hereditament to which the proposal relates unless that person(s) is no longer the occupier and the valuation officer is unable, having taken reasonable steps, to ascertain that person's whereabouts;
- the ratepayer (at the date of agreement) in relation to any hereditament to which the proposal relates;
- any 'interested person' who would have had the right to make the proposal and has served notice on the valuation officer within two months of the proposal being made stating the person wished to be a party to any proceedings unless such a person who gave notice cannot be contacted at the address supplied to the valuation officer. It is rare for such a notice to be given, but it allows former occupiers affected by the proposal to become parties to the appeal;
- the proposer, occupier at the date of proposal and ratepayer at the date of the agreement will, of course, very often be the same person.

15.14.7 Disagreement

Where a proposal is not withdrawn, the valuation officer does not consider it well founded and an agreement is not reached, the valuation officer must serve notice of his or her decision on the proposer, the ratepayer (if not the proposer), any interested person who has served notice (i.e. any 'opted in' party) and the billing authority if the authority has served notice that it wishes to receive a notice of the decision.

The valuation officer's notice must include:

- A statement that the valuation officer does not consider the proposal well founded and the valuation officer disagrees with the proposed alteration and has decided not to alter the list at all or not in accordance with the proposal.
- The reasons for the decision including a statement of the evidence and information used to make the decision.
- A statement in relation to each ground setting out why the valuation officer does not consider the ground made out. It includes a requirement to include a summary of any particulars of the grounds of the proposal with which the valuation officer does not agree. This presumably means that the valuation officer needs to explain in some detail why the ground is not made out.
- The proposer's rights to appeal against the decision.

This is likely to be a lengthy and comprehensive document.

The billing authority who has served notice and any opted in former IP who Is not an IP when the decision notice is served receives a shortened version of the decision notice.

15.14.8 *Appeal to the VTE*

The third part of the system of Check, Challenge and Appeal is 'Appeal'. The proposer does not need to go on to this stage but may wish to appeal to the VTE.

An appeal is submitted direct to the Valuation Tribunal for England via the VTE portal.

The proposer can appeal either because the valuation officer has not altered the list and, in the proposer's view, it remains inaccurate, or because the alteration the valuation officer has made is otherwise than in accordance with the proposal and the list remains inaccurate. An appeal has to be made within four months of the date of the valuation officer's decision notice. The regulations also permit an appeal if the valuation officer has not issued a decision notice within 18 months of the date of submission of the proposal. Again, the appeal has to be submitted within 4 months of the expiry of the 18-month period.

The grounds of an appeal can be either or both of the following:

- the valuation for the hereditament is not reasonable;
- the list is inaccurate in relation to the hereditament (other than in relation to the valuation).

There is the option for the proposer and valuation officer to agree in writing to extend the 18-month period in order to allow the valuation officer to make a decision. This may be preferable where the valuation officer has not been able to attend to the proposal due to pressure of work and the proposer would rather see the valuation officer's decision than go straight on to the Appeal stage.

To initiate the appeal the proposer must serve a notice of appeal on the VTE. This needs to be received within the time limit though there is provision for requesting an extension of time. Under Regulation 13C (3) the notice of appeal *must* be accompanied by:

- a copy of the valuation officer's decision, if made;
- a copy of the proposal and any evidence forming part of the proposal provided by the proposer;
- any information provided to the proposer by the valuation officer under the Challenge procedure;
- the fee, if any;
- if out of time, a request for an extension of time and the reason why the notice of appeal was not served in time.

The appeal has to be made by serving a notice of appeal either through the VTE's electronic portal (www.appeals.valuationtribunal.gov.uk) or in another manner agreed with the VTE. The VTE requires only two documents to be uploaded:

1   the VO's Challenge decision notice,[1] and
2   a supporting evidence statement highlighting the matters in dispute and the key points that will be relied upon (without embedded documents within the submission).

The fee is £150 for a smaller proposer, £300 otherwise. There is no fee if the valuation officer has not made a decision at the Challenge stage and the proposer is appealing after the 18-month period. Should the appeal be decided without a hearing a third is refunded. Should the valuation officer alter the list in accordance with the proposal after the appeal has been made or be ordered to alter the list the fee is refunded in full.

The complication of what is a smaller proposer is dealt with in Regulation 3A. Broadly it is a proposer who satisfies the definition of 'micro business' in s.33(3) of the Small Business, Enterprise and Employment Act 2015, but because non-domestic rates do not just cover business occupiers, they also cover an individual who is not within the definition. Broadly the definition of micro business is one with a headcount of staff of less than ten, and a turnover of less than two million pounds.

Appeal fees are refunded if the VTE decides one or more of the grounds of the appeal are made out or a consent order is made.

15.15 Effective dates

15.15.1 Compiled list entries

Rating lists are not simply lists of addresses and rateable values; they also contain the date each entry becomes effective. This enables the billing authority to calculate rate liability. Where a rateable value is altered, the billing authority will send the ratepayer a revised rates demand showing the change in liability from the effective date of the alteration. For example, if the rateable value of a hereditament increases due to an extension being built with an effective date of 2 July 2024, the increased rates will be payable from 2 July 2024.

For the originally compiled 2023 lists the effective date for all entries was 1 April 2023, which was the date the lists came into force. Subsequent alterations to the lists show the date from which those changes are effective. This may be 1 April 2023 if the alteration replaces the original entry or it may be some later date.

The Regulations for effective dates have changed several times over the years and rating lists. This chapter covers the provisions for the 2023 rating lists from 1 April 2023.

---

1 This will also need to Incorporate the requirements of Regulation 13C(3) and include the proposal document and all evidence submitted at challenge by both parties if not already included in the decision notice itself.

15.15.2 *Effective dates for alterations to lists*

Regulation 14 of the 2009 Regulations makes 1 April 2023 the effective date for an alteration to an original 1 April 2023 compiled list assessment whenever the actual alteration to the list is made. For other alterations the effective date is the day on which the circumstances giving rise to the alteration first occurred. There are a few situations where these simple rules do not apply. These are corrections to wrong assessments which increase the rateable value which are not the fault of the ratepayer, where a completion notice applies and where the actual day on which a change occurred cannot be determined.

Corrections to the compiled list entry that increase the rateable value or corrections to later alterations which also increase the rateable value have, as their effective date, the day the list is altered. This is because it is considered unfair that ratepayers should have a backdated increased liability because of the valuation officer's error in under-valuing their hereditaments. This restriction does not apply where it is the ratepayer's fault that the rateable value has been under-assessed. In this circumstance the increase can be made effective from the normal effective date. An example of under-assessment would be where a ratepayer extended a property in February 2023 before the local 2023 list was compiled, but the valuation officer only learnt of the extension in July 2023. The compiled list valuation will not show the extension because the valuation officer was not aware of it. When the valuation officer corrects the compiled list entry to include the extension the effective date will be the date the valuation officer alters the list and not 1 April 2023. This limitation applies only where the increase is to correct a compiled list inaccuracy or a later incorrect alteration. The limit does not apply to alterations simply to correct the list or MCCs, e.g. the effective date of an increased rateable value to reflect the building of an extension after 1 April 2023 would have as its effective date the date when the work was completed and not the date of the list alteration.

It should be noted that, where an alteration to an entry in the list involves both a MCC (or some other reason) and a correction of an inaccuracy in the originally compiled list which showed too low a figure, the appropriate effective date is not restricted. The effective date should be the date of the MCC (or other change) and should not be restricted to the date the list is altered (see *Goulborn* v *Wynn Cowell (VO)* [2010] UKUT 417 (LC)).

Where an alteration is made to give effect to a completion notice, the effective date is the date specified in the notice or any other date agreed or determined under the completion notice procedure (see Chapter 17).

The draughtsman of the Regulations appreciated that it is not always possible for the actual day on which a change occurred to be discovered or determined. Regulation 14(5) provides that where the alteration results from a proposal, the day is the date on which the proposal was served on the valuation officer. In any other case, e.g. where the valuation officer alters the list, the day is the day on which the alteration is entered in the list.

If the alteration is to insert a new entry in the rating list the effective date will be taken back to the date of event, e.g. if a property is not shown in the compiled list but existed prior to the list coming into force then this is a new entry and not an error in the rating list. See *BMC Properties* v *Jackson 2015* EWCA Civ1306.

15.15.3 *Effective dates determined by tribunals*

Generally, the normal rules for effective dates apply to tribunal decisions but two special provisions apply to decisions of the VTE and the Upper Tribunal.

15.15.4 *Where tribunals increase a rateable value*

The tribunals are empowered to determine a rateable value that is greater than the entry shown in the rating list and greater than the rateable value sought in the proposal giving rise to the appeal. This could result in a ratepayer being faced with a backdated rates bill greater than that before the proposal was made. Therefore, where the VTE increases a rateable value beyond the existing entry in the list and beyond the amount contended for in the proposal, the new higher rateable value will be effective only from the day on which the valuation tribunal's decision was given (Regulation 38(5) of the Valuation Tribunal for England Procedural Regulations 2009 (2009/2269)). This restriction does not apply to splits and mergers.

15.15.5 *Circumstances have ceased*

Sometimes, by the time the VTE hears an appeal, the circumstances justifying an alteration, e.g. a substantial nuisance, will have ceased. To prevent alterations continuing for longer than is in fact justified, valuation tribunals may order the alteration to have effect only for the period which justified the alteration, i.e. giving an effective date for the change and an effective date for restoring the assessment. Two alterations will be required to the list in consequence (Regulation 38(7)).

15.16 Appeals

15.16.1 *The Valuation Tribunal for England*

The VTE was established on 1 October 2009 under Part 13 of the Local Government and Public Involvement in Health Act 2007 to be the first appeal court for rating and Council Tax appeals. It replaced 56 separate local valuation tribunals that undertook the same function up until October 2009.

The VTE consists of the President, one or more Vice-Presidents, members of a panel of persons to act as chairmen (senior members) and other persons appointed as members of the Tribunal. The appointments are made by the Lord Chancellor.

The President must make tribunal business arrangements to provide for the selection of the member or members of the Tribunal to deal with any appeal made to the Tribunal.

The Local Government Act 2003 set up the Valuation Tribunal Service as an independent body to undertake a number of functions in relation to valuation tribunals in England. The Valuation Tribunal Service now provides the accommodation, staff, equipment and training for the VTE, as well as giving general advice about procedures in relation to proceedings before the tribunals.

The VTE is not an expert tribunal as it mostly comprises lay members, though many may be solicitors or valuers by profession. The President, however, is legally qualified and the Vice-Presidents are either legally qualified or valuers.

15.16.2 *Types of appeal*

The VTE is empowered to determine a number of different types of appeal. These are set out in the Regulations:

- rating appeals arising under the Check, Challenge and Appeal system, as described in sections 15.5, 15.8 and 15.9;
- Council Tax appeals;
- appeals against valuation officer's certification of rateable values for transitional relief;
- appeals against completion notices;
- appeals against the imposition of a penalty for failure to supply information to a valuation officer or where a person provides the valuation officer with information in connection with the proposal which is false in a material particular.

15.16.3 *Objective*

The Tribunal sees its overriding objective as being to deal with cases fairly and justly and will:

- deal with appeals proportionately to the importance of the case, the complexity of the issues, the costs of the case and the resources of the parties including the Tribunal;
- avoid unnecessary formality and be flexible in the proceedings;
- deal with the parties fairly allowing them to participate so far as is practical in the proceedings;
- use the special expertise of the Tribunal effectively; and
- avoid delay, so far as it is compatible with the proper consideration of the issues.

15.16.4 *Case management powers*

The Regulations give the VTE power to regulate its own procedure. In particular, it is empowered to do the following:

- Alter the time for complying with any regulation or direction. The statutory time limits can therefore be changed by the VTE.
- Consolidate or hear together appeals raising a common issue.
- Treat an appeal as a lead appeal.
- Permit or require a party or another person to provide documents, evidence, information or submissions to the VTE or a party. As there are no powers given to the VTE to support this power, it is difficult to see how it can be enforced against someone who is not a party and has no interest in the proceedings.
- Deal with an issue as a preliminary issue.
- Hold case management hearings such as pre-hearing reviews.
- Decide the form of any hearing.
- Adjourn or postpone any hearing.
- Require a party to produce a bundle of documents at a hearing.
- Stay proceedings.
- Suspend the effect of its own decision pending the determination of the Upper Tribunal on another appeal.

In practice the general arrangements for valuation tribunal hearings are very substantially a matter for the President of the VTE who is given power in Regulations 5 and 6 of the Valuation Tribunal for England (Council Tax and Rating Appeals) (Procedure) Regulations 2009 (2009/2269), to make arrangements for the organisation of the Tribunal. These arrangements and standard directions are set out in a Consolidated Practice Statement (effective from 1 April 2022 at the time of publication).

It covers the following issues:

PS1 – Extensions of time for making appeals
PS2 – Non-domestic rating appeals
PS3 – Complex cases
PS4 – Postponements and adjournments
PS5 – Decisions without a hearing (2010 or earlier rating list appeals)
PS6 – Transitional relief appeals
PS7 – Lead appeals
PS8 – Model procedure
PS9 – Role of the clerk
PS10 – Discretionary reductions in council tax (liability appeals)
PS11 – Disclosure in all council tax and completion notice appeals
PS12 – Private or closed hearings
PS13 – Reviews and setting aside decisions

PS14 – Reinstatements and lifting of a bar
PS15 – Publication of decisions
PS16 – Consent orders
PS17 – Proceedings in other appeal types

The Consolidated Practice Statement can be downloaded from the VTE's website.

15.16.5 *Notice of hearings*

The Regulations require reasonable notice and at least 14 days' notice to be given by the VTE to parties of the date for a valuation tribunal hearing. Shorter notice can be given with the parties' consent or in urgent or exceptional circumstances. Usually the period of notice will be greater than the 14-day minimum and the VTE aims to give 10 weeks' notice in accordance with its standard directions under the Consolidated Practice Statement.

Appeals for the same hereditament are required to be listed for hearing in a set order. This is determined by establishing the order in which the alterations to the rating list would have taken effect but for the appeals. While appeals on the same hereditament have to be heard in a set order, they can be heard in that order on a single day. However, if there is an initial question of validity to be decided and the tribunal decides the proposal was validly made, then it cannot immediately proceed to determine the appeal arising from the proposal unless all parties agree. This provision is designed to save the parties spending abortive time preparing a case on the subject matter of an appeal when it may be invalid and, therefore, there will be no case to be heard.

15.16.6 *Withdrawal*

Written notice of withdrawal has to be served on the VTE or, providing the VTE panel consents, can be made orally at the hearing.

The proposer can request reinstatement of a withdrawn appeal providing the request is received within one month by the VTE (Regulation 19 of the Valuation Tribunal for England (Council Tax and Rating Appeals) (Procedure) Regulations 2009 (2009/2269), (2009/2269)).

15.16.7 *Pre-hearing*

For some cases it will assist the parties if a case management hearing is held as a preliminary proceeding under the VTE's case management powers.

This will most often be needed with complex cases.

15.16.8 Complex cases

A case is classed as 'complex' if it has one or more of the following:

- national implications;
- lengthy or complex evidence;
- novel, important or contentious points of law; or
- any other factor that makes it desirable to manage the case- and issue-specific directions, including the selection of a lead appeal (see section 15.16.10).

A party can apply for a case to be treated as complex or it can be designated as complex by the Registrar to the VTE. This may well mean it needs to be listed before the President, a Vice-President or a panel picked to contain members with experience of complex issues.

15.16.9 Lead appeals

Sometimes it will appear that the decision on a particular case will be of such significance that it is likely to have an impact upon other appeals awaiting a decision. In these circumstances the case will be designated a 'lead appeal' and the other appeals will be 'stayed' and continue to be stayed for one month (or such other period as the VTE may direct) after the VTE's decision has been issued.

The parties to the related appeals are informed of the designation and can seek a direction that the case is not a lead appeal for their own appeal.

If the lead appeal is appealed to the Upper Tribunal the VTE will direct that the related appeals will continue to be stayed until the appeal is determined.

15.16.10 Conduct of a hearing

Valuation tribunal panels in the past usually consisted of three members. However, the Consolidated Practice Statement says that the hearing of an appeal (including a review of a decision) will normally take place before a panel of two members, unless otherwise directed by the President. At least one of the members must be a senior member. The President, Vice-Presidents and nominated senior members may sit alone. The traditional advantage of a three-member tribunal is that you self-evidently get a majority decision. Under the current arrangements if a panel with an even number of members is unable to agree, a completely new panel will be selected to hear the appeal afresh.

The panel will be assisted by a clerk who is responsible for advising the panel on the law, practice and procedure and any relevant decisions of the superior courts and tribunals. The clerk is required to keep an accurate note of evidence and submissions and may ask questions to clarify evidence and issues. It is not the clerk's role to take part in the actual decision making. The clerk is also under a duty to assist unrepresented parties. The clerk is responsible for recording the panel's reasons, though does not take part in the discussions leading to the decision, and drafting the written decision for approval.

Hearings are in public unless the valuation tribunal decides otherwise. This might be where a party's interests could be prejudicially affected, e.g. by the disclosure in public of business accounts. The Press can and do attend valuation tribunal hearings on occasion.

Valuation tribunal panels can hear and determine appeals in a party's absence where it is satisfied the party has been notified of the hearing, or reasonable steps have been taken to notify and it considers it is in the interests of justice to proceed.

The Regulations (Regulation 3 of the Valuation Tribunal for England (Council Tax and Rating Appeals) (Procedure) Regulations 2009 (2009/2269), (2009/2269)) specifically require valuation tribunals to avoid formality in proceedings as it is recognised that unrepresented ratepayers can be intimidated by over-formal proceedings. This does not mean that normal courtesies such as standing when the tribunal members enter should be dispensed with by the parties. However, since the COVID-19 pandemic, the legislation has been changed and the VTE now normally opts for virtual hearings instead of face to face hearings.

The VTE is to have regard to dealing with appeals in ways which are proportionate to the importance of the appeal, the complexity of the issues, the anticipated costs and the resources of the parties. It is to ensure that so far as practicable the parties are able to fully participate in the proceedings.

The order in which the parties speak is at the discretion of the valuation tribunal. The usual order in a case is for the appellant, most likely seeking a reduction in rateable value, to speak first and then to be cross examined, if appropriate, by the valuation officer. The valuation officer will then present the valuation officer's case and there is an opportunity for the appellant to cross examine the valuation officer on the evidence he or she has given. The appellant would then sum up his or her case but cannot introduce new facts. The arrangements will be similar if other parties appear.

Any party may call witnesses, including expert witnesses, and be represented.

The chairman and panel members will usually also ask questions to clarify the facts and points being raised.

Valuation tribunal panels have no powers to award costs.

15.16.11 *Advocacy and expert evidence*

The Royal Institution of Chartered Surveyors (RICS) has published best practice advice for surveyors when undertaking rating appeals before valuation tribunals, *Rating Appeals Guidance Note* (RICS fourth edition, 2017). This supplements the practice note and guidance, *Surveyors Acting as Expert Witnesses* (RICS fourth edition, 2014) and *Surveyors Acting as Advocates* (RICS Professional Statement second edition, Guidance Note third edition, Guidance Note 2017). Together they require chartered surveyors to include wording along the following lines at the end of their documentation or written report for the tribunal:

i   I confirm that my report has drawn attention to all material facts which are relevant and have affected my professional opinion.
ii I confirm that I understand and have complied with my duty to the [specify the tribunal[2] as an expert witness which overrides any duty to those instructing or paying me, that I have given my evidence impartially and objectively, and that I will continue to comply with that duty as required.
iii I confirm that I am not instructed under any conditional or other success-based fee arrangement.
iv I confirm that I have no conflicts of interest.
v   I confirm that I am aware of and have complied with the requirements of the rules, protocols and directions of the [specify the tribunal].
vi I confirm that my report complies with the requirements of the RICS – Royal Institution of Chartered Surveyors, as set down in the RICS practice statement Surveyors acting as expert witnesses.

The advice stresses the importance of keeping separate the roles of advocate and expert witness. It is usual for surveyors attending valuation tribunals to undertake both roles rather than having a barrister or solicitor undertaking the advocacy and the surveyor just providing the expert evidence on rents, assessments and valuations. In either role it is the duty of the surveyor to assist the tribunal. Witnesses must provide their evidence openly and honestly and must not seek to conceal any relevant matter. As an advocate, surveyors may emphasise a point in a particular way but not to such an extent as actually to mislead the tribunal.

The case of *Banks* v *Speight (VO)* [2005] RA 61 illustrates the importance of surveyors keeping in mind their duty to the tribunal and how surveyor experts should seek to present themselves and their evidence. While the tribunal did not at all doubt the expertise of the ratepayer's surveyor as a mineral valuer, it considered in his evidence that he was 'essentially putting a case rather than giving objective expert evidence'. In contrast, the answers of the valuation officer's expert were wholly objective. He naturally defended the decisions he and his colleagues made when valuing mineral hereditaments for the 1990 lists but did so objectively. The tribunal found that his short, common sense answers showed a wide knowledge and experience of minerals and were helpful and convincing.

In *Abbey National Plc* v *O'Hara (VO)* [2005] RA 247, the Lands Tribunal noted both experts had included in their reports a declaration that they had complied with the requirements of the RICS as set out in *Surveyors Acting as*

2 The reference used may vary, as appropriate to the particular forum.

Expert Witnesses: Practice Statement. However, the Tribunal concluded that the ratepayers' surveyor had only paid lip service to his duty to give impartial evidence to the Tribunal. He was selective in his choice of evidence and had adopted a cavalier attitude to the decision of the valuation tribunal by providing no explanation of a paradox in his valuation methodology identified by the lower tribunal. The Tribunal commented that this was not what it expected from an independent expert whose object should be to try and assist the Tribunal.

The Upper Tribunal has also been concerned about possible bias or the appearance of bias when experts are instructed under a success related fee basis giving them an interest in the outcome of the case. It is clearly important that experts' opinion evidence is free from real or supposed bias with the expert's duty wholly to provide the tribunal with honest and true opinion without favour to the client. In *Gardiner Theobald* v *Jackson (VO) [2017]* the Tribunal said:

> *It remains wholly unacceptable for an expert witness, or the practice for which he or she works, to enter into a conditional fee arrangement, without that fact being declared (and in sufficient detail) to the Tribunal and any other party to the proceedings from the very outset of their involvement in the case. The Tribunal will treat such a failure as a serious matter.*

The Upper Tribunal will no longer accept evidence from experts remunerated under a success related fee basis conditional upon the outcome of the case. 'An expert who is to be paid on a contingency basis may not give evidence' (Practice Directions UT (Lands Chamber) para 18.27 19 Oct 2020).

In *Senova* v *Sykes (VO)* [2019] UKUT 0275 (LC) the Upper Tribunal's displeasure at the ratepayer's surveyor failure to disclose the conditional fee basis of his remuneration, following the Gardiner Theobald case, was evident. The tribunal completely excluded the surveyor's evidence because 'he should have disclosed his conditional fee at the outset (rather than at the last minute and under protest). His failure to make that disclosure at the outset is an abuse of process'. The decision concludes with a statement intended for wider readership:

> *The vast majority of experts who appear before the Tribunal are fully aware of and comply with their obligations both to the Tribunal and to their professional bodies. Most are manifestly professional in their conduct. We see many examples of good and constructive practice and we do not wish to give the impression that abuses are common or that expert witnesses are regarded with suspicion by the Tribunal. On the contrary, the Tribunal depends heavily on expert witnesses in its work and relies on their knowledge and integrity. It is for that reason that it places such importance on the preservation of experts' objectivity.*

15.16.12 *Evidence and the use of forms of return*

The Regulations recommend an informal style of hearing and specifically prevent valuation tribunals being bound by the rules of evidence. Hearsay evidence is, therefore, admissible but it is, of course, for the tribunal to decide what weight to attribute to such evidence.

The 2017 amendments to the procedure regulations restrict the VTE in what evidence it may consider to, essentially, that provided during the Challenge stage. Regulation 17A(1) of the Valuation Tribunal for England (Council Tax and Rating Appeals) (Procedure) Regulations 2009 (2009/2269), restricts the evidence to that 'included in the notice of appeal or any document accompanying the notice of appeal'. There are limited exceptions if the new evidence:

- is provided by a party to the appeal;
- relates to the ground on which the proposal was made; and
- was not known to the party and could not reasonably have been acquired by the party before the proposal was determined;

or all the parties to the appeal agree in writing to the new evidence being provided.

If this new evidence is allowed then the valuation tribunal needs to admit other new evidence relating to it provided by the other parties.

The idea is to restrict the whole evidential process so that there are 'no surprises' at the hearing. All parties should have fully disclosed their evidence and case at the Challenge stage so that it could be properly considered by the other parties. The VTE hearing should be a last resort when there is a definite difference of known opinion based on known facts to be resolved by the tribunal.

Valuation officers are empowered under paragraph 5 of Schedule 9 to the Local Government Finance Act 1988 to serve notices known as forms of return on owners and occupiers requesting the supply of information specified in the notice and 'which the officer reasonably believes will assist him in carrying out functions conferred or imposed on him' by the Act. Failure to return them within 56 days can result in a £100 penalty with a further £100 penalty if the information is still not returned after a further 21 days. A £20 per day penalty follows this. Valuation officers served these forms during 2021–2022 to gather evidence for the 2023 Revaluation. They are no longer paper forms and online submission is required. The evidence of rents from those returns forms the basis of valuation officers' valuations.

Should valuation officers wish to use these returns as evidence at a hearing then Regulation 17 of the Valuation Tribunal for England (Council Tax and Rating Appeals) (Procedure) Regulations 2009 (2009/2269) requires them to give the other parties at least two weeks' notice, specifying the returns they intend to use and the hereditaments to which they relate. This period has been varied by the VTE President's Practice Directions and currently stands at eight weeks.

Having served a Regulation 17 notice, valuation officers must allow any party to inspect and take extracts from the returns having been given a minimum of 24 hours' notice.

It should be noted that only the valuation officer can decide to use returns at a valuation tribunal hearing; another party cannot initiate their use. However, once valuation officers have served their notice of intention to use returns, any party may serve, in effect, a counter notice. This counter notice may specify a number of hereditaments which are comparable in character to the appeal hereditament. However, this number may not exceed four or, if greater, the number of hereditaments in the valuation officer's notice. The valuation officer must then allow any returns for the specified hereditaments to be inspected and also produce them at the valuation tribunal hearing if requested.

15.16.13 *Decisions*

Having heard the evidence, submissions and summing up, the tribunal members usually retire to consider their decision. They may then give their decision orally or reserve the decision to a later date, possibly after an inspection. A written decision notice with reasons must, however, be given.

Having made its decision, the valuation tribunal will, if necessary, order the valuation officer to alter the list in accordance with the decision. The valuation officer is required to do this within two weeks. It may also confirm the list entry and dismiss the appeal.

Valuation tribunals may determine a rateable value for a hereditament which is greater than the existing list entry and that proposed in the proposal.

Valuation tribunals can be requested on the written application of a party to 'review or set aside any decision made' if:

- the decision was wrongly made as a result of clerical error;
- a party did not appear and can show reasonable cause why not;
- the decision is affected by a decision of the High Court or the Upper Tribunal Lands Chamber in respect of the hereditament which was the subject of the valuation tribunal's decision.

If at a hearing any party does not appear then the tribunal can hear and determine the appeal in that party's absence. If the valuation officer is the only party to appear then the valuation tribunal can dismiss the appeal rather than decide the appeal. An appellant can make written submissions rather than appear in person to present the case orally.

The Valuation Tribunal Service has a website (https://valuationtribunal.gov. uk) which lists individual tribunals and future hearings and makes available decisions of the tribunals.

15.16.14 *Appeals from the valuation tribunal*

Appeals may be made from valuation tribunal decisions to the Upper Tribunal (Lands Chamber) ('UTLC') which was previously known as the Lands Tribunal. The UTLC is still commonly referred to as the 'Lands Tribunal' although, strictly, it is a new and different body.

Appeals may be dismissed if not made within four weeks of the date of the valuation tribunal decision. An appeal to the UTLC cannot be made by a party which did not appear at the valuation tribunal hearing. The hearing is a fresh hearing which allows new witnesses, arguments, etc., to be put. Costs are usually awarded and can be very substantial. As the President said in *H&B Foods* v *Johnson (VO)* [2014] UKUT 0458 (LC), the Tribunal 'must, deal with [the] appeal as a "de novo" consideration of the matters decided by the valuation tribunal'. The UTLC President, Deputy President and judges are all members of the Judiciary appointed by the Lord Chancellor or the Lord Chief Justice. There are also three members who are chartered surveyors rather than judges, who sit in a judicial capacity and hear appeals requiring specialist valuation expertise.

The UTLC President's role is to sit as a judge to hear appeals falling within the jurisdiction of the Chamber (particularly disputes connected with the valuation of land) but the President's responsibilities also include giving practice and procedure directions, career development of judges and members, maintaining and building high standards of case management and decision making, and ensuring the Chamber is run efficiently and effectively. Cases are allocated to members (judges and surveyor members) by the President according to their content. For the most part, members sit individually, but major cases are often heard by two members.

In *Stock Auto Breakers* v *Sykes (VO)* [2020] UKUT 0052 (LC) both parties introduced new evidence that was not available to the Valuation Tribunal for England and the question was raised whether, under the new 'Check, Challenge and Appeal' procedures, the UTLC was entitled to take such new evidence into account or whether it was bound by regulation 17A (admission of new evidence) of the Valuation Tribunal for England (Council Tax and Rating Appeals) (Procedure) Regulations 2009 (2009/2269) (see 15.16.12 above). In its judgement, the Tribunal considered it was not so restricted because the 2009 procedure regulations only applied to the VTE and not the UTLC which was governed by its own procedure rules. An Upper Tribunal hearing, as mentioned in H&B Foods above, is also a de novo hearing.

The Upper Tribunal has a searchable website of its decisions: https://landschamber.decisions.tribunals.gov.uk.

The UTLC decision is final on questions of valuation and fact but an appeal on a point of law may be made to the Court of Appeal. An appeal can be made only with the permission of the Chamber or, if refused, by the Court of Appeal, and permission will be given only where the proposed appeal raises some important point of principle or practice, or there is some other compelling reason. An application for leave to appeal must be made to the UTLC within one month of the decision being disputed (Rule 55(2) of The Tribunal Procedure (Upper Tribunal) (Lands Chamber) Rules 2010 (2010/2600)). Appeal may be possible beyond the Court of Appeal to the Supreme Court (formerly to the House of Lords) – the leave of the Court of Appeal needs to be obtained or, if refused, that of the Supreme Court needs to be obtained.

15.16.15 *Written representation and arbitration*

Regulation 29 of the Valuation Tribunal for England (Council Tax and Rating Appeals) (Procedure) Regulations 2009 (2009/2269), (2009/2269 allows valuation tribunals to dispose of appeals on the basis of written representations if all parties agree. This provision is rarely used in practice.

Regulation 4 of the Valuation Tribunal for England (Council Tax and Rating Appeals) (Procedure) Regulations 2009 (2009/2269), provides for the parties to refer disputes to arbitration rather than the valuation tribunal if they wish. An equivalent provision has existed for many years but has been little used.

15.17 Wales

For the 2023 Revaluation the system in Wales is very closely allied to that in England. For Wales this was brought in by The Non-Domestic Rating (Alteration of Lists and Appeals) (Wales) Regulations 2023. The Welsh Government has confirmed that 'Check' 'Challenge' 'Appeal' in Welsh will be 'Gwirio', 'Herio' ac 'Apelio'

The differences are quite small:

As in England penalties may be imposed by the valuation officer for knowingly, recklessly or carelessly providing the valuation officer with information in connection with a proposal which is false in a material particular. A single penalty figure of £200 applies rather than a lower and higher figure. See section 15.14.3 Valuation Officer's response.

There is no fee for an appeal to the Valuation Tribunal for Wales unlike England. See section 15.14.8 Appeals to the VTE.

The system only applies from 1 April 2023. A description of the system applying up to then appears in the fourth edition of this book at page 319.

Appeals are to the Valuation Tribunal for Wales.

The VTW's website is www.valuationtribunal.wales.

16 Rating administration, collection and enforcement

16.1 Rating administration

There are three main parties involved in the general day-to-day administration of the rating system, as follows:

- Billing authorities are local authorities with a duty to collect and enforce the collection of non-domestic rates and in this capacity they act on behalf of the Government. The rest of this chapter deals with the collection and enforcement issues which are the responsibility of the billing authorities.
- Valuation officers of the Valuation Office Agency (VOA), a government body, have the duty to prepare and maintain the rating lists.
- The Upper Tribunal and valuation tribunals for England and Wales are mainly concerned with determining appeals against the rating lists.

The main statutory provisions affecting the collection and enforcement of rates are contained in the Local Government Finance Act 1988 and the Non-Domestic Rating (Collection and Enforcement) (Local Lists) Regulations 1989 (SI 1989/1058), as amended.

Many billing authorities have contracted out the billing, collection and enforcement services to private organisations. In such cases the Local Authorities (Contracting Out of Tax Billing, Collection and Enforcement Functions) Order 1996 (1996/1880) authorises the contractor to exercise certain functions of the billing authority.

16.2 The annual rates bill

To be liable for rates, a person must be the occupier (or owner if the property is empty) of all or part of the hereditament which is shown for that day in a rating list. In addition, a ratepayer is not under a duty to pay rates until such a time as a rates bill has been properly issued by the billing authority.

A rates bill must be served on every ratepayer who is liable to pay rates for that year and a separate bill must be issued for each year. Where there is more

DOI: 10.1201/9781003297369-16

than one occupier of a property they will be jointly and severally liable for the bill. This means that each party is as equally liable as the other. However, where a bill is sent to more than one person it must be made clear that other bills have been sent to other persons and that they are all jointly and severally liable.

In *Atos IT Services Ltd* v *Fylde BC [2020]* EWHC 647 (QB) the High Court had to consider whether s.43 LGFA 1988 required a ratepayer to pay rates on a hereditament when it was only in part occupation, notwithstanding other parties occupied other parts. The liability in s.43 includes the requirement: 'on the day the ratepayer is in occupation of all or part of the hereditament.'

Liability was rejected by the court because, for liability, occupation had to be exclusive and could not be met where parts of the hereditament were let and occupied by a third party. It was necessary to find one occupier or a number of joint occupiers of the relevant hereditament before rateable occupation of the whole could arise. The Court drew support from *Verrall* v *Hackney LBC* [1983] QB 445 (CA) where it was said:

> *The occupation of land can be joint, and it is important to distinguish the case of a building in the hands of joint occupiers from that of a building of which the parts are let separately to several persons, each of whom is the occupier of the part let to him, and of that part only … If the whole building is rated, under one entry in the valuation list, and in the rate, as one indivisible rateable hereditament, no one tenant is liable for the rate on the whole, because he is not the occupier of the whole, nor can he be compelled to pay the rate on the part which he occupies, because there is nothing in the rate, or in the valuation list on which it is based, to show what is the value of that part.*

For liability to arise there needs to be a match between occupation and the description in the list. If the occupation is greater then there is a match, but if it is lesser then it could not be said there is occupation of premises which 'fulfilled the description in the list'. If a ratepayer occupies floors 2 and 3 of an office building but the assessment in the list is floors 2–4 then it cannot be required to pay because there is no entry in the list matching occupation. If, on the other hand, the entries in the list were three separate entries for floors 2, 3 and 4 then the company would be liable for the two entries on floors 2 and 3 because it did occupy them notwithstanding they should properly be shewn as a single entry in the rating list. The solution, as always, is accurate entries in the rating list.

16.3 Form and content of bills

The form of the rates bill or 'demand notice' is not laid down in the Local Government Finance Act 1988, but rather the contents of such a bill are prescribed by regulations.

The bill must include the following information:

- the name of the ratepayer;
- the amount of the bill (chargeable amount);
- details of the facility to pay by instalments;
- the address and description of the hereditament;
- the rateable value of the hereditament;
- the amount of the non-domestic rating multiplier for the year;
- where applicable, the period that the ratepayer was liable for unoccupied rates;
- where applicable, the period the ratepayer was eligible for mandatory or discretionary relief.

16.4 Service of notices

In section 16.2, reference was made to the fact that the rate bill must be 'served' on the ratepayer, and it is important that the correct procedures for service are adopted by the billing authority, otherwise they may not be able to enforce the bill.

The provisions for the service of notices on a ratepayer are contained in section 233 of the Local Government Act 1972.

The bill and other notices are properly served:

In the case of an individual ratepayer:

- by delivering it to the ratepayer; or
- by sending it by post to the ratepayer at his/her last known address, or at an address designated for receipt of the bill or other notice concerned; or
- by leaving it at any such address.

In the case of a body corporate:

- by delivering it to the body corporate or the secretary or clerk of that body; or
- by sending it by post to the body corporate or secretary or clerk of that body at its registered office, or at its principal office in the United Kingdom, or at an address designated for the receipt of the bill or other notice; or
- by leaving it at any such address.

In the case of a partnership:

- by delivering it to the partnership or to a partner or person having control or management of the partnership business; or
- by sending it by post to the partnership or person having control or management of the partnership business body at its principal office in the United Kingdom, or at an address designated for the receipt of the bill or other notice; or
- by leaving it at any such address.

In the case of a trust:

- by serving it on one of the trustees; or
- by delivering the bill to some person on the premises by virtue of which the ratepayer is subject to the rate, or fixing it on some conspicuous part of the premises.

Section 7 of the Interpretation Act 1978 applies to notices which are served by post and prescribes the time when such notices will be deemed to have been received by the ratepayers, regardless of whether they have actually received the bill or not.

The case of *Encon Insulation Ltd* v *Nottingham City Council* [1999] RA 382 highlighted the importance of following the correct procedures for the service of demand notices. The requirements include service of demand notices being as 'soon as practicable in the relevant year'. In this case the authority delayed serving the notice and was in breach of the requirement and as a consequence no rates were payable for that rate year.

16.5 Methods of payment

While ratepayers and the billing authorities can agree in writing to alter the dates when payments are due, most ratepayers will pay under a statutory instalment plan.

This plan will call for:

- the payment of the bill in ten uninterrupted monthly instalments;
- the months of payment to be specified in the bill and the date in the month for payment; usually the instalments will commence in April with no payments being made in February and March;
- the payment of equal monthly instalments;
- provision is made for the number of instalments to be varied where the ratepayer commences occupation during the course of the rate year.

The rate bill should be sent at least 14 days before the first payment is due, but failure to do so will not invalidate the bill.

It is important that the correct procedure is carefully followed. In the case of *Evans* v *Caterham and Warlingham Urban District Council* [1974] RA 65, a distress warrant was quashed because a defective instalment notice had been issued by the rating authority.

Where a ratepayer misses an instalment, a reminder will be issued. If payment is not made within seven days of such a notice the right to pay by instalments is lost and the ratepayer will have to pay the outstanding amount due. If the ratepayer misses a further instalment in that year the right to pay by instalments is automatically forfeited and the full amount becomes due.

16.5.1 Refunds

Provision is made in the Non-Domestic Rating (Collection and Enforcement) (Local Lists) Regulations 1989 (SI 1989/1058)), Regulation 9, for refunds of rates to be made where an overpayment has been made. A 'demand notice' is made at the start of the year and is based on the assumption that the circumstances that exist at that date will continue to do so throughout the remainder of the financial year. Obviously, circumstances can change and this Regulation allows for any refund that the ratepayer is due.

In *AEM (Avon) Ltd* v *Bristol City Council* [1998] RA 98, the billing authority made a refund to the company in the belief that the company had overpaid its rates. As it turned out they were mistaken in this belief and attempted to reclaim the refund and issued a liability notice in an attempt to recover the refund. The High Court quashed the order on the grounds that the notices demanding the return of the overpayment could not be issued under the Regulations and the only way of securing the return of the money was to issue civil proceedings.

16.6 Enforcement

The main means by which authorities enforce the payment of rates is through the magistrates' courts, as described in the following section. However, where any sum has become payable but has not been paid, the authority may seek to recover that payment in any court of competent jurisdiction: Regulation 20(2) and 20(1) of the Non-Domestic Rating (Collection and Enforcement) (Local Lists) Regulations 1989 (SI 1989/1058)).

16.6.1 Demand for payment

The first requirement for enforcement is that an appropriate demand has been made for payment. Following the provisions of the Limitations Act 1980, the action must be commenced within six years of the amount becoming payable. Regulation 5 in Part III to the Non-Domestic Rating (Collection and Enforcement) (Local Lists) Regulations (SI 1989/1058) sets down certain provisions in respect to timing:

> **a demand notice shall be served on or as soon as practicable after … 1st April in the relevant year**

What 'as soon as practicable' means and what happens if the billing authority delays was considered in North Somerset DC v Honda Motor Europe Ltd [2010] EWHC 1505 (QB). It seems late service does not result in automatic invalidity:

> *a failure to serve a Regulation 5 notice as soon as practicable does not result in automatic invalidity. Rather, the court determining any issue resulting from*

such a failure will have regard to the length of delay and the impact of that delay upon the ratepayer, in the context of the public interest in collecting outstanding rates. The greater the prejudice to the ratepayer flowing from the delay, the more likely will be the conclusion that Parliament intended invalidity to follow.

Late service needs to have resulted in prejudice to the ratepayer that is '*substantial and certainly not technical or contrived*' having regard to the countervailing public interest in the collection of rates, the interests of other ratepayers and the revenues of the local authority concerned.

16.6.2 Liability orders

Where a ratepayer fails to pay rates, it is normal for the billing authority to issue a reminder and advise the ratepayer that if the payment is not made within seven days, they will apply to the magistrates for a 'liability order'. The billing authority makes a complaint to the magistrates' court requesting a summons be issued, requiring the ratepayer to appear before the court to show why the outstanding amount has not been paid.

16.6.3 Summons

The summons can be served on the person by:

- serving it on the person;
- leaving it at the person's usual or last known place of abode or, in the case of a company, its registered office;
- sending it to the person at the usual or last known place of abode or, in the case of a company, its registered office;
- leaving it at or sending it by post to the person at the person's place of business;
- leaving it at or sending it by post to the person at an address given by the person as an address at which service of the summons will be accepted.

In *Ralux NV/SA* v *Spencer Martin* [1989] TLR 18/5/89, it was upheld that service of notices by fax met the appropriate requirements for services of notices and, in *R* v *Liverpool Justices, ex parte Greaves* [1979] RA 119, it was upheld that if the authority followed the prescribed procedure it was confirmed that the defence of not having received the notices was not available.

If the debtor fails to attend court, the application may be determined in the debtor's absence.

The magistrates, if they are satisfied that the rates are payable and that they have not been paid, will be obliged to grant the liability order. The order will include the costs of the billing authority.

The billing authority must satisfy the court that:

- an entry appears in the rating list;
- the sums have been calculated, demanded or notified in accordance with the statutory provisions;
- full payment of the amount due has not been made by the due date;
- a second notice has been issued;
- the sum has not been paid within seven days of the second notice;
- the summons has been duly served for the remaining years rates which are outstanding;
- the full sum has not been paid.

The defence available to the ratepayer includes:

- that the property in respect of which the amount is due did not appear for the relevant period on the rating list;
- that the amount due has not been demanded or notified in accordance with the statutory provisions;
- that the amount has been paid;
- that the amount has not been calculated in accordance with the correct statutory provisions, or is otherwise incorrect;
- that the person who is alleged to be jointly or severally liable was not in that relationship with the defaulting ratepayer at the time the debt was incurred.

In *Evans* v *Brook* (1959) 52 R&IT 321, it was held that the justices have the powers to consider whether the amount claimed is due and whether there is an entitlement to any reliefs. The justices cannot enquire into any matter which could be the subject of an appeal against the assessment, and consequently they cannot enquire into matters of valuation and value (see *County & Nimbus Estates Ltd* v *Ealing London Borough Council* [1978] RA 93). In *Hackney Borough Council* v *Mott and Fairman* [1994] RA 381, it was confirmed that the magistrates' court had no jurisdiction to determine whether an entry in the rating list was invalid and thus this was not available as a defence.

A liability order is an all-purpose order obtained in a magistrates' court by the billing authority against the ratepayer and provides the authority with distress or insolvency powers.

16.7 Distress

The amount specified in the liability order can be recovered by 'distress' of any of the debtor's property in England and Wales but not Scotland. This means that the authority may seize goods belonging to the ratepayer and sell them in order to pay the outstanding debt. If the debtor offers to pay prior to goods being seized then this money must be accepted and proceedings ceased.

Regulations govern the goods that can be distrained, which generally include only goods owned by the person concerned. They specifically exclude items such as 'clothing, bedding, furniture, household equipment and provisions as are necessary for satisfying the basic domestic needs of the debtor and his family'.

The debtor has the right to appeal against the exercise of distress to the magistrates' court.

16.8 Commitment to prison

The billing authority may apply for a warrant to commit the debtor to prison, but this can be applied for only if they have first tried to levy distress and there were insufficient goods with which to discharge the debt.

The court may issue a warrant only if it is satisfied that the failure to pay is due to the debtor's wilful refusal to pay or culpable neglect.

The maximum period of imprisonment is three months.

16.9 Rate liability

The amount of rate liability will depend on whether:

- the property is occupied;
- any exemptions or reliefs apply;
- transitional relief applies.

16.9.1 The Uniform Business Rate

The Uniform Business Rate (UBR) is a single tax rate that applies to all properties in England and Wales, with each country having its own individual tax rates as shown in Table 16.1.

Table 16.1 UBRs for England and Wales

Year	England	Wales	England small hereditament
2010/11	41.4p	40.9p	Reduced by 0.7p for small hereditaments
2011/12	43.3p	42.8p	Reduced by 0.7p for small hereditaments
2012/13	45.8p	45.2p	Reduced by 0.8p for small hereditaments
2013/14	47.1p	46.4p	Reduced by 0.9p for small hereditaments
2014/15	48.2p	47.3p	Reduced by 1.1p for small hereditaments
2015/16	49.3p	48.2p	Reduced by 1.3p for small hereditaments
2016/17	49.7p	48.6p	Reduced by 1.3p for small hereditaments
2017/18	47.9p	49.9p	Reduced by 1.3p for small hereditaments
2018/19	49.3p	51.4p	Reduced by 1.3p for small hereditaments
2019/20	50.4p	52.6p	Reduced by 1.3p for small hereditaments
2020/21	51.2p	53.5p	Reduced by 1.3p for small hereditaments
2021/22	51.2p	53.5p	Reduced by 1.3p for small hereditaments
2022/23	51.2p	53.5p	Reduced by 1.3p for small hereditaments
2023/24	51.2p	53.5p	Reduced by 1.3p for small hereditaments

The rate is set each year by a statutory instrument and the amount it can increase in any one year is restricted by statute.

16.10 Occupied property

The calculation of the chargeable amount or amount payable is by use of the following formula:

$$\frac{(A \times B)}{C}$$

where:

- *A* is the rateable value.
- *B* is the non-domestic rating multiplier (UBR).
- *C* is the number of days in the financial year.

This will give the liability for the chargeable day. This is then multiplied by the number of chargeable days that the ratepayer was in occupation of the property.

However, if the property is occupied by a charity then the calculation is:

$$\frac{(A \times B)}{(C \times 5)}$$

For the definition of a charity, see section 67(10) of the Local Government Finance Act 1988. The effect is to make the charity liable for 20% of the normal rate.

The above calculation is for those properties and instances where the transitional relief phasing provisions do not apply.

Where a property is eligible for reliefs (see section 5.3) the method of calculation of the rate liability will be affected by the provisions relevant to that relief.

16.11 Transitional arrangements

16.11.1 Transitional arrangements 2023 – England

In order to protect businesses from large changes in rate liability between revaluations, in 1990 the Government introduced a system for phasing in changes to a ratepayer's rate liability from one year to another. Prior to 2023 the fundamental principles of all the transitional arrangements schemes has been that they are meant to be self-financing schemes, that is, part of the amount which would be saved by persons who have a decreased liability is used to finance the phasing in of the rates bills for those persons who have an increased liability.

From 2023 the scheme has changed in that:

- Where a hereditaments rate liability has decreased then that reduced liability will be payable with effect from 1 April 2023.
- Hereditaments where the rate lability has increased then that increase in lability will be phased in under the provisions described below.

The rules for the calculations are contained in the Non-Domestic Rating (Chargeable Amounts) (England) Regulations 2022 (2022/1403). These regulations contain provisions based on the 2023 rating lists being in force for a *five*-year period and in order to comply with current legislation special provision has been made for an additional amount to be payable in the final year of the 2023 rating lists. However, at the time of writing the Government has confirmed its intention that the next revaluation will take place on 1 April 2026, which would render the above provision inoperative. This will require primary legislation which is to be introduced.

16.11.2 *Application of relief*

Transitional relief is only available to those hereditaments and in those instances specified in the Non-Domestic Rating (Chargeable Amounts) (England) Regulations 2022 (2022/1403).

These hereditaments are termed 'defined hereditaments'.

16.11.3 *Defined hereditament (Regulation 5)*

To be a defined hereditament:

- a property must be shown in a local non-domestic rating list on 31 March 2023 at a rateable value of more than zero; and
- the 2023 rateable value must be greater than zero; and
- for each day after the 31 March 2023 and before the relevant day a property must have a rateable value of more than zero.

Special provisions apply for properties altered after 1 April 2022 and for properties that are split or merged (see of the Regulations).

16.11.4 *Notional Chargeable Amount (Regulation 7)*

Where the property is a defined hereditament then the property is subject to 'transitional relief'. The NCA which is the amount of rates which would be payable in the current year if transitional relief did not apply, needs to be calculated as follows:

$$A \times D$$

where:

- *A* is the rateable value shown for the hereditament for 1 April 2023 in the list.
- *D* is the small business non-domestic rating multiplier.

16.11.5 Base Liability (Regulations 8 and 9)

16.11.5.1 For the year 2022/2023 (Regulation 8)

$$Y \times Z$$

where:

- *Y* is, in the case of a hereditament shown in a local list for 31 March 2023, the rateable value shown for the hereditament for that date in a local list; and
- *Z* is the small business non-domestic rating multiplier for the financial year beginning on 1 April 2022.
- For all other years (Regulation 9).

$$BL \times AF$$

where:

- *BL* is the Base Liability for the hereditament for the relevant year immediately preceding the year concerned; and
- *AF* is the Appropriate Fraction, as found in accordance with Regulation 10, for the relevant year immediately preceding the year concerned.

16.11.6 The Appropriate Fraction (Regulation 10)

The AF is the mechanism which controls the rate of increase or decrease in liability. It is made up of two components: a percentage change in the rate liability for each year, *Q*, and *X* which will vary depending on the location of the property and its rateable value.

The AF is found by multiplying the two components together and dividing by 100 as shown in the following formula:

$$(X \times Q)/100$$

16.11.6.1 'X' (Regulation 10)

In the value of the AF, X is dependent on a number of different factors:

1   Where the NCA for the hereditament for the relevant year exceeds the base liability for the hereditament for the year, and the hereditament is shown in a list and the rateable value shown in that list for that hereditament for 1 April 2023 is more than £100,000, then:

    a   for the relevant year beginning on 1 April 2023, X is 130;
    b   for the relevant year beginning on 1 April 2024, X is 140;
    c   for the relevant year beginning on 1 April 2025, X is 155;

2   Where the NCA for the hereditament for the relevant year exceeds the base liability for the hereditament for the year, and:

    a   the hereditament is shown in a local list in Greater London and the rateable value shown in that list for that hereditament for 1 April 2023 is more than £28,000 and less than or equal to £100,000; or
    b   the hereditament is shown in the central list or in a local list outside Greater London and the rateable value shown in that list for that hereditament for 1 April 2023 is more than £20,000 and less than or equal to £100,000; then

        a   for the relevant year beginning on 1 April 2023, X is 115;
        b   for the relevant year beginning on 1 April 2024, X is 125;
        c   for the relevant year beginning on 1 April 2025, X is 140;

3   Where the NCA for the hereditament for the relevant year exceeds the base liability for the hereditament for the year, and:

    a   the hereditament is shown in a local list in Greater London and the rateable value shown in that list for that hereditament for 1 April 2023 is less than or equal to £28,000; or
    b   the hereditament is shown in the central list or in a local list outside Greater London and the rateable value shown in that list for that hereditament for 1 April 2023 is less than or equal to £20,000.

For the purposes of this regulation, for a hereditament mentioned in paragraph (5):

    a   for the relevant year beginning on 1 April 2023, X is 105;
    b   for the relevant year beginning on 1 April 2024, X is 110;
    c   for the relevant year beginning on 1 April 2025, X is 125.

'Q' (Regulation 10)
For the rate year commencing 1 April 2023, Q is 1.0 but for subsequent relevant years in the relevant period, Q is the amount found by applying the formula:

$$\frac{D(1)}{D(2)}$$

where:

- $D(1)$ is the small business non-domestic rating multiplier for the relevant year concerned.
- $D(2)$ is the small business non-domestic rating multiplier for the financial year which precedes the relevant year concerned.
- Q, if not a whole number, shall be calculated to three decimal places only:
  - adding one thousandth where there would be more than five ten-thousandths;
  - ignoring ten-thousandths where there would be five, or less than five, ten-thousandths.

16.11.7 *The chargeable amount*

The amount of rates payable is now calculated by multiplying the BL by the AF and dividing the result by the number of days in the year. The formula is shown as:

$$\frac{(BL \times AF)}{C} + U$$

where: U is the amount found by applying the formula:

$$\frac{(B-D) \times N}{C}$$

- B is the non-domestic rating multiplier for the relevant year in which the chargeable day falls.
- C is the number of days in the relevant year.
- D is the small business non-domestic rating multiplier for the relevant year in which the relevant day falls.
- N is the rateable value shown in a list for the hereditament for the day.

Special provisions under Regulation 10 apply where the hereditament is eligible for certain reliefs.

16.12 Wales

Transitional Arrangements for Wales are contained in the Non-Domestic Rating (Chargeable Amounts) (Wales) Regulations 2022 (2022/1350) which came into force on 31 December 2022.

The transitional scheme in Wales will last for two years and will apply to all ratepayers where their bill has increased by more than £300 due to the revaluation.

The Regulation starts by defining some common terms used within the Regulations.

16.12.1 *Defined hereditament (Regulation 3)*

To be classed as a 'defined hereditament' the hereditament must be shown in the rating list:

- on 31 March 2023;
- on the relevant day;
- every day between 31 March 2023 and the relevant day.

16.12.2 *Base liability (Regulation 4)*

The base liability depends on whether the property has the benefit of Small Business Relief or not. Where on the 31 March 2023, sections 43(4B)(b) (SBR) applied to the defined hereditament BL is defined as:

$$\frac{A \times 0.535}{E}$$

where:

- *A* is the rateable value shown in the list on 31 March 2023.
- 0.535 is the NNDR multiplier for the year 2022/23.
- *E* is the amount of '*E*' that applies to the defined hereditament on 31 March 2023 in accordance with the Non-Domestic Rating (Small Business Relief) (Wales) Order 2017 (2017/1229) as amended .

In all other cases BL is defined as:

$$A \times 0.535$$

16.12.3 *Notional Chargeable Amount (NCA) (Regulations 6 and 7)*

Where on 1 April 2023 the property was entitled to SBRR, the NCA is:

$$\frac{A \times B}{E}$$

where:

- *A* is the rateable value shown in the list on 1 April 2023.
- *B* is the non-domestic rate multiplier for the year beginning 1 April 2023.
- *E* is the amount of '*E*' that applies to the defined hereditament on 1 April 2023 in accordance with the Non-Domestic Rating (Small Business Relief) (Wales) Order 2017 (2017/1229) as amended.

In all other cases the NCA is calculated by the formula:

$$A \times B$$

16.12.4 *Hereditaments to which the rules apply (Regulation 9)*

The rules apply to a 'defined hereditament' which was shown in the rating list on 31 March 2023, *and* where the NCA is greater than:

$$BL + 300$$

and on the relevant day the occupier was the same person as the occupier on 31 March 2023.

16.12.5 *Hereditaments to which the rules DO NOT apply (Regulation 10)*

- a 'defined hereditament' where the chargeable amount would have been £0; or
- an apportionment has been made for a partly occupied hereditament.

16.12.6 *Rules for finding the chargeable amount (Regulations 11–15)*

To calculate the rates payable the formulas below are applied for the relevant year:

For the year commencing 1 April 2023:

$$\frac{(NCA - BL) \times 0.67}{365}$$

For the year commencing 1 April 2024:

$$\frac{(NCA-BL)\times 0.34}{365}$$

16.12.7 *The amount of 'E'*

Regard must be had to the Non-Domestic Rating (Small Business Relief) (Wales) Order 2017 (2017/1229) as amended by the Non-Domestic Rating (Small Business Relief) (Wales) (Amendment) Order 2018 (2018/1192) which lays down the amount of '*E*' depending on the use to which the property is put and its rateable value.

Table 16.2 The amount of '*E*'

(1)	(2)	(3)	
Article	Rateable value £	Amount of E	
7 (rateable value)	0 to 6,000		5,000,000
7 (rateable value)	6,001 to 12,000	Calculated in accordance with paragraph (3) of this article	
8 (child care)	0 to 100,000	5,000,000	
9 (post office)	0 to 9,000		5,000,000
9 (post office)	9,001 to 12,000	2	

16.13 Examples

Worked examples for England and Wales are contained in Appendix 1

17   The empty property rate

17.1 Introduction

Since 1601, rating has been a tax on the occupation of land. This had a degree of simplicity – if there was an occupier then rates could and should be sought from that person. If the hereditament was unoccupied then there was no liability.

Indeed, the very idea of an unoccupied rate can be seen as anomalous and beyond the very basis of the tax.

This non-liability remained the case until 1966 when the Local Government Act 1966 introduced an unoccupied rate though the City of London did levy its own empty rate under the City of London and Sewers Act 1848 with half the General Rate payable when property became empty.

While the Local Government Finance Act 1988 provides for a non-domestic rate to be payable by owners of unoccupied property, this does not mean liability arises whenever a property becomes empty. Owners of certain types and categories of hereditament are not liable to this rate and there is a period of grace, normally of three months, to allow for normal periods of vacancy between occupiers.

The provisions for liability and exemption are a little complicated. Since 1966 a body of case law has developed to clarify liability. Legislation has also been enacted to determine when newly completed, but empty, buildings are truly complete and can therefore be regarded as vacant hereditaments for the purposes of the unoccupied rating provisions.

17.2 Who is liable?

Section 45(1) of the Local Government Finance Act 1988 provides the basic requirements for liability. Liability is placed on the owner of vacant hereditaments. Clearly, the occupier cannot be liable as is normally the case in rating because, by definition, a vacant hereditament does not have an occupier. 'Owner' does not necessarily mean the freeholder, as section 65(1) defines 'owner' as 'the person entitled to possession' and this may well be a lessee deriving title from the freeholder. A not uncommon situation is for a tenant

DOI: 10.1201/9781003297369-17

who has had to close its business and vacate the premises and finds themselves liable for unoccupied rates rather than the landlord because the lease has not ended and it is the tenant who is still entitled to possession.

In *Kingston upon Thames London Borough Council* v *Marlow* [1995] RA 87, a landlord was held to be liable for unoccupied rates as the person entitled to possession where, following a dispute with his tenant, the tenant had accepted the forfeiture of the lease and vacated the property.

Section 65(5) prevents premises from being regarded as occupied merely because there is plant, machinery or equipment such as furniture on the premises which were previously used in the hereditament, when it was last used, or are intended for future use in the hereditament.

Such ratepayer owners are liable providing certain conditions are fulfilled:

1. No part of the hereditament is occupied. If part is occupied then following the long-established rating maxim that 'occupation of part is occupation of the whole', the billing authority would regard the whole hereditament to be occupied.
2. The ratepayer owner is the 'owner' of the whole hereditament.
3. The hereditament is shown in the local rating list. This is, of course, the same requirement as for occupied rating.
4. The hereditament falls within a description prescribed by the Secretary of State. In practice this has been done in the Non-Domestic Rating (Unoccupied Property) (England) Regulations 2008 (SI 2008/386) and for Wales the Non-Domestic Rating (Unoccupied Property) (Wales) Regulations 2008 (SI 2008/2499) (W.217).

These Regulations provide a twofold test:

1. To be subject to unoccupied rating, the hereditament must be a 'relevant non-domestic hereditament' as defined in the Regulations. The use of the phrase 'relevant non-domestic hereditaments' has been described as an unfortunate piece of drafting as exactly the same phrase is used in defining what hereditaments should appear in rating lists (section 42(1)(b)) but the meaning is different.
2. The Regulations do not list classes of hereditament whose owners are liable to the unoccupied property rate. Instead, in Regulation 4, the Regulation sets out conditions which, if any are satisfied, will take a hereditament outside the ambit of the unoccupied property rate. This means that if one of the conditions is satisfied and the hereditament is vacant, no unoccupied property rate is payable.

The following conditions ((a)–(m), set out in Regulation 4) outline who is liable, namely any hereditament:

(a) which ... has been unoccupied for a continuous period not exceeding three months (26 weeks in Wales);

A period of grace of three months' unoccupancy has always been allowed before the unoccupied rate becomes payable. This is intended to allow a reasonable period in which to complete a sale or new letting and for the new occupier to move in. A longer period of six months is allowed for factories and warehouses (see condition (b)).

The Regulations provide various rules to prevent owners getting around the Regulations by briefly re-occupying and then vacating or re-arranging the property into different hereditaments. Regulation 5 prevents owners from occupying for one day at the end of the three months in order to start another three months running and thereby avoid the unoccupied rate. It provides that any period of re-occupancy of less than six weeks is to be ignored when calculating the expiry of the three-month period.

A new three-month void period does not run merely because of a change of owner ratepayer. The courts have also held that where a vacant hereditament is divided after the three-month 'void period' without structural alterations into two or more hereditaments, no new three-month void period is allowed. Unoccupied rates become payable straight away if the original three-month 'void period' has already been allowed to the undivided hereditament: *Brent London Borough Council* v *Ladbroke Rentals* [1981] RA 153; (1980) 258 EG 857; (CA).

In *Makro Properties Ltd and another* v *Nuneaton and Bedworth Borough Council* [2012] EWHC 2250 (Admin), the High Court considered whether the storage of a few pallets of documents in a vacant cash and carry warehouse, deliberately done with a view to creating six weeks of rateable occupation thereby allowing a further six-month void period, was indeed rateable occupation. The documents comprised paperwork that the appellant was obliged by law to retain and took up only 0.2% of the premises' floor space. After an extensive review of the authorities, the Court said the proper approach was to consider both use and intention. Evidence of intention, together with only slight use, could lead to an inference of occupation.

Similarly in *Sunderland City Council* v *Stirling Investment Properties LLP* *[2013]* All ER 325 the High Court decided that the use by tenants of simply a Wi-Fi transmission box sending marketing and advertising texts to mobile phones was sufficient to constitute rateable occupation and permit a six-week period of occupation to be established.

In *R. on the application of Principled Offsite Logistics Ltd* v *Trafford Council* *[2018]* EWHC 1687; [2018] RA 499 the High Court appear to have taken the Makro case rather further, dispensing with a need for stored documents or other material to have a value in themselves. The company, in the case, was in the business of providing a rates relief service by taking leases and occupying premises for the necessary six-week period to enable a further three-month empty rates period to be obtained. Trafford argued the goods stored by POL were not stored for any business purpose, beyond creating occupation to avoid rates and POL was not really in occupation as there was no benefit gained. The High Court, accepting that ethics and morality needed to be excluded from consideration as it had to act as a court of law not morals, considered there was no requirement to have a benefit derived from storing goods, because the thing of value to the possessor could simply be occupancy as such. 'I find no concept within the meaning of the word requiring a purpose or motive beyond that of the occupation itself.' In essence the four tenets of rateable occupation were present.

In *R. on the application of the Secretary of State For Health & Social Care on behalf of Public Health England* v *Harlow District Council* [2021] EWHC 909, the High Court considered a further case on whether premises were in occupation for the 6 weeks necessary to give another three months void period for the EPR. It very much followed Principled Logistics and went so far as to set out a series of legal propositions in annex A to the judgement to cover such situations and avoid further cases.

The owners had stored various boxes that the billing authority said contained documents of no value which could be disposed of rather than valuable records needing to be retained. The High Court was not much interested in this, being content there was value in occupying for rates avoidance and the owners were undoubtedly storing something it wished to store. It did not consider there was a requirement for the use to be of significant value to the possessor independently of the gaining of a rates exemption. *'It does not matter if the storage is whimsical or eccentric, for example storage of a collector's items or of redundant items.'*

The High Court accepted counsel for the owners' view that, 'actual use of the property, even minimal use as in this case, combined with an intention to occupy it is sufficient for occupation, whether the motive is rates mitigation or any other motive'.

It can be seen from the numerous cases mentioned above that owners have found it worthwhile to arrange short periods of re-occupancy in order to secure a further three month void period when empty property rates are not payable. At the time of preparing this edition the government had announced a consultation to explore the causes of, and potential measures to combat, avoidance, evasion, and what it describes as poor rating agent behaviour within the NDR system particularly in relation to this practice. Its purpose was to:

- consult on specific measures to reform Empty Property Rating, to address known avoidance schemes;
- gather evidence on wider avoidance and evasion practises within NDR, and seek views on whether BAs had sufficient powers and information to combat them;
- gather evidence on 'rogue' rating agent behaviour and seek views on how the government could address any problems.

Possible reforms suggested include:

- Changing the existing 6 week void period to 3 or 6 months
- Repeated short periods of re-occupation not allowing further 3 months periods before Empty Property Rates become payable

- Altering what constitutes occupation for Empty Property Rate purposes
- Altering the 'when next in use' charity exemption from Empty Property Rates.

At the time of writing it is not known what eventual changes, if any, will be made to the legislation.

A hereditament that has not previously been occupied is treated as becoming unoccupied on the day determined under Schedule 4A to the Local Government Finance Act 1988 (completion notice procedure), or where this does not apply, on the day for which the hereditament is first shown in the local rating list.

(b) which is a qualifying industrial hereditament that ... has been unoccupied for a continuous period not exceeding six months;

Up until 1 April 2008, empty property rates were not chargeable on qualifying industrial hereditaments. Since that date they have been brought within the ambit of the empty property rate but are given an extra three months void period compared to other properties, i.e. a six-month period.

Qualifying industrial hereditaments are not just factories but also include warehouses though not retail warehouses. The definition is complicated.

A 'qualifying industrial hereditament' is any hereditament other than a retail hereditament in relation to which all the buildings comprised in the hereditament are:

a constructed or adapted for use in the course of a trade or business; and
b constructed or adapted for use for one or more of the following purposes, or one or more such purpose and one or more purposes ancillary thereto:

i   the manufacture, repair or adaptation of goods or materials, or the subjection of goods or materials to any process;
ii   storage (including the storage or handling of goods in the course of their distribution);
iii the working or processing of minerals;
iv   the generation of electricity.

'Retail hereditament' means any hereditament where any building or part of building comprised in the hereditament is constructed or adapted for the purpose of the retail provision of:

- goods; or
- services, other than storage for distribution services, where the services are to be provided on or from the hereditament.

The case of *Brent London Borough Council* v *Ladbroke Rentals* [1981] RA 153; (1980) 258 EG 857; (CA) examined the definition of 'relevant non-domestic hereditament' and the term 'storage'. The property concerned had been a bus garage and the owners claimed that, as such, it was used for the 'storage' of buses over-night and therefore exempt from unoccupied rates. The High Court rejected this argument on the grounds that storage did not include the parking of buses as this was a 'normal incident of their every day use' in running a bus service.

In *Southwark London Borough Council* v *Bellaway Homes and the Post Office Ltd* [2000] RA 437, a vacant post office had been acquired by a company for redevelopment of the site. The Court held that this was a qualifying industrial hereditament as under the Regulations:

- 'goods' included 'mail';
- the handling of mail was the same as 'subjection of goods or materials to any process'; and
- the post office 'handled goods in the course of their distribution'.

It is important to note there are two limbs to what is a qualifying industrial hereditament. It must both have been used for trade or business and have been constructed or adapted for one of the mentioned purposes.

It is a matter for the billing authority whether a property is a qualifying industrial hereditament and the description in the local rating list is not conclusive of this.

(c) whose owner is prohibited by law from occupying it or allowing it to be occupied;

There are a number of reasons why an owner may be prevented from occupying a property. If the owner is prevented then the unoccupied property rate cannot be levied.

One example is where there is not a satisfactory means of escape in the event of fire. In *Tower Hamlets* v *St Katherine by the Tower* [1982] RA 261, the property being considered did not have a fire escape. Under the London Building Acts (Amendment) Act 1939 a building was required to have a means of escape in case of fire. It was held there was no liability to unoccupied rates as the occupation of the property was prohibited by law.

In *Regent Lion Properties Ltd* v *Westminster CC* [1989] RA 190; [1990] RA 121; (CA); (1989) 154 JP 49; [1990] 04 EG 131, the appellants were prohibited from continuing refurbishment works until various remedial measures had been undertaken relating to loose brown asbestos found in the atmosphere of the building. The prohibition notice was issued under s.22 of the Health and Safety at Work Act 1974. 'The effect of this notice was, as a matter of common sense and for all practical purposes, a prohibition against any beneficial occupation of the building.'

The lack of planning permission for hereditaments previously used as offices and awaiting refurbishment for residential use was found by the House of Lords in *Hailbury Investments Ltd* v *Westminster City Council* [1986] RA 187 not to be a prohibition against occupation. While the premises were described in the list as 'offices' it was accepted they could have legally been used for residential purposes. The requirement for exemption was that any occupation of the 'hereditament' was prevented and not merely occupation for the particular mode and category of use indicated by the description in the rating list was prohibited. The ratepayers were not prohibited from occupying the hereditaments in question but were only prohibited from using them as offices.

In *Pall Mall Investments (London) Ltd* v *Gloucester City Council* [2014] EWHC (Admin) 2247, the owners argued that occupation of two office buildings was prohibited by law because the interiors were in poor condition and occupation as offices would have put the owners in breach of the Health and Safety at Work Act. The Court decided there was no actual prohibition of occupation. The owners could take occupation to repair or for purposes that did not breach health and safety requirements.

(d) which is kept vacant by reason of action taken by or on behalf of the Crown or any local or public authority with a view to prohibiting the occupation of the hereditament or to acquiring it;

This includes the compulsory acquisition of property.

A planning restriction, however, does not necessarily prevent occupation. For example in *Hailbury Investments Ltd* v *Westminster City Council* [1986] RA 187, planning restrictions prevented the use for office purposes of four empty hereditaments described in the list as offices. The hereditaments were not exempt from unoccupied rates since occupation was not prevented as such; it was merely limited by the planning restriction.

The expiry of a planning permission would also not seem to be a prohibition on occupation. In *Regent Lion Properties Ltd* v *Westminster CC* [1989] RA 190; [1990] RA 121; (CA); (1989) 154 JP 49; [1990] 04 EG 131, the judge observed, without deciding the case on this ground, that it would have been possible to put the premises to some commercial use and even if an enforcement notice was served its use could have legally continued until determination. So use was not actually prevented by planning. The judge also suggested it was inconceivable that the premises would not be granted planning permission for some commercial use and, indeed, it ultimately received permission for use as a snooker hall.

In the same case, *Regent Lion Properties Ltd* v *Westminster CC* [1989] RA 190; [1990] RA 121; (CA); (1989) 154 JP 49; [1990] 04 EG 131, the appellants were prohibited by the local authority from continuing refurbishment works until various remedial measures had been undertaken relating to loose brown asbestos found in the atmosphere of the building. The effect of the notice was a prohibition against any beneficial occupation of the building and therefore it was kept vacant by reason of action of the local authority. Exemption was therefore found both under condition (d) and, as mentioned, under condition (c).

(e) **which is the subject of a building preservation notice within the meaning of the Planning (Listed Buildings and Conservation Areas) Act 1990 or is included in a list compiled under section 1 of that Act;**

This is an important exemption. It applies just as much to a listed office building modernised to the very highest standard as it does to a fourteenth century unmodernised museum building. It is important to note that if only part of the hereditament is listed then the whole will be liable to unoccupied rates. In *Debenhams Plc* v *Westminster City Council* [1987] AC 396, the ratepayers argued that a building joined by a tunnel and bridge to a building which was listed was also included in the listing as an 'object or structure fixed to the building'. The House of Lords did not construe 'structure' as including a complete building and was intended to be limited to structures ancillary to the building itself. It seems from this that an extension can, if subsidiary and ancillary, be treated as listed with the main building. However, in *Richardson Developments Ltd* v *Birmingham CC* [1999] All ER (D) 34, the High Court did not accept extensions which were larger than the original building qualified. It took the view that the listed building had, in effect, become ancillary to the newer structures.

Rather surprisingly the wording of the exemption suggests that a hereditament can be 'listed' but as is clear from the planning acts only a 'building' can be listed. This difficulty has been resolved by the courts, perhaps most succinctly by Mr Justice Irwin in *G E Bowra Group Ltd* v *Thanet DC* [2007] All ER (D) 184 (Apr), 'The Regulation means that where a hereditament is wholly included within a building which is listed, exemption should follow.'

Listing can also apply to 'any object or structure within the curtilage of a building which ... forms part of the land' but this applies only where they have done so since before 1 July 1948. So it appears that if the owner of a large listed building takes over an adjoining post-war small building which was not listed and uses it with the large building in such a way that they together constitute a single hereditament then the listed building exemption from the empty property rate is unlikely to apply in the event of vacation.

(f) **which is included in the schedule of monuments compiled under section 1 of the Ancient Monuments and Archaeological Act 1979;**
(g) **where its rateable value is less than a set RV figure;**

To save local authorities chasing comparatively trivial sums, the Regulations have always provided for empty property rates not to be liable below a rateable value threshold. For the 2017 and 2023 rating lists this is £2,900.

(h) where the owner is entitled to possession only in his capacity as the personal representative of a deceased person;

(i) where, in respect of the owner's estate, there subsists a bankruptcy order within the meaning of section 381(2) of the Insolvency Act 1986;

(j) whose owner is entitled to possession of the hereditament in his capacity as trustee under a deed of arrangement to which the Deeds of Arrangement Act 1914 applies;

(k) whose owner is a company which is subject to a winding-up order made under the Insolvency Act 1986 or which is being wound up voluntarily under that Act;

In *Hurstwood Properties (A) Ltd* v *Rossendale Borough Council [2021]* UKSC 16 the owners let properties to non-trading special purpose vehicles (SPVs) that lacked assets or liabilities which were then put into liquidation. Because of the exemption from empty property rates where the owner was insolvent no rates would be payable if the device succeeded

The local authorities sought to establish either the leases to the SPVs were ineffective to make the SPV the 'owner' within the meaning of the legislation; or, alternatively, the separate legal personality of the SPV should itself be ignored by 'piercing the corporate veil'.

The Supreme Court, adopting a purposive approach, decided identifying 'the person entitled to possession' in s.65(1) of the 1988 Act as the person with the immediate legal right to possession would defeat the purpose of the legislation. The schemes were designed to ensure the SPV lessee had no real or practical control over whether the property was occupied or not and such control remained at all times with the landlord.

> *Parliament cannot sensibly be taken to have intended that 'the person entitled to possession' of an unoccupied property on whom the liability for rates is imposed should encompass a company which has no real or practical ability to exercise its legal right to possession and on which that legal right has been conferred for no purpose other than the avoidance of liability for rates.*

The landlord was therefore the owner for EPR not the SPV.

The Supreme Court did not consider the corporate veil should be pierced and upheld the Court of Appeal on this point.

(l) whose owner is a company in administration within the meaning of paragraph 1 of Schedule B1 to the Insolvency Act 1986 or is subject to an administration order made under the former administration provisions within the meaning of article 3 of the Enterprise Act 2002 (Transitional Provisions) (Insolvency) Order 2003;

(m) whose owner is entitled to possession of the hereditament in his capacity as liquidator by virtue of an order made under section 112 or section 145 of the Insolvency Act 1986.

17.2.1 *Charities and amateur sports clubs*

A further exemption, in effect, was introduced by the Rating (Empty Properties) Act 2007 inserting section 45A into the Local Government Finance Act 1988. This provides for hereditaments to be 'zero rated' notwithstanding they would otherwise come within the ambit of empty property rates and are not exempt providing:

- The ratepayer is a charity/trustees for a charity or a registered community amateur sports club and it appears when next in use the hereditament will be wholly or mainly used for charitable purposes or the purposes of a community amateur sports club, respectively.
- In the case of a charity, the likely future use can be by other charities and in the case of sports clubs it can be that the future use includes use by other such registered clubs as well as the ratepayer club.

The effect of this relief is that if it is established that the owner is a charity and the next likely use is charitable by a charity, then no empty property rates are payable.

What 'charitable purposes' actually are has been considered by the courts in relation to the 80% mandatory and 20% discretionary relief from occupied rates. In *Oxfam* v *City of Birmingham District Council* [1976] AC 126, the House of Lords considered they were 'those purposes or objects the pursuit of which make it a charity – that is to say in this case the relief of poverty, suffering and distress'. It was content that relief also extended to ancillary uses but not to simply raising money for a charity because this was not a charitable purpose even though it facilitated the work of a charity. Charity shops used mostly to sell donated goods are treated as being used for charitable purposes by section 64(10) of the Local Government Finance Act 1988.

In *Kenya Aid Programme* v *Sheffield City Council* [2013] RA 75, the High Court rejected the ratepayer's (a charity) contention that once it was established, the charity was in rateable occupation and it necessarily followed that the only use to which the premises were being put was charitable. The High Court held the 'wholly or mainly' phrase did not refer simply to the purpose of the use but that the magistrates' court was entitled to look at the extent or amount of actual use. However, it was not correct to look at how efficiently the space was being used.

In *Public Safety Charitable Trust* v *Milton Keynes* [2013] EWHC 1237 (Admin), the charity took leases of premises at nominal rents and installed small broadcasting transmitters providing free Wi-Fi access and broadcasting public safety Bluetooth messages. The charity invited the judge to disagree with the ruling in the *Kenya Aid Programme* case. The judge declined to do so, considering the extent of use interpretation correct. The charity suggested such an interpretation would render the word 'wholly' as meaningless as the extent of use interpretation would always resolve into a question about whether it was 'mainly used' for charitable purposes since it would be very rare for any building to be wholly used all the time. The judge disagreed, explaining that a 'building may fairly and properly be described as being wholly or mainly used for a particular purpose even though not every square metre of floor space is in constant use all of the time'.

The valuation officer had decided the use of part of the premises (very small parts) for the Wi-Fi equipment constituted a separate use justifying separate hereditaments. The councils had granted relief for these occupations but not the main hereditaments comprising the business premises. It was the main hereditaments that were the subject of the appeals.

In *My Community Space* v *Ipswich BC [2018]* EWHC 3313 (Admin) It was argued that otherwise vacant office space was being used for an exhibition. To visit, a prior appointment was necessary as there was nobody in attendance; it was found very few members of the public visited; there was no signage or promotional material and the floors were actually empty except for promotional posters on easels. The High Court held it was not wholly or mainly used for the charitable object of encouraging volunteering.

In considering such cases it seems courts will need to evaluate the 'extent of use' to which premises are put by a charity but excluding irrelevant considerations such as necessity of use and efficiency in the way the space is used. Where the court finds the activity relied upon by the charity to justify mandatory charitable relief lacks the appearance, purpose or intent of the use it was supposed to possess according to the charity, then the factors which are relevant changes and factors that were prohibited from consideration as irrelevant may become 'highly relevant'.

17.2.2 Reliefs

Billing authorities are able to offer discretionary or hardship relief in the same way as for occupied hereditaments.

Small business relief is not available for empty property. A ratepayer in receipt of Small Business Rates Relief (SBRR) will lose the relief when the property becomes vacant.

17.3 Partly occupied hereditaments

While empty property rating strictly applies only to hereditaments that are completely unoccupied, billing authorities are given discretion to apply the empty property rate provisions to part of a hereditament if the part is likely to be vacant for a short time only. The way this is done is by the billing authority requiring a certificate from the valuation officer apportioning the rateable value of the hereditament between the occupied and unoccupied parts.

The wording in section 44A of the Local Government Finance Act 1988 is:

Where a hereditament is shown in a billing authority's local non-domestic rating list and it appears to the authority that part of the hereditament is unoccupied but will remain so for a short time only the authority may require the valuation officer for the authority to apportion the rateable value of the hereditament between the occupied and unoccupied parts of the hereditament and to certify the apportionment to the authority.

Valuation officers are obliged to provide a certificate when it is requested and will need a clear explanation of what parts are to be treated as vacant in order to prepare the apportionment of the existing rateable value.

What is 'a short time only' is not defined and, in practice, different local authorities take different views on how long a 'short time' is.

When unoccupied rates were at 50% of the occupied charge, the provision enabled a lower charge on the unoccupied part. Section 44A certificates now allow the three- or six-month exemption period before the empty property rate starts to be applied, although after that period the charge will be at the same level as if it was occupied. Of course, for hereditaments not subject to the empty property rate, there will be no charge on the unoccupied part at all while the certificate is in force.

17.4 Completion notices

The provisions for completion notices are in section 46A and Schedule 4A to the Local Government Finance Act 1988. They require billing authorities to serve completion notices where they are aware that the work remaining on a new building is such that it can reasonably be expected to be completed within three months. The authority is required to serve the completion notice on the owner. The notice has to specify the building and state the day it proposes as the 'completion day' when the hereditament is deemed to be completed. The completion day may not be more than three months from the day the notice is served, and it cannot be earlier than the date of service.

In an Upper Tribunal case *Hermes Property Unit Trust* v *Roberts (VO) and Trafford Council [2021]* UKUT 308, the billing authority had not allowed any time for service, both issuing the notice and specifying the completion date on the same date. The ratepayers argued it did not comply with the statutory requirements in para. 2(3) to Schedule 4A LGFA1988:

Where at the time a completion notice is served it appears to the authority that the building to which the notice relates is completed, the authority shall propose as the completion day the day on which the notice is served.

The date used was some six days earlier than service.

While the VTE did not consider this failure automatically invoked invalidity; regarding the notice as substantially compliant with Schedule 4A, the Upper Tribunal disagreed taking the view that Parliament cannot have intended substantial compliance to be good enough as that would result both in uncertainty and in retrospectivity. The UT suggested if six days was good enough, 'what about 10 days? Or a couple of months.'

This case appears to confirm the long-held view that a completion day cannot be earlier than the service of the completion notice.

In *UKI (Kingsway) Ltd* v *Westminster City Council [2018]* UKSC 67 the Supreme Court, reversing the decision of the Court of Appeal, decided that a completion notice addressed to 'the owner' but which had merely been left at the premises with the receptionist (an employee of the facilities management company, who was neither the owner nor an agent of the owner) and which was then passed on to the owner in electronic form was properly served. The Upper Tribunal had considered the indirect transmission was sufficient as it had considered what mattered was the eventual receipt by the owner.

The Supreme Court noted it was accepted by all that the owner did receive the completion notice. It saw the real issue being whether the authority had 'caused' the notice to be received and considered it had. The means by which a notice arrived at its destination was not wholly immaterial. Not all situations where ultimately the intended recipient came to know of the contents of a notice would satisfy, as there had to be both actual receipt of the notice and a sufficient causal link with the actions of the council.

It saw no reason why the service via the receptionist, who had done no more than would reasonably be expected of a responsible employee in her position, should not be treated as effective service under ordinary principles of causation. The situation was no different from a notice mistakenly delivered to a neighbouring address being passed to the correct recipient by a friendly neighbour. The Supreme Court also examined the question of whether electronic receipt – via e-mail – was sufficient. The court had no difficulty with this, holding there was no good reason suggested for distinguishing transmission by fax, that had been accepted as valid transmission of notices in earlier cases, from transmission by email.

Notwithstanding the UKI (Kingsway) case, it is clearly wise to take care in serving notices and following clear and proper means of service. If the authority considers the building complete then it will adopt the date of service as the completion day.

Completion notices can be served for just part of a building.

Completion notices not only are used for brand new buildings but can also be served for converted or refurbished buildings, providing the work is sufficiently substantial. The term 'new building' is defined in section 46A(6) as including a building produced by the structural alterations of an existing building where the existing building is comprised in a hereditament which, by virtue of the alteration, becomes, or becomes part of, a different hereditament or hereditaments.

In *Tull Properties Ltd* v *South Gloucestershire Council [2014]* RA 180, the President of the Valuation Tribunal for England accepted alterations to a building (Beluga House) could be significant enough to create what could be described as a 'new building'. But, in the particular case, the building did not constitute a new building within the terms of section 46A(6) as the alterations carried out to it were not significant enough. He also considered the fact that it had been made capable of occupation in two parts did not make it a new building as it remained a single building which was just as suitable for single occupancy as multiple. The President posed two tests:

(i)   whether the works done have produced a 'new building' in the ordinary sense of that expression or
(ii) whether they constitute a structural alteration to the hereditament which in consequence becomes a different hereditament or hereditaments.

This question about what constituted a new building within the regulations was considered again by the Valuation Tribunal for England in *Delph Property Group Ltd* v *Alexander (VO) and Leicester CC 2018 (unreported)* concerning several floors of an existing office building that had been refurbished and had had completion notices served on individual floors. It was accepted without completion notices the floors were not sufficiently complete to form rateable hereditaments. The new President adopted the tests posed in Tull Properties.

He did not think it was a 'new building', noting that while there had been a number of significant alterations to it, the question was whether they were so significant as to transform it into something else? With the exception of new cladding and the addition of balconies, the building was basically the same as regards skyline, footprint and size. He suggested there would need to be a clear and significant change to the shape or size of the building or to its character to create a new building out of an existing one. He did not think the change of use from full non-domestic to part non-domestic and domestic was sufficient to change its character.

Looking at each floor as it had its own completion notice he did not think two of them had had a sufficient degree of structural alteration but with the second floor a large atrium in the ceiling had been removed and in-filled increasing the size of the hereditament and creating a roof garden above. This significant element of structural work meant it satisfied the criteria of being a different hereditament resulting from structural alterations and permitting the service of a completion notice.

A Non-Domestic Rating Bill had its first reading in the House of Commons on 28 March 2023, with the aim of legislating for various reforms to the rating system. Clause 7 of the bill, as originally drafted, aims to amend the definition of a 'new building' so that a completion notice may be served in England for a building that, although not new in itself, was deleted from the rating list due to refurbishment/alteration and the works are now capable of being completed within the usual three-month period. This is seen by the government as closing a lacuna in the current legislation, which does not permit completion notices on altered buildings unless satisfying the tests in the Delph Property case. It will allow the existing completion notice procedure, as currently used for completely new or significantly structurally altered buildings, to be used for such refurbished buildings.

The time 'reasonably required for carrying out the work' does not include the time needed to find a tenant to occupy the property (*J and L Investments Ltd* v *Sandwell DC* [1977] RA 78).

17.4.1 Appeals

The time limit for appealing against a completion notice is tight. It needs to be made within four weeks of the owner receiving the completion notice.

If an owner disagrees with the completion date given in a completion notice then the owner and the billing authority can agree a different completion day from that given in the notice. Alternatively, the owner can appeal to the valuation tribunal within four weeks of receiving a completion notice on the grounds that the building cannot reasonably be expected to be completed by the day specified as the completion day in the notice (Schedule 4A of the Local Government Finance Act 1988, paragraph 4). The detailed regulations are contained in the Non-Domestic Rating (Alteration of Lists and Appeals) (England) Regulations 2009 (2009/2268), regulation 19. The notice of appeal has to be accompanied by a copy of the completion notice, and a statement of the grounds on which the appeal is made.

The 2009 Appeals Regulations do not make the four-week time period an absolute requirement as regulation 19(3) permits a request that the Valuation Tribunal for England should allow an extension of time for appeal. Clearly there is a considerable risk that this will not be allowed without a good reason and appeals should be made within the four-week time limit.

There is no provision in Wales for the four-week period to be extended.

Local authorities can withdraw completion notices but only by serving a replacement notice. If an appeal is made then the consent of the owner is required to a new notice being served.

In *Spears Brothers* v *Rushmoor Borough Council* [2006] RA 86, the Lands Tribunal quashed a completion notice on a small workshop because the workshop could not reasonably have been expected to be completed within three months of the service of the completion notice. The Tribunal considered that the state of the joinery could have been remedied in a matter of days and the fact that block walls needed plastering and painting did not make the property incapable of immediate occupation as a workshop, but the lack of electrical wiring, lighting and a fire alarm system, together with the absence of any prospect of an independent electricity supply being provided to the appeal property within the statutory three-month period, meant that the completion notice failed.

In *London Borough of Newham* v *Rad Phase 1 Type B Property Co. No.1 Limited (2020)* RA55/2019 the Upper Tribunal considered whether, as in *Spears Brothers* v *Rushmoor* above, the Valuation Tribunal for England was correct to quash a completion notice where the works could not reasonably be completed by the date in the completion notice. The Tribunal said, 'On my reading of Schedule 4A, there is nothing to prevent the parties agreeing, or the VTE determining, a date outside of the three-month period specified for the purposes of the initial notice.' It was clearly influenced by the absurdity of the local authority having to start again if, say, the work would take but a week longer than that specified in the notice perhaps three months and one week. This is a perhaps surprising decision as it would appear to render the three months requirement of much less importance.

It would seem, following the Valuation Tribunal for England's decision in *Delph Property Group Ltd.* v *Alexander (VO) and Leicester CC 2018 (unreported)*, which was unappealed, that if a ratepayer is of the opinion a completion notice was not validly served it may be possible to challenge this by way of proposal. In the case, the VO had argued that without a completion notice appeal the completion notice stood and the only way to challenge this was by judicial review. The President decided that the VTE could consider a proposal under Reg. 4(1)(h) 'a hereditament shown in the list ought not to be shown in that list' involving consideration of whether the completion notice was valid.

17.4.2 Hereditaments divided or amalgamated after the service of a completion notice(s)

Where a completion notice is served in respect of a whole building but the building is subsequently assessed as a number of separately rateable hereditaments, the completion notice is effective for all the hereditaments (*Camden London Borough* v *Post Office* [1978] RA 57).

17.4.3 Works customarily done after a building has been substantially completed

The legislation has also made provision for situations where it is normal practice to partly finish the works on a new property before marketing is undertaken, the final works being carried out only after an occupier has been found, e.g. where shops are only completed to a 'shell state' before marketing to allow new tenants to complete to their own specification, or where large office floors are left without partitioning, a full power supply and even lighting. Without special provisions these properties would not fall within the ambit of unoccupied property rating unless less than three months' worth of fitting out remained to be done.

The wording in paragraph 9, Schedule 4A, of the Local Government Finance Act 1988 is not overly clear in its meaning:

9.-(1) This paragraph applies in the case of a building to which work remains to be done which is customarily done to a building of the type in question after the building has been substantially completed.

(2) It shall be assumed for the purposes of this Schedule that the building has been or can reasonably be expected to be completed at the end of such period beginning with the date of its completion apart from the work as is reasonably required for carrying out the work.

The meaning of the wording (or rather its earlier enactment in Schedule 1, paragraph 9, to the General Rate Act 1967) was considered by the Court of Appeal in *Graylaw Investments Ltd* v *Ipswich BC* [1979] RA 111. The case concerned a newly erected block of offices that was unoccupied and available to let. It was accepted that work remained to be done which is customarily done to a building of this type after the building had been substantially completed. The floors were large and partitioning and electrical wiring works were required. Six months had been determined as the likely time required to finish the works.

Had this work been work required to substantially complete the building, then a completion notice would not have been effective because a greater period than three months was required (Schedule 4A, paragraph 1(1), and its predecessor legislation allowing completion notices to be served only when completion 'can reasonably be expected within three months'). The special provisions now in Schedule 4A, paragraph 9, however, allow a longer period where the works are those customarily done after substantial completion.

The question decided in the case was whether the six months should run from the actual date of substantial completion, which happened to be a year before the notice, or the date of the notice – should the six months run properly from the date of substantial completion then the finishing works could already be deemed to have been done at the date of the notice.

If the date was the date of the notice then *Graylaw Investments Ltd* effectively gained an additional six months before the unoccupied rate became payable. The Court of Appeal decided certain words in the paragraph would be 'meaningless unless they make the date of substantial completion the starting date' and found for the local authority.

Paragraph 9 applies only after a building has been substantially completed. A local authority cannot look back a year on an incomplete building and say it could have taken three months to achieve substantial completion, a further six months for fitting out works, and since more than the nine months had now elapsed a completion notice can now be served. It can do this only if substantial completion has actually been achieved.

In *London Merchant Securities Plc* v *Islington LBC* [1988] AC 303, the House of Lords decided the date of substantial completion was the date a building was fully complete apart from the customary works, not when there were still some minor works to be finished or when a building had reached such a stage in its construction that it would be practical for the fitting out work to commence.

In the *London Merchant Securities* case the completion notice was served on 1 June 1983, stating 1 September 1983 as the completion date. The House of Lords decided that the building was not actually complete apart from the customary works until 31 August 1983 when various works had been completed. The completion notice therefore failed because it was served (1 June 1983) before the building was complete apart from the customary finishing works.

17.4.4 Valuation and entry into the rating list

The completion notice procedure allows a local billing authority to deem complete a new building that is not actually complete. The valuation officer can then enter the notionally completed hereditament into the rating list.

Note that if the hereditament, though vacant, is in fact complete a completion notice is not required and the valuation officer can enter the actually complete hereditament into the list. This was confirmed by the Lands Chamber of the Upper Tribunal in *Porter (VO)* v *Gladman SIPPS* [2011] RA 337. The case, however, emphasised the importance of the building actually being complete, with the Tribunal deciding that it was not sufficient for a new building to be nearly complete or 'even very nearly' complete in order for a valuation officer to enter it into the list without a completion notice but that it should actually be complete.

The case confirms that if a building lacks features that:

- will have to be provided before it can be occupied for the purpose for which it is designed to be occupied; and
- when provided, will form part of the occupied hereditament and form the basis of its valuation;

then it does not constitute a hereditament and should not be shown in a rating list.

The case of *Porter (VO)* v *Gladman SIPPS* [2011] RA 337 concerned a number of newly built office units on two to three floors with areas of 216.7 m^2, 566 m^2 and 934.6 m^2. All units had plastered walls, suspended ceilings, fluorescent lighting, raised floors and WCs. They were lacking various minor works as there was only one double gang power point on each floor, no full height partitioning, no tea point (except in the smaller units) and no data cabling or telephone equipment. The Lands Chamber was:

entirely satisfied on the evidence that a potential occupier of any of the appeal properties would have required small power (a ring main and power points), tea points and at least some power points to be installed before occupying them as offices.

A more recent case, *Aviva Investors* v *Whitby (VO)* [2013] RA/3/2011, cleared up a question left over from the decision in the *Gladman* case about what to do with buildings which are not actually completed but similar buildings are occupied in an incomplete state nearby.

In the case the warehouse buildings were accepted by the valuation officer not to be fully finished, but she said similar buildings nearby were occupied in an unfinished state. This showed that the buildings did not lack features that would have to be provided before they could be occupied. The ratepayers argued that evidence that a very small minority of occupiers with unusual or atypical requirements were prepared to occupy premises without undertaking otherwise essential works was not a reliable guide to the test from *Porter (VO)* v *Gladman SIPPS* [2011] RA 337 of whether a building was ready for occupation.

The Tribunal strongly rejected the valuation officer's arguments. The Tribunal considered that the examples provided by the valuation officer did not represent use of the properties in the manner for which they were designed. The Tribunal found that additional lighting and power distribution in the warehouse areas of the appeal properties was required before they were ready for, or capable of, occupation.

It seems it is necessary to decide how a particular building would typically be completed to be used in the manner for which it was designed. If it is not fully finished to that standard then it will not be ready to be included in a rating list unless a completion notice has been served or it is, in fact, occupied.

In *Cyxtera Technology UK Limited.* v *Ricketts (VO)* [2021] UKUT 0265 (LC) the question arose as to whether parts of a new data centre were in beneficial occupation and formed part of the new hereditament when they were completed as 'white space' but not adapted to customer requirements ('White space' is an industry recognised term and refers to a standard of data hall fitting out with raised floor and suspended ceiling together with the Power Distribution Units ('PDU') and Computer Room Air Handling Units ('CRAH'). It can be contrasted with 'Fallow space' where nothing has been done and 'Customised White Space' which has been adapted to a customer's requirements.)

The Tribunal looked back to *Post Office* v *Nottingham City Council [1976]* 1 WLR 624 where the question had been whether a building was ready for occupation as a telephone exchange. The Upper Tribunal noted Cyxtera Technology had actively maintained and constantly adapted the white space since it took occupation of the whole data centre. The Tribunal regarded the ratepayer's arguments about the extent to which the white space was ready for customers as a 'red herring'. It was the company's own beneficial occupation of the white space that mattered. The Tribunal found the white space was capable of occupation for the intended purpose of the building, i.e. as a data centre. The alteration work to the white space did not involve any alteration to the building itself and was akin to the addition of furniture. And, in any case, none of the adaptions to the white space were necessary for beneficial occupation as a data centre.

An interesting question arose in a VTE hearing before the President as to whether the Tribunal could consider the validity of a completion notice in valuation proceedings (*Prudential Assurance* v *VO for Weybridge* [2011] RA 490) concerning The Heights, Brooklands Road, Weybridge. The valuation officer argued that the Tribunal could not consider validity as the facts behind the completion notice's service were not known to him as he was not responsible for service and there was a separate appeals route for completion notices. The President found that the Tribunal did have jurisdiction on the basis that if the completion notice was not valid then the entry in the list could not be made, so it was an essential question. The completion notice appeal procedure was limited to the question of completion date only and not wider matters and, if necessary, the billing authority could be added as a party to the VTE proceedings so that evidence of service could be provided.

Having found that the original completion notice was not valid because it had been served on the wrong person, the VTE then went on to find that a later agreement on completion day between the billing authority and the ratepayers was either a completion notice itself or, in the alternative, binding.

17.5 Wales

The rating of unoccupied property in Wales is governed by the following Regulations:

- Non-Domestic Rating (Unoccupied Property) (Wales) (Amendment) Regulations 2009 (SI 2009/255);
- Non-Domestic Rating (Unoccupied Property) (Wales) (Amendment) Regulations 2010 (2010/272);
- Non-Domestic Rating (Unoccupied Property) (Wales) Regulations 2008 (SI 2008/2499).
- The Non-Domestic Rating (Unoccupied Property) (Wales) (Amendment) Regulations 2021 (2021/118);
- The Non-Domestic Rating (Unoccupied Property) (Wales) (Amendment) Regulations 2011 (2011/197).

18 The Council Tax

18.1 Introduction

The Council Tax was proposed in a Government consultation paper, *A New Tax for Local Government*, in April 1991. It replaced the short-lived experiment of the Community Charge or 'poll tax' which in its turn had replaced domestic rates in April 1990 but had not proved successful.

The principle act is the Local Government Finance Act 1992, but most of the detailed statutory provisions are contained in regulations made under powers contained in this Act.

The tax came into force on 1 April 1993 in England, Wales and Scotland.

18.2 The basic concept

The use of capital values to replace notional rental values, used for the old domestic rating system, had been discussed for many years. The lack of open market rents upon which to base a domestic rating revaluation suggested an alternative basis should be adopted. Because of the limited time available to introduce the new tax, a practical alternative to allocating precise capital values was identified. This was to allocate each dwelling a particular level of tax if it fell within a given range or band of values.

The Council Tax uses very broad bands of capital value. Each dwelling, rather than being given a value, is assigned a letter denoting the band into which its value falls. Originally, eight value bands were used, A–H, although in the 2005 re-banding exercise in Wales a ninth band, Band I, was added for Wales .

A standard level of Council Tax is set by the local billing authority for Band D dwellings with the amounts for the other bands falling into precise ratios to Band D. These ratios are set out in the legislation. These result in the Council Tax payable for the highest band (Band H in England) being three times that of the lowest (Band A). The set proportions cannot be varied by the individual billing authorities.

DOI: 10.1201/9781003297369-18

Table 18.1 Value bands for England and properties (1993 Revaluation) (Sept 2022)

Valuation bands	Range of values (£)	Number of properties by band
A	up to 40,000	6,039,152
B	40,001–52,000	4,924,539
C	52,001–68,000	5,522,567
D	68,001–88,000	3,937,056
E	88,001–120,000	2,457,264
F	120,001–160,000	1,304,484
G	160,001–320,000	890,403
H	320,000 upwards	150,016
Total		25,225,481

Table 18.2 Value bands and properties for Wales (2005 Revaluation) (Sept 2022)

Valuation bands	Range of values (£)	Number of properties by band
A	up to 44,000	211,570
B	44,001–65,000	303,850
C	65,001–91,000	316,690
D	91,001–123,000	235,630
E	123,001–162,000	195,790
F	162,001–223,000	118,940
G	223,001–324,000	53,970
H	324,001–424,000	12,930
I	424,001 upwards	5,610
Total		1,454,980

18.3 Liability

Liability to the tax is set out in an order of possible persons starting with residents, and descending down a list to a non-occupying owner if there is no one in residence. A resident freeholder is first liable; if the freeholder is not resident then the liability falls on a resident leaseholder, followed by resident statutory tenants, residential licensees, residents and lastly the owner. If more than one person, such as a husband and wife, are resident freeholders together or fall in one of the other liable categories, then they are jointly and severally liable.

In some situations the billing authority can find difficulties in ensuring the tax is collected or would find it easier if the owner was liable for the Council Tax. In these cases the billing authority can determine that liability will fall on owners of all that class of property in its area.

The classes are set out in the Council Tax (Liability of Owners) Regulations 1992 (SI 1992/551) (as amended) and, in general terms, are:

- Class A – Residential care homes, nursing homes and the like.
- Class B – Dwellings inhabited by religious communities such as nunneries.
- Class C – Houses in multiple occupation. These are dwellings constructed or adapted for use by persons who do not form a single household, e.g. a house adapted as bedsits.
- Class D – Dwellings where there are live-in staff in domestic service.
- Class E – Dwellings occupied by a minister of any religion and from which the minister's duties are performed.

18.4 Discounts

Discounts apply in certain circumstances.

Single person's discount: if there is only one resident then a 25% discount is given. This applies also when there are other residents but they fall to be disregarded. Such persons are children, persons who are severely mentally impaired or are full-time students.

No resident: if the property does not have anyone resident then the billing authority has discretion to give a discount. Many authorities offer no discount or make it available only for a short period, perhaps a month. Originally a 50% discount was automatically applied but this ended in 2013. The rules are a little complicated and provide different rules for various classes of property:

- Class A – Second homes where continuous occupancy for 28 days or more is prohibited. The billing authority may offer discounts from 0% to 50%. Second homes are properties which are not the sole or main residence of an individual.
- Class B – Second homes where continuous occupancy for 28 days or more is not prohibited. The billing authority may offer discounts from 0% to 50%.
- Class C – Properties which are 'unoccupied' and substantially unfurnished. The billing authority may offer discounts from 0% to 100%.
- Class D – Vacant properties undergoing 'major repair work' or 'structural alteration'. The billing authority may offer discounts from 0% to 100% for up to 12 months.

The single person discount applies only while there is a single resident. If the property becomes vacant it is treated the same as any other vacant dwelling and if any exemption period finished before re-occupation the liability will rise to 100%.

Sometimes people own or occupy more than one property such as a main house and a holiday home. Establishing which property is the person's sole or main residence may require the consideration of several factors, including where most of the person's belongings are kept, where the person's spouse or children live and where the person sleeps.

Relief may be available for a physically disabled person under the Council Tax (Reduction for disabilities) Regulations 1992 (SI 1992/554) (as amended). The relief is given by notionally reducing the banding of the person's dwelling by a formula equivalent to one band.

Where a person has a 'second home' but that second home is 'job-related' then if the main residence is vacant a mandatory 50% relief applies. Job-related dwellings are those provided for the proper performance of the duties of the employment, where provided for the better performance of the duties and it is one of those employments where it is customary for employers to provide dwellings or there are special threats to the employee's security and the employee resides in the job-related dwelling as part of the security arrangements.

Where an annex or other part of a property has been 'disaggregated' because there is more than one self-contained unit in a dwelling (see later in this chapter) and that dwelling is being used by the resident or members of the family but no person is actually resident in the disaggregated part then a mandatory 50% relief applies.

18.4.1 Empty homes premium

Billing authorities have the power to increase Council Tax on dwellings that have been 'unoccupied and substantially unfurnished' for a long period of time. The authority can choose to charge up to 150% of the normal charge.

18.4.2 Exempt dwellings

Certain occupied, unoccupied or vacant properties listed in the Council Tax (Exempt Dwellings) Order 1992 (SI 1992/558) are exempt from Council Tax. This Regulation treats a dwelling as unoccupied if no one is living there. If it is substantially unfurnished as well as unoccupied then it is regarded as vacant. These exempt classes of occupied, unoccupied and vacant properties apply to dwellings that are:

- Class B – Owned by a charity and has been unoccupied for less than six months;
- Class D – Unoccupied and the sole or main residence of a person detained in prison or for mental care;
- Class E – Unoccupied and the home of a person in hospital or residential care;
- Class F – Unoccupied following the death of the occupier providing it was the occupier's main residence, no one else's main residence and the deceased was sole owner, lessee or a tenant where the executor continues to pay rent and either probate has not been granted or less than six months have elapsed since its grant;
- Class G – Occupation is prohibited;

- Class H – Kept for occupation by a minister of religion;
- Class I – Unoccupied but previously the home of someone whose permanent residence is now elsewhere in order to be cared for due to old age, disablement, drug misuse, etc.;
- Class J – Home of a person who is resident elsewhere to take care of others;
- Class K – Home of a person who is resident elsewhere for the purposes of study;
- Class L – Properties in the possession of a mortgagee;
- Class M – Hall of residence provided predominantly for the occupation of students and owned or managed by an educational establishment;
- Class N – Wholly occupied by students;
- Class O – Part of British armed forces accommodation;
- Class P – Where at least one person occupying is a member of the visiting armed forces (as defined);
- Class Q – In the hands of a trustee in bankruptcy;
- Class R – Pitches or moorings where a caravan or boat is not present;
- Class S – Occupied only by persons aged under 18;
- Class T – Unoccupied granny annexes and other disaggregated dwellings which cannot be separately let without breaching planning control;
- Class U – Dwellings occupied only by severely mentally impaired persons or such persons and students;
- Class V – Occupied by certain diplomats.

18.4.3 Class W – Disaggregated dwellings occupied as the sole or main residence of dependant relatives of a person who is resident elsewhere in the property

Table 18.3 shows the number of exempt properties per class.

Table 18.3 Number of exemptions per class as at September 2022

Exemption class		
B	Unoccupied dwellings owned by a charity (up to six months)	25,453
D	A dwelling left unoccupied by people who are detained, e.g. in prison	1,838
E	An unoccupied dwelling which was previously the sole or main residence of a person who has moved into a hospital or care home	32,786
F	Dwellings left empty by deceased persons	121,900
G	An unoccupied dwelling where the occupation is prohibited by law	7,961
H	Unoccupied clergy dwellings	1,054
I	An unoccupied dwelling which was previously the sole or main residence of a person who is the owner or tenant and has moved to receive personal care	2,729

Table 18.3 (Continued)

Exemption class		
J	An unoccupied dwelling which was previously the sole or main residence of a person who is the owner or tenant and has moved to provide personal care to another person	809
K	An unoccupied dwelling where the owner is a student who last lived in the dwelling as their main home	206
L	An unoccupied dwelling which has been taken into possession by a mortgage lender	2,154
M	A hall of residence provided predominantly for the accommodation of students	73,337
N	A dwelling which is occupied only by students, the foreign spouses of students, or school and college leavers	73,337
O	Armed forces' accommodation. (Included as exempt dwellings rather than chargeable dwellings because contributions in lieu will be paid in respect of them, allowed for in line 20 of the CTB form.)	44,049
P	A dwelling where at least one person who would otherwise be liable has a relevant association with a Visiting Force	10,056
Q	An unoccupied dwelling where the person who would otherwise be liable is a trustee in bankruptcy	486
R	Empty caravan pitches and boat moorings	6,370
S	A dwelling occupied only by a person, or persons, aged under 18	6,034
T	A dwelling which forms part of a single property which includes another dwelling and may not be let separately from that dwelling, without a breach of planning control	9,774
U	A dwelling occupied only by a person, or persons, who is or are severely mentally impaired who would otherwise be liable to pay the Council Tax or only by a one or more severely mentally impaired persons and one or more students, students' foreign spouses and school and college leavers	101,326
V	A dwelling in which at least one person who would otherwise be liable is a diplomat	4,029
W	A dwelling which forms part of a single property including at least one other dwelling and which is the sole or main residence of a dependent relative of a person who is resident in the other dwelling	9,621
Total		692,206

18.5 Valuation lists

The responsibility for preparing and maintaining a list of bands for the local billing authority falls upon the listing officer. Like valuation officers these are chartered surveyors or valuers working for the Valuation Office Agency (VOA) who have been appointed to this statutory role. The lists are called valuation lists and are required to show each dwelling situated in the local billing authority's area.

Originally, the Government did not envisage any revaluation of the valuation bands. However, the Local Government Act 2003 introduced a requirement for a minimum of a ten-year cycle of revaluations commencing in Wales with new lists being compiled on 1 April 2005. The planned revaluation for England in 2007 was postponed in September 2005 and the Act amended to make future revaluations a matter for ministerial order (s.22B Local Government Finance Act 1992). The Act also permitted a variation in the number of bands from the original scheme of eight bands. This has been exercised in Wales and nine value bands were used for the 2005 Revaluation.

Listing officers are required to send a draft valuation list to the local billing authority not later than 1 September prior to the date on which the list is to be compiled.

The Local Government Act 2003 also provided for a transitional relief scheme to smooth the changes in Council Tax liability resulting from the coming into force of a revaluation.

The band letter, but not a value, appears alongside the address for each dwelling in a valuation list. The lists also show a reference number, the effective date of the alteration shown, whether the dwelling is composite and whether the alteration shown was made by order of a valuation tribunal or the High Court. The lists are amended and kept up to date by the listing officer making deletions and alterations.

Council tax valuation lists can be viewed on the VOA website at www.voa. gov.uk.

18.6 The banding process for the 1993 valuation lists

The task of project managing the original banding exercise involving some 21 million dwellings was given to the VOA, an executive agency of HM Revenue & Customs. Just over half of the banding work was undertaken by private surveyors under contract to, and supervised by, the VOA. In Scotland, all banding was undertaken by the local assessors under the direction of the VOA.

18.7 The dwelling

For a taxpayer to have a Council Tax liability, a dwelling has to appear in a valuation list. A dwelling is defined in section 3 of the Local Government Finance Act 1992. Essentially a dwelling is any property which would have been a hereditament had domestic rating continued but does not actually appear in a rating list or is exempt from rating. Additionally, the following minor properties are not treated as dwellings under section 3(4) unless they form part of a larger property, such as a house, which is itself a dwelling:

- a yard, garden, outhouse or other appurtenance belonging to or enjoyed with property used wholly for the purposes of living accommodation;
- a private garage which either has a floor area of not more than 25 m^2 or is used wholly or mainly for the accommodation of a private motor vehicle;

- private storage premises used wholly or mainly for the storage of articles of domestic use.

While Council Tax has its own valuation rules and procedures, the concept of the dwelling as the unit of assessment for Council Tax is firmly based on the concept of the rateable hereditament. Council Tax can therefore be seen as the son or daughter of rating. The starting point when considering whether a property is a dwelling is to decide if it would have formed a hereditament for the purposes of domestic rating that existed prior to 1990. The rules as to what constitutes a hereditament are set out in Chapter 3.

Section 3(3) of the 1992 Act provides that a composite hereditament is also a dwelling. A 'composite hereditament' is defined in the Local Government Finance Act 1988, which provides that a 'hereditament is composite if part only of it consists of domestic property'.

Usually, composite dwellings are shown both in the council tax valuation list and the non-domestic rating list. However, where the non-domestic part of a dwelling is exempt from rating, the dwelling is shown in the council tax valuation list alone. For example, most farms are wholly exempt from rating but the farmhouse and any farm cottages, forming part of the farm, appear in the valuation list.

18.8 Basis of valuation

While specific values are not assigned to each dwelling for Council Tax, the legislation sets out a series of valuation assumptions to enable a decision to be reached on the valuation band for a property. These valuation assumptions are contained in Regulations 6(1) and (2) of the Council Tax (Situation and Valuation of Dwellings) Regulations 1992 (SI 1992/550). These set out the valuation assumptions in a more formal way than is the case for rating, where the assumptions have often developed from case law. Like rating, Council Tax uses an antecedent valuation date (AVD) that is two years before the valuation lists came into force. The value of any dwelling is taken to be the amount it 'might reasonably have been expected to realise if it had been sold in the open market by a willing vendor on 1 April 1991' subject to the following valuation assumptions. In broad terms these are:

- That the sale was with vacant possession.
- That the interest sold was the freehold or, in the case of a flat, a lease for 99 years at a nominal rent.
- That the dwelling was sold free from any rent charge or other encumbrance. These are items of a financial nature in the nature of a claim or liability rather than something like a restrictive covenant (see *Coll (LO)* v *Walters* [2016] EWHC (Admin) 831).
- That the size, layout and character of the dwelling, and the physical state of its locality, were the same as they were at the relevant date. This was 1 April 1993 for the original valuation lists.

- That the dwelling was in a state of reasonable repair.
- In the case of a dwelling, the owner or occupier of which is entitled to use common parts, that those parts were in a reasonable state of repair and the purchaser would be liable to contribute towards the cost of keeping them in such a state.
- Fixtures designed to make a dwelling suitable for use by a physically disabled person which add to the dwelling's value are assumed not to be included in the dwelling.
- That the use of the dwelling would be permanently restricted to use as a private dwelling.
- That the dwelling had no development value other than value attributable to permitted development.

The rating concept of *rebus sic stantibus* does not apply to Council Tax. Instead the assumptions as to use, physical changes which can be envisaged and the assumed state of repair are set out in the Regulations. A dwelling has to be valued on the basis that it will always be restricted in its use to a private dwelling. The potential to make physical changes to the dwelling can be taken into account only for the sort of minor alterations permitted by the Town and Country Planning (General Permitted Development Order) (England) 2015 (2015/595), as amended, without the need to obtain planning permission. Permitted development includes the enlargement of, improvement of or other alteration to a dwelling-house within set tolerances, the construction of a porch (subject to size limits) and the construction of outbuildings or a swimming pool within the curtilage of a dwelling-house (again subject to limits on size).

The need to assume a 'state of reasonable repair' does not mean that all properties are assumed to be in a 'good' state of repair. What has to be decided is the state it would be reasonable to expect for such a dwelling, having regard to its age, locality and character. This may not be the actual condition, as dwellings will be banded in the typical state of repair of similar dwellings in the neighbourhood. If a dwelling is unmodernised and its fixtures and fittings very dated, whereas the model for other houses in the street reflects improvements and modernisation, it may be appropriate to regard its character as different from the others in the locality and treat its actual state as the basis for its banding. If, however, a dwelling is in such a poor state of repair that it can no longer be regarded as a dwelling then it should be removed from the valuation list. In *Wilson* v *Coll (LO)* [2012] RA 45, the High Court considered the assumptions as to repair contained in Council Tax. The judge accepted that the assumption did not contain any reference to what was or was not economic as was the case with rating under the definition of rateable value – such a test is absent from the Council Tax legislation. The test is therefore different:

> *I accept that as a general matter of law the crucial distinction for the purposes of deciding whether there is, or continues to be, a hereditament should focus upon whether a property is capable of being rendered suitable for occupation*

(… as a dwelling) by undertaking a reasonable amount of repair works. The distinction, which was correctly drawn by the respondent … is between a truly derelict property, which is incapable of being repaired to make it suitable for its intended purpose, and repair which would render it capable again of being occupied for the purpose for which it is intended.

A dwelling that is capable of repair remains a hereditament even if it would not necessarily be economic to carry out the repair works, though there may come a point at which a property is so derelict as to be incapable of repair. The important distinction is not between economic and uneconomic repair but between repair and reconstruction works. In the rating case of *Newbigin (VO)* v *Monk [2015] EWCA Civ 78* the Supreme Court held that a property in course of reconstruction was no longer a hereditament. The same will apply for Council Tax given the basis of the 'dwelling' for council tax is the concept of the hereditament. So, *if there is a scheme of works*, then Listing Officers will need to consider the evidence and *judge whether the works make a building uninhabitable*. This will only apply to vacant property and where there is major renovation and extension underway. It is unlikely to apply to works involving merely replacing kitchen or bathroom fittings.

18.9 Altering bands

There are restrictions on when council taxpayers may appeal by proposing a different band for their properties, and there are also restrictions on when the listing officer may alter a band in a valuation list. In designing the Council Tax, the Government considered it unfair for taxpayers who improved their homes by adding an extension then to have their bands increased to reflect the added value. The Regulations prevent this by allowing a band to be increased for physical changes to a dwelling only when the taxpayer who undertook the extension sells the property. The physical change is known as a material increase and the sale a relevant transaction.

Material increase means:

> **any increase in the value of a dwelling which is caused (in whole or in part) by any building, engineering or other operation carried out in relation to the dwelling, whether or not constituting development for which planning permission is required.**
>
> (Section 24(10) of the Local Government Finance Act 1992)

Relevant transaction means:

> **a transfer on sale of the fee simple, a grant of a lease for a term of seven years or more or a transfer on sale of such a lease.**
>
> (Section 24(10) of the Local Government Finance Act 1992)

It should be noted that changes to a dwelling's locality, such as improved road links, that increase a dwelling's value are not included in the definition of a material increase and even if the dwelling is sold, its banding cannot be reviewed to reflect the improved value unless there have also been improvements to the property which do constitute a material increase.

Alterations can also be made to bands for a 'material reduction' in a dwelling's value. Material reduction is quite tightly defined to cover quite limited circumstances. These are in relation to the value of the dwelling.

Material reduction is any reduction which is caused (in whole or in part) by:

- the demolition of any part of the dwelling;
- any change in the physical state of the dwelling's locality;
- any adaptation of the dwelling to make it suitable for use by a physically disabled person.

In *Chilton-Merryweather (LO)* v *Hunt* [2008] EWCA Civ 1025, the Court of Appeal had to consider whether a change in the volume of traffic, with its attendant noise and pollution, was a change in the physical state of the locality of a house next to a motorway and therefore a material reduction, allowing the Council Tax band to be reduced. Rix LJ said:

> the listing officer is properly concerned only with the essential fabric and character of house and locality, but not with other matters which go to their enjoyment, use, occupation or activity, such as, I would suggest, the particular degree of traffic to be met on a particular date.

He accepted that the expression 'physical state' could embrace traffic and its physical consequences such as noise and pollution, but the context showed the wording was limited only to looking at changes to the physical fabric. It was, however, a different matter if the fabric of a local road changed, such as where a motorway or any other road had an extra lane fitted or was itself altered in some other way. This would be a change to the physical fabric. It might also be different if the character of the road changed, such as where a road is re-categorised and signage, or the stoppage up of other roads, directed traffic into a previous backwater.

The Local Government Act 1992 fixes the valuation for Council Tax at the AVD and it is the value at that date which is adopted, except to the extent that the AVD value would itself have to reflect the 'material reduction' of a subsequent change in the physical condition of the dwelling or the physical state of the locality.

Changes to bands may also be made to reflect dwellings becoming or ceasing to be composite hereditaments, changes in the balance of domestic and non-domestic use and where the band given to a dwelling is incorrect.

18.10 Valuation date

The valuation date used for Council Tax is, like rating, an AVD. The date in England is 1 April 1991, which was two years before the valuation lists came into operation on 1 April 1993. For the 2005 valuation lists in Wales the AVD is 1 April 2003. In a similar way to rating the Regulations set out the particular days on which the state of the factors affecting the value of a dwelling is taken. Like rating, non-physical matters are taken as they were at the AVD. Again, like rating, the physical state of the property and the physical state of the locality are looked at as they were on different and later days from the AVD. These days vary depending on the type of alteration made and can even be different days for the physical state of the dwelling and the physical state of the locality. Council Tax can therefore go one better than rating and have three different valuation dates for one alteration!

- When considering an original list entry or the insertion of a dwelling which was omitted from an original list, the size, layout and character of the dwelling and physical state of its locality are assumed to be the same as they actually were on the date the valuation lists came into force.
- For a newly built or converted dwelling, all physical factors are taken as they actually were when the dwelling was completed.
- Some properties, such as most public houses and many shops which have living accommodation occupied by the shopkeeper, are both domestic and non-domestic and have a rating assessment and a Council Tax banding. When altering a list because a property becomes, or ceases to be, one of these composite dwellings, or the balance of domestic and non-domestic use in a composite alters, the physical factors are taken as they actually were on the date the change happened.
- If a dwelling has been improved, for example by an extension being added, the band can only be altered to reflect the material increase in value if there is a later relevant transaction. In considering the new band all physical factors are taken as at the date of the sale. So if over a number of years the taxpayer had made several improvements to the property then the value of all of these improvements would be taken into account when the band is reviewed following the sale.
- If part of a dwelling is demolished, for example a garage is demolished or a wing of a house is taken down, then all the physical factors, both the physical state of the dwelling and its locality, are taken to be as they were when the demolition was undertaken. So, if there were physical improvements to the property not included in the original banding, the value of these would be included and might reduce or offset the material reduction in value due to the demolition.

The Regulations do not, however, permit a reduction on account of part demolition, if the work is part of a scheme of improvement to the dwelling, such as replacing a small extension with a large extension. This is to prevent a temporary reduction in value resulting from the demolition of part of a dwelling becoming a permanent reduction in banding. This would otherwise have happened because of the restriction on increasing bands on account of a material increase until there is a subsequent sale. This restriction is contained in Regulation 3(3) of the Council Tax (Alteration of Lists and Appeals) (England) Regulations 2009 (2009/2270) and for Wales Regulation 3(2) of the original Non-Domestic Rating (Alteration of Lists and Appeals) Regulations 1993 (SI 1993/290).

- Where a material reduction in the value of a dwelling is caused wholly by the demolition of any part of the dwelling, the valuation band shall not be altered if the works of demolition are part of, or connected with, a building, engineering or other operation carried out, in progress or proposed to be carried out in relation to the dwelling.
- If there is a material reduction in value due to adapting the dwelling to make it suitable for a physically disabled person then in reviewing the band all physical factors are taken as they actually were when the adaptation was made.
- If there is a material reduction in value caused by a change to a dwelling's locality, for example due to a new road or other development being built close to it, the physical state of the dwelling and the physical state of the locality are taken at different dates. The physical state of the locality is looked at as it was on the date of the physical change, but the physical state of the dwelling is viewed for banding as it was at the most recent of the last alterations to the list entry, the last relevant transaction that did not result in an alteration or 1 April 1993. This means that if taxpayers suffer a reduction in the value of their homes due to a change to the locality, the full effect of this is taken into account in the banding and it is not offset by any improvements to their properties unless there has been a relevant transaction.
- Where the alteration is to correct an earlier error, the physical state of the dwelling and the physical state of the property are taken at the dates that applied to the earlier incorrect alteration.

18.11 Disaggregation – treatment of self-contained units

While the concept of the dwelling for Council Tax follows the rating concept of the hereditament, there is also a requirement that where there is more than one self-contained unit within a dwelling then each unit has to be treated as a separate dwelling and given its own band. The most common example of this is where a taxpayer adds a granny annex to an existing house for an elderly relative. If the annex is really a separate hereditament in its own right, because of the rules of what constitutes a separate hereditament, then as a separate hereditament it will have its own band. However, if in reality the taxpayer is in control (paramount occupation) of the annex as well as the main house, perhaps because of the infirmity of the relative, then the requirement to treat self-contained units as separate dwellings will apply and the annex given a separate band despite being a single hereditament with the main house. This is termed disaggregation.

The Council Tax (Chargeable Dwellings) Order 1992 (1992/549) (as amended) requires, in article 3, that:

where a single property contains more than one self contained unit … the property shall be treated as comprising as many dwellings as there are such units included in it and each such unit shall be treated as a dwelling.

A single property means 'property which would apart from this Order, be one dwelling within the meaning of section 3 of the Act' (i.e. it would be a single hereditament/dwelling).

Self-contained means 'a building or part of a building which has been constructed or adapted for use as separate living accommodation'.

The legislation does not set out what facilities are needed for a unit to be self-contained, but clearly living, sleeping, cooking, washing and sanitary facilities are required. A number of High Court cases have considered the wording of article 3. The test of whether a unit is self-contained does not depend on the actual use made of it but whether it was in fact constructed or adapted to be capable of use as separate living accommodation. The degree of communal living which will often occur in granny annex cases does not make an annex not self-contained when it has all the features needed for separate living. In *Batty (LO)* v *Burfoot* [1995] RA 299, Ognall J said:

It will be obvious that the purpose behind the great majority of these annexes is to furnish separate accommodation for an older generation in such a way as to allow of mutual privacy, whilst at the same time allowing for the degree of community which gives peace of mind to both parts of the family. It follows that in most cases the degree of communal living will be, or probably will be, significant. But that cannot assist, in my judgement, in answering the question as to whether the annexe in question was constructed or adapted for use as a separate dwelling.

The accommodation does not need to be physically separate from the main property, and indeed most disaggregated dwellings will not be physically separate. The sharing of services such as a common electricity meter does not indicate the unit is not self-contained. Article 3 does not require the self-contained unit to be capable of being separately sold, although clearly if separate sale is difficult or impracticable this will affect the value band of the dwelling. It may, as a consequence, be valued in the lowest band (Band A). Self-contained units will not be spread over different parts of the main property but will form an entity with its rooms all together. In *McColl* v *Subacchi (LO)* [2001] RA 242, the High Court confirmed that both a house and a flat with its own lockable door but reached by passing through the hall, stairs and landing of the house, were separate self-contained units and should have separate bands:

> *In relation to the test of whether a unit is 'really separate', I would myself say that both the flat and the house are to be considered 'really separate', in that they both respectively contain all that is necessary for separate living accommodation and are intended to be lived in separately. The flat has the locked door which makes it separate, and the house is separate from the street and the flat. The fact that, as I have said, there is a licence to pass through the house to get to the flat is not, on the way I construe the Order, a decisive consideration.*

Article 3 refers to 'buildings constructed or adapted for use'. At first sight this seems to imply a test of what the original developer or person who made the adaptions intended or how it is actually used. This is not the case. In *Jorgensen (LO)* v *Gomperts* [2006] RA 269, the High Court said, 'The test is an objective bricks and mortar test. Intention and use, actual or prospective, are not relevant.' What has to be considered is whether the property does actually comprise more than one self-contained unit.

In *Jorgensen (LO)* v *Gomperts* [2006] RA 269, it was argued that a lockable door was necessary to achieve the separateness sufficient for self-contained units. The High Court considered it wrong to treat the presence of a lockable door between the parts as a necessary condition:

> *there may well be particular circumstances where some degree of privacy could reasonably be regarded as a material element in determining whether something constituted separate living accommodation and the Tribunal must then decide whether having regard to the physical characteristics, the part of the building in question should be so treated. I stress again that the test is an objective one. It is emphatically not a question whether any particular occupant might believe that privacy required a locked door, and without it there was no separate living accommodation. It is for the Tribunal, looking at the physical characteristics of the building, to determine on an objective basis whether those characteristics are sufficient to make part of the building separate living accommodation.*

In a later case the High Court (*Ramdhun* v *Coll* (LO) [2014] EWHC 946 (Admin)) was unwilling to disturb a valuation tribunal decision that not even a door was necessary between two floors to make them self-contained. The Court was content that the tribunal had given careful consideration to privacy and separation and formed a view. It noted questions of privacy were ones of degree and balance and as there were fully functioning kitchens and bathrooms on each floor it was a rational decision for the tribunal to make that the floors were self-contained in the circumstance.

Not all disaggregated dwellings are granny annexes or converted flats in former houses. Article 3 applies to all types of property. *Beasley (LO)* v *National Council for YMCAs* [2000] RA 429 concerned ten flats owned by the National Council for YMCAs operated as a 'foyer housing scheme' for homeless young people and purpose built in 1995. It was accepted that the YMCA was in paramount occupation and the whole housing scheme was a single hereditament, but it was disputed that each of the flats was a separate self-contained unit. The flats were bed-sitting rooms with a kitchenette area and an en suite shower room with a lavatory. Communal facilities included two meeting rooms, laundry, disabled lavatory, bicycle store, refuse store and a kitchen. Sullivan J decided that each flat should be disaggregated:

> *When looking at Articles 2 and 3 of the 1992 Order, one focuses not upon the use that is actually made of the building, but upon whether it has been constructed for use as separate living accommodation ... Whether Pinder House was controlled by one body, and whether that body had criteria for residency, had nothing to do with whether the flats had been constructed for use as separate living accommodation ... I would be prepared to accept that in deciding whether a particular flat has or has not been constructed for use as separate living accommodation within a larger building, it will often be relevant to consider the extent of the communal facilities which have been provided in the flat and the extent of the communal facilities which have been provided in the remainder of the building.*

There is a clear difference between, for example, traditional student accommodation in a student hostel where a student's room is simply a study bedroom with the student having to use communal facilities such as the dining hall and bathrooms, compared to a flat with kitchen, bathroom and sanitary facilities where there might also be communal facilities in addition. The former is not self-contained whereas the latter, like the flatlets in the YMCA case, is self-contained.

Care homes, within the meaning of the Care Standards Act 2000, are excluded from disaggregation by article 3A. They are treated for banding as comprising, as a single dwelling, the care home itself, irrespective of how many self-contained units it comprises, plus as a separate dwelling any self-contained unit occupied, or if currently unoccupied, provided for the purpose of accommodating the warden or other person registered as in charge of the care home.

18.12 Aggregation

Provision is made for listing officers to treat, as single dwellings, properties that would otherwise be separately banded. The provisions are to aid Council Tax collection on buildings occupied as a number of separate but not self-contained units such as bedsits or tenement blocks, where there is a high turnover of occupiers. If a bedsit, for example, without its own washing or cooking facilities, is separately occupied, then the requirement for a separate rateable hereditament and therefore Council Tax dwelling may be satisfied. The Regulations do not simply work as the reverse of the disaggregation provisions by requiring all non-self-contained units in a building to be aggregated. Instead, they give discretion to the listing officer to aggregate where there is a self-contained multiple property occupied as more than one unit of separate living accommodation. Article 4 of the Council Tax (Chargeable Dwellings) Order 1992 (SI 1992/549) provides:

4(1)   Where a multiple property –

(a)   consists of a single self contained unit, or such a unit together with or containing premises constructed or adapted for non-domestic purposes; and

(b)   is occupied as more than one such unit of separate living accommodation, the listing officer may, if he thinks fit, subject to paragraph (2) below, treat the property as one dwelling.

(2)   In exercising his discretion in paragraph (1) above, the listing officer shall have regard to all the circumstances of the case, including the extent, if any, to which the parts of the property separately occupied have been structurally altered.

Multiple property means property which would, apart from the order, be two or more dwellings within the meaning of section 3 of the Local Government Finance Act 1992 (i.e. two or more separate hereditaments/dwellings).

The listing officer's discretion is informed by article 4(2). The whole multiple property has to be a self-contained unit. If units within it are self-contained they cannot be aggregated. Therefore, a block of self-contained flats cannot be aggregated. Nor can aggregation be used to reverse the effect of disaggregation because, in order to be disaggregated, units must, in the first place, be self-contained. The listing officer has to have regard to all the circumstances of the case, including the amount of structural alteration. The frequency of tenant turnover will also be a factor for consideration.

18.13 Appeals process

The system for challenging Council Tax bandings and for listing officers to amend the valuation lists is similar to the rating system. The Regulations for England are contained in the Council Tax (Alteration of Lists and Appeals) (England) Regulations 2009 (2009/2270).

If taxpayers consider their bandings to be incorrect they may be able to challenge them by means of making proposals which offer an alteration to the valuation list. The right to make proposals extends beyond the current taxpayer to interested persons, a term defined to include the owner if different from the taxpayer. Billing authorities may also make proposals in many of the same circumstances as taxpayers. Listing officers usually encourage taxpayers to make less formal enquiries to see if it is possible for their concerns to be satisfied without undertaking the formal appeals process of making a proposal; the availability of the proposal route remains should they not be satisfied.

As already explained, there are limitations as to what alterations may be made to a valuation list. For example, proposals or listing officer alterations may only be made to reflect improvements to dwellings when there is a sale, and improvements to the locality cannot be reflected until there is a national revaluation. Within these limits interested persons and billing authorities may make proposals under Regulation 5 when they consider the points set out below:

- A property is wrongly included in the valuation list. This includes disaggregated dwellings.
- A property should be included in the list as a dwelling.
- The valuation band is wrong. Generally this ground was available only for the first eight months of the life of the valuation lists, i.e. in England up until the end of November 1993. The exception is for people moving into a dwelling and becoming taxpayers for it for the first time. They are allowed six months from the date they become taxpayers to make a proposal. However, they can do this only if a proposal on the same facts for the same dwelling has not already been determined by a valuation tribunal or the High Court.
- There has been a material increase in the value of the dwelling coupled with a subsequent sale.
- There has been a material reduction in the value of the dwelling.
- The dwelling has become or ceased to be a composite hereditament, or there is a change in the balance of domestic and non-domestic use.
- A valuation list entry should be altered to take account of a relevant decision by a valuation tribunal or the High Court. A proposal on this ground must be made within six months of the decision.
- An alteration to the valuation list made by the listing officer is incorrect. A proposal on this ground must be made within six months of the service of the listing officer's notice of the alteration.

The Regulations provide detailed requirements to be satisfied for proposals to be validly made. They include the following points:

- They must be in writing.
- They must be served on the listing officer for the billing authority containing the dwelling. It will not be validly made if served on the wrong listing officer.
- They must give the name and address of the person making the proposal.
- They must state the capacity of the proposer, e.g. taxpayer or owner.
- They must identify the dwelling/dwellings. It may be for more than one dwelling if it proposes the deletion of the dwellings or, in other cases, if

the dwellings are within the same building or curtilage and the proposer is making the proposal in the same capacity for each of the dwellings, e.g. as owner.

The proposal needs to include a statement of the reasons why it is believed the valuation list is inaccurate. If the proposal is on the basis of a material increase coupled with a sale, or a material reduction, or the property has become or ceased to be a composite hereditament, or there has been a change in the balance of domestic/non-domestic use then the proposal must:

- State the reasons for believing the event occurred.
- State the date the change is believed to have occurred.
- If the proposal is challenging an alteration of the list by the listing officer, it must identify the alteration challenged. This can be done by stating the day on which it occurred. If disputing an effective date, then the substitute date must be stated if the proposal relies on a decision of a local valuation tribunal or the High Court; then the property concerned must be identified, giving the date of the decision and stating whether the decision was by a local valuation tribunal or the High Court.
- If the proposal is made on the basis that the proposer is a new taxpayer then the proposal needs to state the date the proposer became the taxpayer.
- If the proposal is made to challenge an alteration to the valuation list made by the listing officer then the proposal needs to state the date of the listing officer's notice.
- If the proposal disputes the effective date of an alteration the proposal needs to state the date proposed in substitution.

The Regulations provide for procedures to be followed if a listing officer considers that a form or letter he or she has received purporting to be a proposal is not valid. On receiving an invalid proposal the listing officer may, within four weeks of its service, notify the sender of the reasons for considering the proposal not to be validly made and advise the sender of the sender's right either to appeal to the valuation tribunal within four weeks of receiving the invalidity notice or to make a fresh proposal rectifying the defect(s) in the original proposal.

Having received a valid proposal, the listing officer has to acknowledge it within four weeks of receipt. The listing officer is allowed four months to consider whether its contentions are correct and alter the list treating the proposal as 'well founded', discuss and negotiate to reach an agreement on what alteration should be made or decide that the proposal is not well founded. The listing officer then sends a 'decision notice' to the proposer, any other taxpayer for the dwelling(s) covered by the proposal and any 'competent person' (meaning anyone else who had the right to make the proposal).

Having received the decision notice the proposer or 'competent person' either can accept the decision or has three months in which to 'appeal direct' to the Valuation Tribunal for England (VTE). The appeal has to be in writing and must give reasons for the appeal and other information.

The Council Tax appeals system provides for the VTE to decide appeals in a similar manner to rating appeals. However, unlike rating appeals, an appeal from the VTE lies to the High Court but only on a point of law. The VTE's decision on a value band is therefore final unless some error of law is established.

18.14 Effective dates

The Regulations provide for changes to bandings to become effective for billing purposes from set dates. These are contained in Council Tax (Alteration of Lists and Appeals) (England) Regulations 2009 (2009/2270) Regulation 11 and vary depending on the type of alteration:

- When a new dwelling is constructed or converted from an existing property, the effective date of the new entry is from the date the dwelling is finished and comes into existence. This does not apply where the alteration is to disaggregate a single dwelling because it comprises more than one self-contained unit and treats it as being a number of dwellings. In the case of disaggregation the date is the date the alteration is entered in the list.
- The completion notice procedure used in rating (see Chapter 17) also applies to Council Tax. Where a notice has been served, a new dwelling will be treated as coming into existence on the completion day specified in the notice.
- Where there is a material increase in value due to, for example, an extension, and there is a subsequent sale allowing the value of the improvement to be included in the value for banding, the effective date is the date the alteration is entered in the list.
- If the banding is reduced because there is a material reduction in value then the effective date is the date of the physical change which caused the reduction.
- If the alteration is to reflect a property becoming or ceasing to be a composite dwelling or there is an increase or decrease in the domestic use of a composite dwelling then the effective date is the date this happens.
- Sometimes it will be realised many years after Council Tax came into force that the original banding was incorrect. If the correction reduces the band then the Regulations allow the backdating to be taken back to 1 April 1993. On the other hand, so that council taxpayers do not receive backdated bills for an error not of their own making, where the banding was set too low in the original compiled list, or the dwelling should have been disaggregated from 1 April 1993, the correction is made effective only from the day the list is altered.
- In most cases where a list is altered and the alteration is later found to be incorrect, the amending alteration has effect from when the original alteration had or should have had effect. The exception is where the originally

altered figure was too low or the dwelling should have been disaggregated. As with similar compiled list inaccuracies the effective date is the date the list is altered.

18.15 Revaluation

The Government announced in the 2000 White Paper *Strong Local Leadership – Quality Public Services* that there would be a cycle of ten-yearly Council Tax revaluations. The Local Government Act 2003 inserted section 22B into the Local Government Finance Act 1992, requiring a statutory ten-yearly revaluation cycle. By order, a revaluation of bands came into force on 1 April 2005 in Wales with an AVD of 1 April 2003. It was intended that England would follow in 2007 with an AVD of 1 April 2005. However, the English revaluation was postponed in September 2005 and subsequently cancelled. The VOA had already been working on valuations for the English revaluation using an Automated Valuation Model derived from work on computer aided mass appraisal (CAMA) in the USA. The exercise was well advanced at the time of its postponement and a 'first pass' at valuation had already been undertaken representing the largest single mass appraisal carried out anywhere in the world. Future revaluations are likely to be based on these techniques, although the 2005 Welsh exercise was undertaken without computer valuation support. The 2003 Act also provided for the numbers of bands to be changed at a revaluation, and this was done in Wales with a ninth band being added as Band I.

The lack of a revaluation in England means Council Tax, remarkably, is still levied on the basis of 1991 property values – values some 32 years out of date! Inevitably, the result for those householders whose property values have not kept pace with the general increase in values is that they are paying too much and for those whose properties have gained in relative value that they are paying too little. It also means those occupying dwellings which have been extended since 1 April 1993 without a subsequent sale are not paying in line with their neighbours on the relative value of their properties.

18.16 Wales

A revaluation was undertaken of property in Wales in 2005 with a valuation date of 1 April 2003.

The value bands were revised in Wales for the 2005 Revaluation and are now different from those in England.

In October 2022 the Welsh Government closed a consultation 'A Fairer Council Tax' on reforms to Council Tax in Wales with both an initial consultation and a later Phase 2 consultation on more detailed proposals.

Views were sought on:

- undertaking a revaluation in 2025 of all 1.5 million dwellings in Wales with an AVD of 1 April 2023;

- adding more council tax bands (for example at the top and bottom ends of the scale) and revising the bands to help make the system more progressive;
- undertaking revaluations on a more regular basis at least every five years in the future, to ensure the tax burden is shared fairly amongst taxpayers;
- whether when dwellings are significantly improved they should be rebanded if the value is band significant, whereas at present this, normally, can only happen on a sale;
- improving the framework of discounts, disregarded persons, exemptions and premiums to ensure the arrangements are better aligned with priorities;
- improving the Council Tax Reduction Scheme which provides support to low-income households.

Under the Regulations which came into force on 1 April 2008 in England, all valid proposals to alter Council Tax lists received by listing officers are to be considered within a period of four months and a decision notice issued. If dissatisfied with the listing officer's decision, the recipient may then 'appeal direct' to the VTE within three months of receiving the notice. In Wales the system prescribed by the 1993 Regulations (as amended) still applies as it also did in England prior to 2008, whereby if the proposal is not settled the listing officer is required to refer the disagreement to the relevant valuation tribunal without the appellant needing to take any action. This constitutes an appeal by the proposer against the listing officer's refusal to alter the list. The reference must be no later than six months from the date the proposal was served.

Properties undergoing repair or structural alteration and that are vacant attract a six-month exemption.

A power exists to charge up to 200% on holiday homes with no permanent resident.

An additional aggregation requirement is contained in the Council Tax (Chargeable Dwellings) (Amendment) (Wales) Order 2014 (2014/2653). This amends the Chargeable Dwellings Order for Wales only and requires any property that satisfies the definition of a 'refuge' to be treated as a single dwelling with a single band. This aggregation provision applies irrespective of whether any part of the refuge is self-contained or would otherwise constitute separate dwellings. A 'refuge' means a building in Wales which is operated by a person otherwise than for profit and is used wholly or mainly for the temporary accommodation of persons who have been subject to any incident or pattern of incidents of controlling, coercive or threatening behaviour, physical violence or abuse from persons to whom they are or were married or with whom they are or were in a civil partnership or co-habiting.

19 Rating in Northern Ireland

19.1 Introduction

Both the Republic of Ireland and Northern Ireland are within the rating family – countries that use some form of rating derived from the system that originated with the 1601 Poor Relief Act of Elizabeth I. The two Irish systems have significant similarities as they originate from the common Valuation of Ireland Act 1826 during the reign of George IV.

Sir Richard Griffith (1784–1878) was appointed as the first Commissioner of Valuation to oversee the implementation of the 1826 Act. This required a survey of all property as well as its valuation in what is known as the 'Townland Valuation'.

The Valuation of Ireland Act 1852 provided for the first general valuation of all tenements in Ireland, a task which was undertaken and completed in 1864 during Sir Richard Griffith's long career. Whilst formally known as the Primary Valuation of Ireland it is more commonly known as 'Griffith's Valuation'.

Completion dates for Griffith's Valuation varied as Griffith's team moved across Ireland. Most in Ulster date from 1858 to 1864. Griffith's Valuation was revised annually, at first under the new Commissioner, Sir John Ball Greene CB. A new General Revaluation of Northern Ireland took place in 1935 followed by subsequent revisions and revaluations with the most recent in 2020.

The basis set down in the 1852 Act for valuation was Net Annual Value (NAV). This was defined in section 11 as:

The rent for which, one year with another, the same might in its actual state be reasonably expected to let from year to year, the probable average annual cost of repairs, insurance, and other expenses (if any) necessary to maintain the hereditament in its actual state, or rates, taxes, or public charges, if any, (except tithe rentcharge) being paid by the tenant.

However, land was to be valued by reference to set standard average prices of agricultural produce contained in the Act.

Originally both domestic and non-domestic properties were subject to the same rating system but since 2007 in Northern Ireland domestic rates have

DOI: 10.1201/9781003297369-19

been assessed on a capital value basis with non-domestic premises staying liable to rates assessed on rental value.

The Rates (Northern Ireland) Order 1977 (1977/2157) is the principal measure setting out the current system of rates in the province of Ulster. It is a comprehensive code, all in one document, covering the rating system in 146 pages.

There are approximately 73,000 non-domestic hereditaments in the province with a total NAV of approximately £1.64 billion. The total receipts from both domestic and non-domestic rates is some £1.33 billion with about half coming from non-domestic rates (2018/19).

Rates remain a local tax with the amount in the pound being set for the District Rate by the individual local councils and for the Regional Rate by the Department of Finance.

19.2 Rateable occupation and the hereditament

There is little differentiation in identifying the rateable occupier between England/Wales and Northern Ireland but there are differences in identifying the hereditament.

'Hereditament' is defined in article 2(2) to the 1977 Order as:

property which is or may become liable to a rate, being a unit of such property which is, or would fall to be, shown as a separate item in a valuation list.

The leading case on the hereditament in Ireland is *Switzer and Co. Ltd* v *The Commissioner of Valuation* [1902] 2 IR 2 75, decided by the King's Bench Division and confirmed by the Court of Appeal. This case lies at the heart of rating law in Northern Ireland. It concerned nine separate leases but with one occupier. It was decided that where property is held under one lease, by one occupier from one immediate lessor it must be entered in the valuation list as a separate entry. The way in which a property is held is therefore very important in identifying the hereditament.

It was decided when a very different definition of hereditament was in force (section 12 of the Valuation of Ireland Act 1852) but recent cases have not considered this alters the rule in *Switzer*.

Given the likely practical difficulties of applying *Switzer* in its purest form the legislature has made provision to ameliorate the effects of that decision (see the following).

19.2.1 *(i) General discretionary power to amalgamate*

Section 8 of the Valuation Acts Amendment Act (Northern Ireland) 1932, now as article 38(3) of the Rates (Northern Ireland) Order 1977 (1977/2157), gives the Commissioner power to amalgamate contiguous hereditaments in the occupation of one and the same occupier as a single hereditament, notwith-standing that they are held under different titles.

38.– (3) Notwithstanding anything contained in paragraph (2), the Commissioner, or the district valuer with the approval of the Commissioner, may, if he thinks it proper to do so having regard to the circumstances of the case,–

(a) value contiguous hereditaments in the occupation of one and the same occupier as a single hereditament, notwithstanding that they are held under different titles;

(b) where a hereditament comprises two or more parts capable of separate occupation, although in the same occupation, value the several parts as separate hereditaments;

and where hereditaments or parts of a hereditament are valued as mentioned in sub-paragraph (a) or (b), they shall be treated as a single hereditament, or, as the case may require, as separate hereditaments, for all the other purposes of this Order.

It is important to note that this is a power of amalgamation and enables the *Switzer* position to be overridden by decision but does not alter the under-lying law.

19.2.2 (ii) Industrial hereditaments

The Rating Valuation (Apportionment) Act (Northern Ireland) 1928 section 4, now Schedule 2, paragraph 3, of the Rates (Northern Ireland) Order 1977 (1977/2157), in relation to industrial hereditaments, provides for industrial hereditaments in the same occupation which are contiguous or within one curtilage to be treated as if they were a single hereditament for the purpose of deciding if the hereditaments together are industrial and entitled to be distinguished in the list as industrial.

Where two or more properties within the same curtilage, or contiguous to one another are in the same occupation and, though treated for any reason as two or more hereditaments for the purposes of valuation and rating, are used as parts of a single mine, quarry or factory, then, for the purposes of determining whether the several hereditaments are industrial hereditaments, they shall be treated as if they formed parts of a single hereditament comprising all those hereditaments.

This paragraph does not alter the actual number of hereditaments – it does not result in amalgamation. It enables a hereditament which would otherwise not be industrial to enjoy industrial derating. So if there were buildings A, B and C occupied together in a curtilage under separate leases and A and B were largely industrial but C was offices then the apportionment would be worked out by treating the three hereditaments as one and considering whether derating applied to the whole. If it did then the amount of non-industrial NAV would be apportioned between the three units as a common percentage.

Effectively the provision only has effect where either contiguous properties held under one lease have not been amalgamated under article 38(3) or the properties are not contiguous but within the same curtilage.

In *Belfast Collar Co. Ltd* v *Commissioner of Valuation* [1959] NI LR 198, the Court of Appeal considered what was meant by the word 'curtilage' in relation to the requirement to treat industrial units as if they were one for the purposes of deciding the extent of industrial distinguishment. In the case, two properties were divided by a highway and cloth was stored on one side of the road and transported to the property on the other side for processing.

The lower court had found the two premises to be one following the English approach to a functional connection but the Court of Appeal emphasised the difference between the Irish and English statutory backgrounds. It reaffirmed the *Switzer* case which had established 'premises had to be entered and valued having regard to the leases on which they were held, and that there had to be separate entries for each hereditament held on a separate lease'. It said the *Switzer* decision was 'founded on a consideration of the provisions of the Irish Valuation code and has ever since been a basic part of the valuation law of this country'.

The Court of Appeal's view was that *Switzer* gave no opportunity for premises either side of a highway to form a single hereditament and rejected the idea that some functional connection between a factory and a store in single occupation on either side of a street might make them one hereditament.

The ratepayers nonetheless argued the two properties were within one curtilage and that allowed them to be considered under Schedule 2, paragraph 3, as if they were one hereditament. Lord MacDermott LCJ rejected this:

> *The definition of 'curtilage' contained in the Shorter Oxford English Dictionary is – 'A small court, yard, or piece of ground attached to a dwelling-house and forming one enclosure with it.' The word 'enclosure' is, perhaps, the key word, signifying a close or enclosure containing one, or even more than one, tenement. I cannot find, however, that these two hereditaments are within the same curtilage, divided as they are by a public highway. There is no authority to indicate that one can have a curtilage bisected in that way by a public highway. Accordingly, that being my view, and it being conceded that the premises are not contiguous, I am of opinion that section 4(3) cannot apply.*

The use of the word 'enclosure' probably extends the definition beyond a small court yard perhaps as far as buildings within the boundary fence of an industrial park.

It is interesting to note that the *Switzer* decision was founded on a very different definition of hereditament in the 1852 Act, yet the 1977 Order definition of hereditament is very similar to the English/Welsh definition. It seems, though, that the *Switzer* requirement to look at the lease in order to establish the hereditament is very strongly embedded in the law and unlikely to be upset by the *Woolway* decision in relation to England, Wales and Scotland.

19.3 Definition of Net Annual Value

19.3.1 Basic definition

Schedule 12 to the Rates (Northern Ireland) Order 1977 (1977/2157) provides the basis of valuation in paragraph 1:

1   Subject to the provisions of this Schedule, for the purposes of this Order the net annual value of a hereditament shall be the rent for which, one year with another, the hereditament might, in its actual state, be reasonably expected to let from year to year, the probable average annual cost of repairs, insurance and other expenses (if any) necessary to maintain the hereditament in its actual state, and all rates, taxes or public charges (if any), being paid by the tenant.

An important proviso to the definition of NAV is contained in paragraph 2:

2(1)   Subject to sub-paragraph (2), in estimating the net annual value of a hereditament for the purposes of any revision of a valuation list, regard shall be had to the net annual values in that list of comparable hereditaments which are in the same state and circumstances as the hereditament whose net annual value is being revised.

This paragraph makes the existing level of value in a valuation list very much a matter to be taken into account when considering the correctness of a valuation in a new list or in deciding on the value for a new or altered hereditament. This does not apply where the valuation is being made by reference to accounts following sub-paragraph 2(2):

(2)   Sub-paragraph (1) shall not apply to any hereditament for whose valuation special provision is made by or under Part IV or any of the succeeding Parts of this Schedule, or to any hereditament whose net annual value falls to be ascertained by reference to the profits of the undertaking or business carried on therein.

For mineral hereditaments such as mines where the output extracted is very relevant, then paragraph 4 requires the probable quantity to be adopted to be that for the first year with respect to which the valuation will be in force.

19.3.2 *Uniformity and tone of the list*

The task of the district valuer and Commissioner in preparing a valuation list is comparatively straightforward and little different from the task of valuation officers in England and Wales. Valuations need to follow Schedule 12, paragraph 1, taking the valuation date as that provided under article 39A – the antecedent valuation date (AVD) currently being 1 April 2018 for the 2020 general revaluation. Like in England and Wales, all factors about the hereditament and its locality and beyond are taken as at this date except those provided as a list of 'matters', which are taken at the date the list comes into force. These are contained in paragraph 39A(1A) and are broadly physical matters about the property and locality (they are covered in more detail in section 19.4). So, in essence, it is a valuation at the AVD but modified so that the physical state of the hereditament and the locality are taken as at the date the list comes into force.

The difference to England and Wales becomes sharply visible when dealing with 'revisions' to the valuation list. An old problem for rating has been in dealing with changing property values. If there was no property price inflation or deflation, rating valuations could always be done at current values. But values do move, and it would be unfair if some hereditaments were valued at one date and others at other valuation dates. The solution has been various ways of establishing a 'tone of the list' whereby any valuations made after a list comes into being aim to be in line with the existing 'tone' of that list. The original English/Welsh way of achieving this was to have a ceiling as a limiting factor on current values set so that the valuation in the list could not exceed the value which would have applied had the current hereditament been valued for the original list though it could fall below it.

The modern English/Welsh approach is to require valuations to be made as at the AVD but modified by various physical matters about the property and locality taken at the current date. So a new hereditament is valued as it is now but imagining it existing back at the time of the AVD. A tone of the list comes about by consideration by ratepayers and the valuation officer as to what the correct level of values should have been at the AVD and ensuring all properties in that class are valued at that level.

Northern Ireland's legislation adopts a different approach. It does this by requiring that in making any revision to the original list (paragraph 2(1)),

regard shall be had to the net annual values of comparable hereditaments in that list which are in the same state and circumstances as the hereditament whose net annual value is being revised.

This means the original NAVs appearing in the valuation list constitute not just the Commissioner's tone but also the actual 'tone of the list' and are to be followed.

At a general revaluation, evidence of rents and turnover are applied in the context of the actual AVD economic state to arrive at valuations. The principles upon which the Commissioner applied at the general revaluation stage cannot be challenged on revision but individual valuations can be challenged at the revision stage. As the Lands Tribunal said in *McKeown Vintners* v *Commissioner of Valuation* VR/9/1985:

> *it is ... absurd to argue that a change in economic or trading conditions can be used at revision stage to attack, or even demolish, principles which were applied at general revaluation stage. The Tribunal does agree however that at revision stage any individual entry, whether relating to the hereditament under revision or not, can be challenged and shown to be incorrect, but this is a very different matter from an attack on the principles applied at a general revaluation. There is no machinery for challenging the validity of the principles in the Lands Tribunal. If, as the Ratepayer contends, they are fundamentally flawed then it may well be that his remedy lies elsewhere.*

As the Tribunal pointed out in the case:

> *As time progresses, if actual rental levels and turnover figures were used for the revision of a particular entry in the valuation list, it would inevitably result in that entry being increased to a level significantly higher than other entries in the list. There must therefore be a limiting factor, and this is provided by paragraph 2(1) which, in essence, produces what is often termed a 'tone of the list', and which ensures fairness and uniformity. It does this by providing that at revision stage regard 'shall be had' to the net annual values in the valuation list of comparable hereditaments ... Suffice to say that the significance of this role increases with the passage of time ... It is a limiting factor designed to maintain uniformity over intervals of five years (between general revaluations).*

The restriction to have regard to other assessments in the valuation list which are similar acts is a limiting factor. It is the Northern Ireland means of ensuring fairness and consistency. All hereditaments are assessed in line with others that are similar. In considering any revision the district valuer, the Commissioner and indeed the Lands Tribunal must use as the primary guideline not rental evidence or other evidence but the NAVs of comparable hereditaments which are in the same state and circumstances as the appeal hereditament. Economic changes are not relevant because the valuation date is fixed at the AVD but nor are other changes which might affect value relevant if they apply equally to a whole class because each will remain assessed comparably to the others. So, for example, if a new building resulted in a doubling of modern factory space in a town this would not affect NAVs. The new factories would be valued in line with the existing comparable factories and the greater supply which very probably in the real world would have reduced current rents would not affect NAVs because all the factories would be in the same state and circumstances as the other factories. Only at a general revaluation could the change in demand be taken into account (see discussion of *Trustees of Glenkeen Orange Hall* case below).

The Tribunal did explore where there were no true comparables. In this situation it seems paragraph 2(1) has no effect and the valuer simply has to determine the NAV under paragraph 1.

A number of further cases have considered the meaning of paragraph 2(1).

In *Trustees of Glenkeen Orange Hall* v *Commissioner of Valuation* [1994] VR/31/ 1993, the question before the Northern Ireland Lands Tribunal was whether the assessment of an Orange hall should be reduced to take account of a change in the relative value of such halls reflecting the changed demand for their use as a result of an increased availability of superior alternative premises and social change in the rural community.

The Tribunal explained the legislation had changed in 1970 from a requirement to take current state and circumstances into account, and value as if these matters had been so at the last general revaluation date and had been replaced by a requirement to refer to comparables in the valuation list. The effect being that new or altered premises had to be valued so as to keep them in line with valuations in the list of others in the same class and/or locality.

Under the earlier scheme, if circumstances affecting a whole class and/or locality had changed and would have affected rental values then the new circumstances could have been taken into account and the valuations of the entire class changed. But, under the current scheme, the subject of revision would be in the same circumstances as others of that class and/or locality in the valuation list and so its valuation must follow the existing tone.

A-Wear v *Commissioner of Valuation* [2001] VR/3/2001 concerned the levels of value in a newly built shopping centre, The Quays, situated within close proximity to an existing but recently improved shopping centre called the Buttercrane.

The Commissioner considered the correct legal position, in having regard to the tone of the comparables, was to look at the situation of the comparables when the new list had been issued and before the improvements to the Buttercrane. He considered that The Quays units were comparable to those in the original Buttercrane prior to the improvements and should be assessed at that level. His reasoning was that under Schedule 12, 'regard shall be had to net annual values in the valuation list of comparable hereditaments which are in the same state and circumstances as the hereditament whose net annual value is being revised'. He considered that the original Buttercrane Centre prior to improvements was in the same state and circumstances.

The Northern Ireland Lands Tribunal disagreed with this interpretation. It considered the comparison should be with the present situation and not the past and that the Schedule did not provide for any assumption that comparables are in a state and circumstances different from those at the relevant time. It concluded that as by this time Buttercrane was superior to The Quays it followed that The Quays should have a lower relative valuation.

The Lands Tribunal pointed to the 'absurdity' that, if the Commissioner was right and had The Quays in fact been as good as the improved Buttercrane, then the result would be that The Quays should be entered at a higher level on the grounds that the units would be superior to the units in the original Buttercrane Centre – which would be unfair.

As mentioned in *McKeown Vintners* v *Commissioner of Valuation* VR/9/1985, the principles which the Commissioner applied at revaluation cannot be challenged at the Lands Tribunal. They can, of course, be challenged by Judicial Review.

In *Belfast International Airport* v *Commissioner of Valuation* [2017] VR/9/2016, the date at which the economic circumstances of the hereditament should be applied for a revision was considered. The appellant said the economic landscape of the AVD of 1 April 2001 should be adopted, as was the case at the revaluation, but the Commissioner contended for 1 April 2003, the date the Fifth General Revaluation List came into effect.

The Tribunal was in no doubt the date was the AVD:

> *The suggestion by the respondent that on a revision, which dealt with the tone of the list, and with the rents of comparables, a date different to the date when the comparable rates were assessed should be taken, makes no sense. As the appellant submits 'the valuer has little chance of having regard to the tone of the list that paragraph 2(1) of the Schedule 12 requires, if, in his revision valuation, he took general economic conditions (including demand) at a date different from the date at which they were taken for the revaluation'.*

and:

> *It seems to us that the purpose of the legislation, namely to achieve fairness and uniformity among ratepayers of different hereditaments, is much better achieved if demand, which dictates value, is assessed at the AVD both for revaluation and revision.*

19.3.3 Disrepair

The definition of NAV states that 'the tenant is responsible for the average annual cost of repairs'. This means that if the property is capable of beneficial occupation the repair assumption to be made is that the property is in its actual state of repair subject only to minor alterations.

The main case considering repair in Northern Ireland is *Alfred Black* v *Commissioner of Valuation VR/4/1998*. In this case an old terraced property used as a dry cleaners was in disrepair. The Lands Tribunal explained Schedule 12 of the 1977 Rates Order meant that:

> *The premises are to be valued in their actual state and put to their most valuable use in that state and, secondly, the hypothetical tenancy is on the assumption*

that the cost of repairs are to be met by the tenant who is obliged to maintain the premises in that actual state, no better or no worse.

In looking at valuation the tribunal preferred the approach of the Commissioner's expert who had adjusted the £ per m^2 placed on the areas where the repair was not as good as that in the comparables.

If the property is in such a poor state of repair that it is no longer capable of commanding a rent then it should be removed from the valuation list. Examples of such hereditaments might include those that have been extensively damaged by bomb, fire or flood. If the property is undergoing major repair works with a planned schedule of works, then it will clearly not be capable of rateable occupation, and should be deleted from the valuation list.

19.4 Revaluation

Article 38(1)(a) of the Rates (Northern Ireland) Order 1977 (1977/2157) gives the Commissioner of Valuation the power to 'conduct ... such general revaluations of hereditaments as are necessary for the preparation of new valuation lists under Article 45'.

Article 45 of the Rates (Northern Ireland) Order 1977 (1977/2157) sets out the legislation for the preparation of new valuation lists.

Article 39A of the Rates (Northern Ireland) Order 1977 (1977/2157) goes on to state that 'the NAV of any hereditament in a new valuation list coming into effect on the 1st of April shall be ascertained by reference to such an earlier time as the department shall determine ...'. This 'such earlier time' is the AVD used by most modern rating systems.

The most recent revaluation, which came into force on 1 April 2020, has an AVD of 1 April 2018. The next revaluation is planned for 1 April 2023 but with an antecedent date of 1 October 2021, an advancement of six months compared to the previous two-year gap between AVD and publication.

The 1977 Order was amended by the Rates (Amendment) Act (Northern Ireland) 2012 to clarify what elements of the state and circumstances of the hereditament and locality should be taken to be as they were at the date the list comes into force rather than at the AVD. The earlier wording in assuming the 'hereditament was in the same state and circumstances as at the time the list came into force' could have been interpreted as requiring many more factors concerning the locality to have been taken as at the time the list came into force. The amended wording now has strong similarities to the English/Welsh differentiation between matters to be taken as at the AVD and those at the time the list comes into force. The article reads:

39A(1)   Any net annual value to be ascribed to any hereditament in a new NAV list coming into force on 1st April in any year shall be ascertained by reference to such earlier time as the Department may by order subject to negative resolution specify, but on the assumption that at the time

specified in the order the matters mentioned in paragraph (1A) were as at the time when the list comes into force.

(1A)   The matters referred to in paragraph (1) are–

- matters affecting the physical state or physical enjoyment of the hereditament;
- the mode or category of occupation of the hereditament;
- matters affecting the physical state of the locality in which the hereditament is situated; and
- the use or occupation of other premises situated in the locality of the hereditament.

In essence the legislation requires the NAV in a new list to be ascertained by reference to the AVD, but on the assumption, first, that at the AVD the mode of use and physicality of the hereditament are as at the date the list comes into force and, second, that the assumption is widened to include the use or occupation of other premises situated in the locality of the subject and matters affecting the physical state of the locality. Economic and other changes between the AVD and the date the list comes into force cannot be taken into account.

As mentioned earlier, Schedule 12, paragraph 4, to the 1977 Order makes the additional assumption for mineral hereditaments that the probable quantity to be assumed is that for the first year with respect to which the valuation will be in force. This is different from hereditaments such as pubs or multi-storey car parks as their NAVs will be fixed by reference to the likely volume of trade at the AVD of the list and not the year the valuation will come into force. This is because trade is not mentioned in the list in 39A(1A).

The Non-domestic Rates Valuations (Coronavirus) Act (Northern Ireland) 2022 varied the application of Article 39A (1A) making the direct or indirect effects of coronavirus to be disregarded in considering physical matters at the time of a new valuation list. Any impact on NAVs of physical matters affecting the property or its surrounding locality at the date of publication of a new list, which are due to coronavirus restrictions, are to be disregarded. This applies both to the circumstances of the 2020 List and subsequent non-domestic lists.

1.–(1)   For the purposes of the application of Article 39A of the Rates Order (valuation of hereditament) to the net annual value list that came into force on 1st April 2020 and to each subsequent net annual value list, the matters mentioned in paragraph (1A) of that Article are to be treated as not including, and as never having included, any matter directly or indirectly attributable to coronavirus.

Article 39A (1A) will apply as originally intended in all other circumstances.

The Act also provides for amendments to the legislation to allow it to apply should variations of the COVID-19 coronavirus infection arise in future.

Clause 2 of the Act outlines the matters that are directly or indirectly attributable to coronavirus but is not limited to the examples mentioned.

These are, anything done by any person:

a) in compliance with statutory provisions to prevent the spread of coronavirus;
b) in compliance with any other statutory provisions relating to the incidence or spread of coronavirus; or
c) in response to advice or guidance given by a public authority relating to the incidence or spread of coronavirus.

While not specified in the legislation, practical examples of the consequences of coronavirus, which may lead to a direct or indirect impact on Article 39A (1A) matters, are likely to include, but are not limited to:

• closure of premises, either the hereditament itself or others in the locality as a result of coronavirus measures;
• restricted/reduced operating times/opening hours;
• restrictions on the activities that can be carried on (seated eating, dancing, singing);
• maximum capacity;
• restrictions on numbers of different households in certain settings;
• social distancing;
• requirements for physical barriers;
• additional costs associated with coronavirus for the provision of cleaning, hand washing, one-way pedestrian systems, signing in etc.;
• vaccine status checks.

The impact of the coronavirus pandemic and other economic forces on the property market will be captured for Revaluation 2023 and will be reflected in NAVs in the 2023 List, to the extent that they have affected the rental market at the new AVD, 1 October 2023.

The legislation will again act to ensure any direct or indirect consequence of coronavirus that may occur between the AVD, 1 October 2021, and the date of publication, 1 April 2023, are disregarded for the purposes of Article 39A (1A). So, for example, new physical restrictions to prevent the spread of coronavirus coming into effect at the date of publication, will be disregarded.

All other physical matters that occur between the AVD and date of publication, leading to consideration of an Article 39A (1A) matter at the effective date of the new list 1 April 2023, are unaffected by the legislation.

19.5 Exemptions, reliefs and distinguishment

There are three ways in which a ratepayer's liability might reduce on account of the ratepayer's business or activities or the characteristics of the premises:

- The property may not be considered a hereditament.
- It may be wholly or partly distinguished.
- A relief may be available.

The provisions contained in the 1977 Order are quite complex, and it is not possible in this textbook to cover all the nuances and detail of the various exemptions and reliefs. The relevant article and associated schedules in the 1977 Order should be consulted when considering a problem.

19.5.1 Not a hereditament

Article 37 provides in Schedule 11 to the 1977 Order a list of various types of property which are not to be treated as hereditaments under Schedule 10. This is a similar list and uses much the same wording as the England and Wales Schedule 5 exemptions, though the list is not the same.

1	Agricultural land
2	Agricultural buildings
3	Livestock or poultry buildings
3A.	Fish farms
3B.	Salmon fishings or eel fishings
3C.	Buildings associated with salmon or eel fishings
4	Turf bogs and turf banks used for making turf mould, for fuel or manure
5	Mines – which have been opened for less than 7 years or are abandoned
6	Lighthouses etc.
7	Sewers, pumping stations etc.
8	Cemeteries and crematoria not yielding a private profit
9	Air raid shelters
11A.	Public parks
10	Property excluded by local acts etc.

Such properties are simply not hereditaments and therefore should not appear in valuation lists.

19.5.2 Distinguishment

Certain hereditaments are to be distinguished in the valuation list either completely or to the whole of the extent they are used for exempting purposes.

   The distinguishment is shown alongside the NAV in the relevant list by way of a code, e.g. industrial (IN), exemption (EX), sport and recreational (S&R), and freight transport (FT).

19.5.2.1 Hereditaments altogether of a public nature

Article 41(2)(a) covers hereditaments altogether of a public nature and occupied and used for the purposes of the public service.

It has long been the case in Ireland that exemption has been granted to hereditaments which are altogether of a public nature and are occupied and used for purposes of the public service. Indeed this sub-paragraph essentially re-enacts section 2 of the Valuation [Ireland] Act 1854 which provided for the distinguishment in the valuation lists of 'all hereditaments and tenements, or portions of the same, of a public nature' and went on to provide for their exemption from rates 'so long as they shall continue to be of a public nature and occupied for the public service'.

Clearly property in which each member of the public can be said to have an interest such as a road dedicated to the public comes within the exemption but so also does property occupied by a local authority for the purpose of providing a service for the benefit of members of the public who are inhabitants of the district, where the service is also in fact available on equal terms to other members of the public, e.g. a swimming pool. Article 41(b) specifically restricts the exemption from applying to hereditaments occupied for the purposes of a public utility or by a body specified in Schedule 13 (the fire and police authorities, certain education bodies and the Board of Trustees of the National Museums and Galleries of Northern Ireland).

In *Commissioner of Valuation* v *Lurgan Borough Council* 1968 NI 104, it was found that 'public purposes' extended beyond government purposes and meant those in which all members of the community were interested as opposed to individuals, a class or a locality on its own. The Court of Appeal held that members of the public in general had an interest in the Lurgan swimming pool and that it was established and used for the benefit of all the public at large, not merely the inhabitants or ratepayers of the borough of Lurgan or even Lurgan and surrounding districts. It was also considered whether charges levied for the use of the swimming pool should deny exemption, but it was held that limited restrictions on the use of the hereditament by members of the general public may be ignored 'if they are purely regulatory'.

19.5.3 *Religious occupations*

Article 41(2)(b) covers churches, chapels and similar buildings, occupied by a religious body and used for the purposes of public religious worship and church and chapel halls or similar buildings and ancillary buildings.

19.5.3.1 *Charitable occupations*

Article 41(2)(c) covers hereditaments occupied by a charitable body and used wholly or mainly for charitable purposes. The test is a double one of both occupation and use. Occupation by a charity is not sufficient; the use must be in connection with the charitable objects. The article provides that any use for profit, unless directly facilitating the carrying out of the purposes of the charity, is not to be treated as a use for charitable purposes.

In *Irish World* v *the Commissioner of Valuation* [1990] RA 263, the Lands Tribunal found that, while a body was recognised by the Inland Revenue as being a charity to promote knowledge and education about Irish culture, its occupation of a shop in Dungannon was not exempt because it was used not wholly or mainly for the charitable objects of the society but rather for administration and other purposes.

19.5.3.2 Hereditaments not established for profit

Article 41(2)(d) covers hereditaments not established or conducted for profit, whose main objectives are charitable or are concerned with science, literature or the fine arts and are used wholly or mainly for the purposes of those main objects.

In *Lisburn Camera Club* v *the Commissioner of Valuation* [VR/105/1978], the Tribunal found that (a) the occupier was a body not established or conducted for profit and (b) that the club's main objects were concerned with the fine arts as shown by its Rule 2, 'The objects of the Club shall be the advancement of the Science and Art of Photography in all its branches.' It was therefore entitled to be distinguished as exempt under article 41(2)(d)(ii).

19.5.3.3 Recreational

Article 41(2)(e) covers hereditaments used wholly or mainly for purposes which are declared to be charitable by the Recreational Charities Act (Northern Ireland) 1958. This requires the provision to be in the interests of social welfare, to be available to members of the public at large and to be of public benefit.

All of the foregoing are provided under article 41. This article provides where a hereditament is not wholly used for exempting purposes the NAV is apportioned between the exempting and non-exempting purposes and the non-exempting apportionment is to be shown as the NAV in the valuation list. Where the whole is exempt it is shown as nil rateable value (Schedule 7).

19.5.3.4 Community halls

Article 41A distinguishes as wholly exempt from rates any hereditament occupied by a listed body which is not established or conducted for profit providing:

- the Commissioner or the district valuer is satisfied it is to a substantial extent used or made available for use for purposes which are declared by the Recreational Charities Act (Northern Ireland) 1958 to be charitable or are otherwise charitable; and
- any charges are not more than necessary to defray reasonable expenses actually incurred; and

- the hereditament is not one to which article 31 (reduction of rates on certain hereditaments used for recreation) or article 41(2)(e) (exemption for recreational charities) applies; and
- it does not have a full liquor licence.

This is aimed at encouraging community halls but does not include those with full liquor licences as it is considered they have the capacity to generate income from the bar to assist with paying rates.

19.5.3.5 *Enterprise zones*

Article 42(1A) distinguishes as wholly exempt properties in enterprise zones not occupied by listed public bodies. There are no designated enterprise zones in Northern Ireland at present.

19.5.3.6 *ATM sites in rural areas*

Article 42(1F) distinguishes in the NAV list as wholly exempt from rates any automatic telling machine (ATM) site which is situated in a rural area during the relevant year.

19.5.3.7 *Industrial distinguishment*

Industrial de-rating still applies in Northern Ireland. For many years it was at full de-rating giving no liability. A decision to phase out industrial de-rating was announced in April 2003 following a public consultation exercise on the Review of Rating Policy in May 2002. Its phased removal began in April 2005 with 15% of the NAV becoming rateable. Originally the phasing out was planned as a gradual increase in liability until full rates were achieved in 2011. However, following a further review of rating policy initiated by the Northern Ireland Assembly, it was decided to cap liability at the then current level of 30%. The decision to complete the phasing out of industrial de-rating remains under review.

Article 43 together with Schedule 14 provide that industrial and freight transport hereditaments are to be distinguished by apportioning the NAV between the use for those and other purposes.

Where the NAV of the whole hereditament does not exceed a set figure (£2,290 from 1 April 2003) or where the use for other purposes does not exceed 10% of the part used for industrial purposes then the hereditament is deemed entirely industrial. If it exceeds 10% then only the element over 10% is treated as not used for industrial purposes.

EXAMPLE 1 – NAV EXCEEDS SET FIGURE (£2,290)

Premises having an NAV of £10,000 where £8,000 is attributable to Industrial uses and £2,000 attributable to other purposes.

NAV £10,000   Industrial £8,000   Other purposes £2,000
Sch 14 apportionment +/- 10%   +£ 800   – £ 800

Whilst the actual amount attributable to other purposes is £2,000 only the excess over 10% of the £8,000 attributable to Industrial uses is treated for apportionment as attributable to other purposes.

NAV in List   £10,000   Industrial £8800   Other purposes £1200

EXAMPLE 2 – NAV LESS THAN SET FIGURE (£2,290)

Premises having an NAV of £2,000 where £1,500 is attributable to Industrial uses and £500 attributable to other purposes.

NAV £2,000   Industrial £1,500   Other purposes £500
Sch 14 apportionment +/– 10%   +£150   – £150

But due to threshold all is treated as Industrial.

NAV in List   £2,000   Industrial £2,000   Other purposes £0

Schedule 7 deals with what happens to the NAV where the hereditament is distinguished as wholly or partially industrial.

If it is wholly industrial the NAV to appear in the valuation list is taken as 30% of the full figure. If it is only partly industrial the industrial part is taken at 30%, any part of the NAV apportioned to transport purposes is taken at 30% and the NAV attributable to its use for other purposes is taken at 100%.

19.5.4 Reliefs

19.5.4.1 Recreation

Article 31 provides for a reduction of rates on certain rateable properties used for 'prescribed recreations'. The level of reduction in such cases is an 80% reduction of the normal rate. This can be increased to 100% by the Department and the 100% applies to Community Amateur Sports Clubs (article 31(5A) and (5B)). Ancillary social facilities, such as bars, restaurants, card rooms, etc., are fully rateable.

'Prescribed recreation' is 'a recreation, whether conducted indoors or outdoors, which in the opinion of the Department demands an appreciable degree of physical effort …'. It follows that games that require a high degree of skill or contribute to intellectual development but require little physical effort do not generally qualify as prescribed recreations though billiards and pigeon racing are included.

It applies to a hereditament, or any part:

- used solely for the purposes of a prescribed recreation;
- which is occupied for the purposes of a club, society or other organisation;
- not established or conducted for profit;
- that does not employ any person to engage in any recreation for reward, except for the instruction of other persons who are themselves engaging or preparing to engage in it otherwise than for reward; and
- that is not distinguished in the list as exempt under article 41(2)(e) (recreational charities).

The wording is somewhat complex. In *Drummond Cricket and Football Club v Commissioner of Valuation* [2003] RA 209, a cricket club employed a professional coach who, as well as undertaking the usual coaching role, was also required to play in all club matches. The district valuer considered exemption did not apply because the exception appeared to permit a person engaged in recreation to be paid if the person was coaching amateur players only rather than engaged in playing for reward. The Lands Tribunal carefully examined the wording of the exception. It seemed to the Tribunal that it clearly contemplated the situation of a professional coach imparting instructions during the course of a competitive match in which the coach also took part as a player at the same time. The coach was not engaging 'in any recreation for reward, except for the instruction of other persons who are themselves engaging or preparing to engage in it otherwise than for reward'.

Together with article 44, article 31 engages in an unusual mix of exemption and relief by requiring an adjustment to the NAV and then providing the 80% relief.

The NAV is unaffected if the hereditament is used solely for the purposes of a prescribed recreation, though it is to be shown in the valuation list as such, but if only part is so used, article 44 requires the NAV is to be apportioned between the parts so used in various ways depending on the proportion:

i   if the amount apportioned to the part used solely for the purposes of a prescribed recreation is less than 20% of the NAV then no part is treated as so used. It is as if the prescribed recreation was not present;

ii if the apportioned amount is between 20% and up to 50% the apportionment is to be shown in the valuation list;

iii if the apportioned amount is between 50% and up to 80% the amount is to be increased by 20% and the amount apportioned to the remainder of the hereditament proportionally reduced and this amended apportionment is to be shown in the valuation list;

iv if the amount apportioned is 80% or more the hereditament is shown in the NAV list as used solely for the purposes of a prescribed recreation.

Having done this, article 31 provides for the rates to be reduced by 80% of the normal level if the hereditament is shown as used solely for the purposes of a prescribed recreation and if there is an apportionment then the amount apportioned to the part used for prescribed recreation has its rates reduced by 80%.

This remarkably complex scheme is designed to give a tapering relief with no relief being given when 20% or less of the NAV is attributable to prescribed recreation. There is special provision for where the proportion of use changes during the year.

19.5.4.2 *Persons with a disability*

Article 31B provides for a rebate for any hereditament which is used wholly or mainly for one or more of the specified qualifying purposes (including purposes ancillary thereto). The purposes are:

a   **the provision of residential accommodation for the care of persons suffering from illness or the after-care of persons who have been suffering from illness;**
b   **the provision of facilities for training or keeping suitably occupied persons suffering from illness or persons who have been suffering from illness;**
c   **the provision of such accommodation or facilities as are mentioned in sub-paragraph (a) or (b) for persons with a disability not falling within that sub-paragraph;**
d   **the provision of personal social services for persons with a disability;**
e   **the provision of facilities under section 15 of the Disabled Persons (Employment) Act (Northern Ireland) 1945.**

The rebate is 100% for wholly used hereditaments or 100% on the qualifying part's NAV if only partly used based on a certified apportionment by the district valuer.

19.5.4.3 *Hardship*

Article 33B permits the Department to reduce or repay where, it appears to the Department, there are exceptional circumstances and without such a reduction or repayment a person would suffer hardship.

19.5.4.4 *Small Business Rates Relief*

A scheme for Small Business Rates Relief (SBRR) came into effect in April 2010 for five years but has been renewed and annually extended. As with the schemes in England and Wales it is targeted at small businesses generally rather than specifically at businesses in need of support. Small businesses can, of course, be highly profitable and this suggests a degree of mis-targeting.

There are three levels of relief:

1 Hereditaments with an NAV of £2,000 or less receive 50% relief.
2 Hereditaments with an NAV of more than £2,000 but not more than £5,000 receive 25% relief.
3 Hereditaments with an NAV of more than £5,000 but not more than £15,000 receive 20% relief.

Various classes of hereditament are excluded from the scheme:

- unoccupied properties;
- partially unoccupied properties;
- ATMs;
- property used for the display of advertisements;
- car parks;
- sewage works;
- telecommunications masts;
- government buildings;
- contiguous properties;
- multiple properties (more than three);
- sports and recreational relief;
- freight and transport relief;
- industrial derating.

19.5.4.5 Rural general stores

Mandatory and discretionary relief is given to qualifying general stores and post offices in rural settlement lists under article 31C and Schedule 9A.

19.6 Administration and appeals

Responsibility for valuation and indeed the collection of rates both domestic and non-domestic lies with Land and Property Services Northern Ireland. This was formed by a merger of the Valuation and Lands Agency, the Ordnance Survey of Northern Ireland, the Rate Collection Agency, and the Land Registry within the Department of Finance on 1 April 2008.

Under article 36, the head of the Department of Finance is required to appoint an officer of the Department as the Commissioner of Valuation for Northern Ireland. The Commissioner is required to divide Northern Ireland into valuation districts and for each district appoint an officer of the Department as the district valuer. Any district valuer with the approval of the Commissioner may act in the stead of any other district valuer. The Commissioner is required to supervise the district valuers and other officers and may assign their functions.

Interestingly, the offices of the Commissioner are to be known as the 'Valuation Office' though this term is not used in practice.

Immediate responsibility for valuation, therefore, lies with the district valuer. There is an appeals system which begins with an application to the district valuer for a revision of a valuation list. It may move on to an appeal first to the Commissioner and then beyond to the Northern Ireland Valuation Tribunal (NIVT) or the Lands Tribunal for Northern Ireland.

Article 49(1) of the 1977 Order provides for any person – it does not need to be someone connected with the property – to serve an application on the district valuer for a 'revision' of a valuation list. The district valuer can make a revision, alter the assessments of other hereditaments or decide not to make an alteration. The district valuer is required to make any revision within three months, though this period can be extended for further three-month periods by serving a notice on the applicant stating the reasons for delay.

Article 54(3) provides that 'any valuation shown in a valuation list with respect to a hereditament shall be deemed to be correct until the contrary is shown'. This means the burden of proof rests on the applicant to show that the list entry is incorrect. This is, of course, very important should the case proceed on appeal to the Lands Tribunal.

If the district valuer amends the list then certificates of the alteration are to be served on various persons. These are the person who made the application (the occupier in most cases), then any person on whom a copy of the application was served and who submitted comments or furnished evidence, and the district council and/or water authority if either of them requested a copy.

If, on the other hand, the district valuer decides not to amend the list then a notice of the decision is to be served on the applicant. Sub-paragraph (6) provides for the district valuer to regard an application as frivolous or vexatious and in that circumstance the district valuer can simply respond with notice to that effect and the applicant has no right to challenge the decision by appeal under article 51.

Article 51 provides that any person, other than the Department, who is aggrieved by the district valuer's decision to alter or not alter a list may, within 28 days of the service on him of the certificate of alteration or notice, appeal to the Commissioner. An appeal is instituted by a notice of appeal, signed by the appellant, stating:

a the alteration desired or objected to;
b the reasons for desiring or objecting to the alteration; and
c where the appellant is not the owner, or is not the occupier, of the hereditament, the name and address of the owner, or, as the case may require, of the occupier or of both.

The appellant is required to serve a copy of the appeal within the 28-day period on the occupier and the owner if different from the appellant.

The appellant may, at any time before the Commissioner's decision on the appeal is issued, abandon the appeal by serving notice on the Commissioner.

Article 52 requires the Commissioner to investigate the subject matter of the appeal and review the district valuer's decision. In doing so the Commissioner:

- shall afford an opportunity for concerned persons to comment and furnish oral or other evidence;
- may obtain information from such persons and make inquiries;
- may call for a report on the hereditament from a suitably qualified officer other than the officer previously employed;
- shall make such decision as appears proper; and
- where appropriate alter the list.

If the Commissioner alters the list a certificate is served on the same persons as the district valuer would serve if making a revision. If the Commissioner dismisses the appeal or alters the valuation list other than as desired by the appellant, a notice of dismissal or a statement of reasons for making the alteration is served on the appellant, the district council if requested by the council to do so and every other person on whom a copy of the notice of appeal was served providing the person submitted comments or furnished evidence.

There is an optional power for the district valuer, with the consent of the applicant, to transfer directly to the Commissioner an application served on the district valuer under article 49. The district valuer's functions are then exercisable by the Commissioner (article 49A).

Any person, other than the Department, who is aggrieved by the decision of the Commissioner may appeal under article 54 to the Valuation Tribunal for NAVs up to £12,000 or the Lands Tribunal for higher values. On appeal the Tribunal may:

a **make any decision that the Commissioner might have made; and**
b **if any alteration in a valuation list is necessary to give effect to the decision, direct that the list be altered accordingly.**

Note: The Valuation Tribunal was primarily established to hear capital value appeals and, initially, rating appeals have not been sent to it but, instead, all have gone directly to the Lands Tribunal. This may change.

An appeal lies from the Valuation Tribunal to the Northern Ireland Lands Tribunal and thence to the Court of Appeal on a point of law.

Alterations to a valuation list arising from an application for revision made within six months of the list coming into force are backdated to the date the list came into force.

Where the alteration is to include a newly constructed hereditament or one which has been out of occupation on account of structural alterations whether remaining in the list or not, the alteration has effect from the date of coming into occupation or where a Material Change of Circumstance (MCC) occurs requiring the change.

Corrections of clerical errors are effective back to the date the error originally had effect.

19.7 Domestic rates

Domestic rates have continued in Northern Ireland. The province has not had the short lived Community Charge, Poll Tax, in the early 1990s or the Council Tax as in the rest of the United Kingdom.

Domestic rates continued to be levied on the basis of rental value until the Rates (Capital Values Etc.) (Northern Ireland) Order 2006 changed the basis on which domestic hereditaments are assessed. This stated that from 1 April 2007, domestic properties would be assessed by reference to their capital value. The capital value being the amount 'the hereditament might reasonably be expected to realise if it had been sold on the open market on the relevant capital valuation date'.

It is therefore a proper capital value based tax rather than the broad brush approach of the Council Tax and reflects the true variations between different hereditaments' values rather than grouping them together in broad bands of values.

Domestic rates work in the same way as those on non-domestic properties with the value multiplied by a poundage (pence in the pound). Each of the 11 council areas sets its own poundage. For example, if the capital value of a house in the valuation list is £80,000 and it is within the Armagh City, Banbridge and Craigavon Council area, the rates calculation for the year 2016 to 2017 would have been:

£80,000 (capital value) × 0.008148 = £651.84 rates payable

The maximum capital valuation for a domestic property is £400,000. For any hereditament valued over £400,000, the additional value is disregarded when calculating rates liability.

Hereditaments with a mixed use, such as a shop with living accommodation over, have the non-domestic part assessed on the basis of its NAV and the domestic part is assessed on the basis of its capital value (see article 39(1B) of the 1977 Order).

Farmhouses are valued in the same way as other domestic hereditaments, but there is a 20% reduction if:

• they are occupied in connection with agricultural land;
• the main occupation of the person who lives in the house is farming that land.

As with rates on non-domestic hereditaments an AVD is adopted – known as the 'relevant capital valuation date' – and is 1 January 2005. To accommodate the change to a capital value basis the 1977 Order was again amended quite extensively.

Article 39(1A) provides that specified hereditaments are to be valued upon 'an estimate of their capital value'. These are:

- any dwelling-house;
- any private garage;
- any private storage premises.

Hereditaments which are used partly as a private dwelling and partly for other purposes, e.g. a shop with living accommodation or a public house with living accommodation, are valued on the basis of an estimate of both its NAV and its capital value.

Article 39(1C) covers how hereditaments which are not in use should be treated. They are to be treated as dwelling-houses, private garages or private storage premises if the district valuer or Commissioner considers they will, when next in use, fall partially or wholly into one of those classes.

Schedule 5 provides the definitions of each of the classes of hereditament to be valued on a capital value basis as follows.

19.7.1 Dwelling-houses

'Dwelling-house' is defined as 'a hereditament used wholly for the purposes of a private dwelling' but this is subject to a few exceptions.

The first is the 'hotel exception' for short stay accommodation such as guest houses and hotels. These will not be treated as private dwellings. This exception is somewhat of a 'belt and braces' job as they would probably not be thought of as constituting private dwellings in any case. The exception makes clear that a hereditament which is wholly or mainly used in the course of a business for the provision of short stay accommodation, that is to say accommodation:

- which is provided for short periods to individuals whose sole or main residence is elsewhere; and
- which is not self-contained self-catering accommodation provided commercially; shall be deemed not to be used for the purposes of a private dwelling.

However, 'bed and breakfast' type accommodation will not fall into this exception and will still be domestic notwithstanding it is accommodation provided for short stay if:

- it is intended that short stay accommodation will not be provided within the hereditament for more than six persons simultaneously;

- the person intending to provide such accommodation intends to have his sole or main residence there; and
- that any use of living accommodation within the hereditament which would, apart from this sub-paragraph, cause any part of it to be deemed not to be used for the purposes of a private dwelling, will be subsidiary to the use of the hereditament for, or in connection with, his sole or main residence.

The second exception covers holiday cottages and the like – the 'self-catering' exception. If these, whilst appearing very much to be private dwellings, are used substantially for holidays then they are not really somewhere people 'dwell'. A hereditament or self-contained part will not be deemed a private dwelling if the owner or occupier intends in the coming year that it is to be available for letting commercially, as self-catering accommodation, for short periods totalling 140 days or more. This will not apply if it is used as the sole or main residence of any person because then it would, indeed, be a private dwelling.

The definition in Schedule 5 provides that merely because the hereditament includes a garage, outhouse, garden, park, pleasure ground, yard, court, forecourt or other appurtenance which is not used, or not used wholly, for the purposes of a private dwelling, that does not result in the hereditament being treated as not being used wholly as a private dwelling.

So having a garage with a house used by the occupier for storing a commercial taxi would not result in the house being treated as other than wholly used as a private dwelling. This treatment of minor non-domestic uses as not changing the status of a dwelling is extended to other parts of the hereditament other than garages, etc., providing the part was not constructed, or had not been adapted, for those other purposes. So using a bedroom as an office would not result in the house being treated as anything other than a private dwelling because it would not have been constructed or adapted as an office. However, if a chiropodist built a room on at the side of the house as a surgery or converted a garage into a surgery then that would have been constructed or adapted, and the hereditament would not be treated as just used wholly as a private dwelling.

19.8 Private garages

'Private garage' is defined as 'a hereditament which is used wholly or mainly for the accommodation of a motor vehicle'. However, specifically excluded from this are hereditaments used for the purposes of a trade or business, or by a charity, a public body or any other body that is not established or conducted for profit.

19.8.1 Private storage premises

'Private storage premises' are defined as 'a hereditament which is used wholly in connection with a dwelling-house or dwelling-houses and so used wholly or mainly for the storage of domestic articles belonging to the residents'.

'Domestic articles' are defined (unlike the equivalent English/Welsh) in legislation as meaning:

a   household stores and other articles for domestic use;
b   light vehicles, whether mechanically-propelled or not.

19.8.2 *Valuation*

Schedule 12 provides the definition of capital value:

> **The amount which, on the assumptions mentioned in paragraphs 9 to 15, the hereditament might reasonably have been expected to realise if it had been sold on the open market by a willing seller on the relevant capital valuation date.**

As mentioned, the relevant capital valuation date is 1 April 2005.

As with the rental value assumption, when making a revision to an existing list regard has to be had to the capital values in the valuation list of comparable hereditaments in the same state and circumstances as the hereditament whose capital value is being revised.

The main valuation assumptions are:

- The sale is with vacant possession.
- The estate sold is the fee simple absolute or, in the case of a flat, a lease for 99 years at a nominal rent.
- The hereditament is sold free from any rent charge or other incumbrance.
- The hereditament is in an average state of internal repair and fit out, having regard to the age and character of the hereditament and its locality.
- The hereditament is otherwise in the state and circumstances in which it might reasonably be expected to be on the relevant date (this is the date a list comes into force, currently 1 April 2007). This assumption applies only in preparing a list because of the requirement in making a revision to have regard to other hereditaments in the same state and circumstances. This effectively means the current state of the hereditament is compared to the current state of other hereditaments.
- The hereditament has no development value other than value attributable to permitted development.
- The hereditament will always remain as a private dwelling, garage or private storage premises or, though not a private dwelling, will always remain partly used for the purposes of a private dwelling.
- There has not been contravention of any statutory provision, or any requirement or obligation, whether arising under a statutory provision, an agreement or otherwise which would affect the capital value.

19.8.3 Appeals

The NIVT was established on 1 April 2007 to hear and determine appeals against new capital values for property in Northern Ireland in respect of domestic rates. It also falls within the NIVT's remit to hear and determine appeals against decisions relating to a number of rating rebates. It considers appeals by domestic ratepayers who have first applied to the district valuer and then sought a review by the Commissioner of Valuation but been dissatisfied.

19.9 Empty property

The rating of unoccupied non-domestic property was introduced in Northern Ireland from 1 April 2004, which was later than on the mainland. The term used is Non-domestic Vacant Rating (NDVR).

Liability is at 50% of the occupied rate.

Liability flows from the Rates (Northern Ireland) Order 1977 (1977/2157) which, in article 25A, provides for a liability on the person entitled to possession to apply for hereditaments specified by Schedule 8A. This Schedule provides for a statutory rule listing classes of hereditament to be prescribed. This list of prescribed classes is contained in the Rates (Unoccupied Hereditaments) Regulations (Northern Ireland) 2011 (2011/36).

As in England and Wales, to be prescribed and a person liable, a hereditament has to consist of 'a building or of part of, any building, together with any land ordinarily used or intended for use for the purposes of the building or part'.

19.9.1 Exclusions

Again as in England and Wales, there is then a list of exclusions from liability where:

1. The hereditament has been unoccupied for a continuous period not exceeding three months.

   This three-month void period is extended to 12 months in two situations:

   a   Vacant and in the developer's hands – where the hereditament was first included in a valuation list after 31 March 2007 and has not previously been occupied, and the person entitled to possession of the hereditament is a developer.

   b   Public funded developments – where the person entitled to possession of the vacant hereditament throughout the vacant period is a person that:

      i   is a company limited by guarantee within the meaning of Part 1 of the Companies Act 2006;

ii   is funded in respect of its capital costs, in whole or in part, by a public body; and

iii provides accommodation and support services for a trade or business.

In the above:

- 'Developer' means a person who has undertaken, arranged or carried out development work and who is also the first owner of the hereditament.
- 'Development work' means any building operations which result in a new building.
- 'Owner' means any person, for the time being, receiving or entitled to receive, on his own account, the rack rent of the hereditament in connection with which the word is used or who, if the hereditament were let at a rack rent, would so receive or be entitled to receive that rent.

2. To the extent (if any) to which it appears that the hereditament, when next in use, will be used for the purposes of a private dwelling.

3. The hereditament is a qualifying industrial hereditament. Here:

- 'qualifying industrial hereditament' means any hereditament other than a retail hereditament in relation to which all buildings comprised in the hereditament are constructed or adapted:

a   in the course of a trade or business; and

b   for one or more of the following purposes, or one or more such purposes and one or more purposes ancillary thereto:

i   the manufacture, repair or adaptation of goods or materials;

ii   the working or processing of minerals;

iii the generation of electricity;

iv   'retail hereditament' means any hereditament where any building or part of a building comprised in the hereditament is constructed or adapted for the purpose of the retail provision of goods or services on or from the hereditament.

4. The person entitled to possession of the hereditament is prohibited by law from occupying it or allowing it to be occupied.

5. The hereditament is kept vacant by reason of action taken by or on behalf of any public body with a view to prohibiting its occupation or to acquiring it.

6. Any building or part of a building comprising the hereditament or part of the hereditament is included in a list compiled under article 42 of the Planning (Northern Ireland) Order 1991 or is the subject of a building preservation notice within the meaning of article 42A of that Order.

7. Any building or part of a building comprising the hereditament or part of the hereditament is included in the Schedule of Monuments compiled under article 3(1) of the Historic Monuments and Archaeological Objects (Northern Ireland) Order 1995.

8. The rateable NAV of the hereditament is less than £2,000.
9. The person entitled to possession of the hereditament is so entitled only in his capacity as the personal representative of a deceased person.
10. There subsists in respect of the estate of the person entitled to possession of the hereditament a bankruptcy order within the meaning of Parts I and VIII to X of the Insolvency (Northern Ireland) Order 1989.
11. The person entitled to possession of the hereditament is so entitled in his capacity as a trustee under a deed of arrangement to which Chapter 1 of Part VIII of the Insolvency (Northern Ireland) Order 1989 applies.
12. The person entitled to possession of the hereditament is:

    a a company in administration within the meaning of paragraph 2 of Schedule B1 to the 1989 Order or is subject to an administration order made under the former administration provisions within the meaning of article 2 of the Insolvency (2005 Order) (Transitional Provisions and Savings) Order (Northern Ireland) 2006; or
    b a company which is subject to a winding-up order made under the Insolvency (Northern Ireland) Order 1989 or which is being wound up voluntarily under that Order.

13. The person entitled to possession of the hereditament is so entitled in his capacity as liquidator by virtue of an order made under article 98 or article 123 of the Insolvency (Northern Ireland) Order 1989 Order.
14. The person entitled to possession of the hereditament is the Foyle, Carlingford and Irish Lights Commission.

19.9.2 Completion notices

A completion notice procedure exists with completion days determined under article 25B and Schedule 8B to the 1977 Order in relation to a building that comprises the whole or part of that hereditament. The Schedule provides that if it appears to the Department of Finance that a building has been completed or the work remaining to be done on a new building is such that the building can reasonably be expected to be completed within three months, the Department may serve a completion notice on the person entitled to possession.

The notice needs to specify the building to which it relates and the day that the Department proposes as the completion day in relation to the building. There is provision for an appeal, which has to be made not later than 28 days from the date of service of the completion notice.

19.9.3 Temporary rebate for previously unoccupied premises

The Rates (Amendment) Act (Northern Ireland) 2012 amended the 1977 Order to provide a temporary 50% rebate for previously vacant retail hereditaments in the new article 31D.

This applied to 1 April 2016 and then, subsequently, was extended to 1 April 2017.

An application now needs to be made by 1 April 2023 and applies to hereditaments that become occupied during the year up to that date but were unoccupied for at least 12 months immediately before re-occupation and either were used for retail purposes or, if never occupied, retail was their most likely prospective use. The rebate applies for 24 months from the date of occupation at 50% of the normal charge and comes to an end if the person ceases to occupy.

Other reliefs and exemptions do not apply if this rebate is obtained.

Appendices

Appendix 1A – England

A1.1 Example 1 – NCA exceeds BL, RV >£20,000 and <£100,000

	Abbreviation	Formula	Value	Regulation	Calculation
NCA exceeds BL, RV >£20,000 and <£100,000					
Property outside Greater London					
2022 RV			£60,000		
2023 RV			£80,000		
Notional Chargeable Amount					
RV in 2023 rating list	A		£80,000	Regulation 7(2)	
NNDR multiplier for relevant year – 2023/24	B		£0.512	Regulation 12	
Small business NNDR multiplier – 2023/24	D		£0.499	Regulation 7(2)	
Notional Chargeable Amount (NCA)	NCA	A × D	£39,920	Regulation 7(1)	£80,000 × £0.499
Base liability					
RV 31 March 2023	y		£60,000	Regulation 8	
Small business multiplier 2022/23	Z		£0.499	Regulation 8	
Base liability – 2022/23	BL	Y × Z	£29,940	Regulation 8	£60,000 × £0.449
Appropriate fraction					
X	X		115	Regulation 10	
Q	Q		1	Regulation 10(7)	

(Continued)

A1.1 (Continued)

	Abbreviation	Formula	Value	Regulation	Calculation
Appropriate fraction	AF	(X × Q)/100	1.15	Regulation 10	(115 × 1) / 100
Chargeable amount					
Number of days in year	E		365	Regulation 12	
The amount prescribed by s.44(9)	N		1	Regulation 13	
RV 2023	U		£80,000	Regulation 14	
		(B − D) × N	£1,040	Regulation 15*	(£0.512 − £0.499) × 100
		(BL × AF) + U	£35,471.00		(£29,940 × 1.15) + £1,040
Chargeable amount			£35,471.00		

* amended to calculate on yearly basis

A1.2 Example 2 – NCA exceeds BL, RV <£20,000

Example 2	Abbreviation	Formula	Value	Regulation	Calculation
NCA exceeds BL, RV <£20,000					
Property outside Greater London					
2022 RV			£15,000		
2023 RV			£19,000		
Notional Chargeable Amount					
RV in 2023 rating list	A		£19,000	Regulation 7(2)	
NNDR multiplier for relevant year – 2023/24	B		£0.512	Regulation 12	
Small business NNDR multiplier – 2023/24	D		£0.499	Regulation 7(2)	
Notional Chargeable Amount (NCA)	NCA	A × D	**£9,481**	Regulation 7(1)	£19,000 × £0.499
Base liability					
RV 31 March 2023	y		£15,000	Regulation 8	
Small business multiplier 2022/23	Z		£0.499	Regulation 8	
Base liability – 2022/23	BL	Y × Z	**£7,485**	Regulation 8	£50,000 × £0.449
Appropriate fraction					
X	X		105	Regulation 10	
Q	Q		1	Regulation 10(7)	
Appropriate fraction	AF	(X × Q)/100	1.05	Regulation 10	(105 × 1) / 100

(*Continued*)

A1.2 (Continued)

Example 2	Abbreviation	Formula	Value	Regulation	Calculation
Chargeable amount					
Number of days in year	E		365	Regulation 12	
The amount prescribed by s.44(9)	N		1	Regulation 13	
RV 2023	U		£19,000	Regulation 14	$(£0.512 - £0.499) \times 100$
		$(B - D) \times N$	£247	Regulation 15*	$(£7,485 \times 1.05) + £1,040$
		$(BL \times AF) + U$	£8,106.25		
Chargeable amount			**£8,106.25**		
				* amended to calculate on yearly basis	

A1.3 Example 3 – NCA exceeds BL, RV>£100,000

Example 3	Abbreviation	Formula	Value	Regulation	Calculation
NCA exceeds BL, RV>£100,000					
Property outside Greater London					
2022 RV			£125,000		
2023 RV			£150,000		
Notional Chargeable Amount					
RV in 2023 rating list	A		£150,000	Regulation 7(2)	
NNDR Multiplier for relevant year – 2023/24	B		£0.512	Regulation 12	
Small business NNDR multiplier – 2023/24	D		£0.499	Regulation 7(2)	
Notional Chargeable Amount (NCA)	NCA	A × D	**£74,850**	Regulation 7(1)	£150,000 × £0.499
Base Liability					
RV 31 March 2023	y		£125,000	Regulation 8	
Small business multiplier 2022/23	Z		£0.499	Regulation 8	
Base Liability – 2022/23	BL	Y × Z	**£62,375**	Regulation 8	£125,000 × £0.449
Appropriate Fraction					
X	X		130	Regulation 10	
Q	Q		1	Regulation 10(7)	
Appropriate fraction	AF	(X × Q)/100	1.3	Regulation 10	(130 × 1) / 100

(*Continued*)

A1.3 (Continued)

Example 3	Abbreviation	Formula	Value	Regulation	Calculation
Chargeable amount					
Number of days in year			365	Regulation 12	
The amount prescribed by s.44(9)			1	Regulation 13	
RV 2023	E		£150,000	Regulation 14	
U	N	(B − D) × N	£1,950	Regulation 15*	(£0.512 − £0.499) × 100
	U	(BL × AF) + U	£83,037.50		(£62,375 × 1.15) + £1,040
Chargeable Amount			**£83,037.50**		

* amended to calculate on yearly basis

Appendix 1B – Wales

The following examples of Transitional Relief calculations can be found on the Welsh Government website at https://businesswales.gov.wales/non-domestic-rates-transitional-rates-relief-2023-revaluation.

Example 1

A hereditament's rateable value increases from £6,000 to £9,000 and SBRR entitlement is reduced.

Determine the base liability

On 31 March 2023, the hereditament is eligible for SBRR:

$$(6{,}000 \times 0.535) / 5{,}000{,}000 = £0$$

Determine the notional chargeable amount

On 1 April 2023, the hereditament is eligible for SBRR:

$$(9{,}000 \times 0.535) / 2 = £2{,}408$$

Compare the base liability to the notional chargeable amount

Check whether the notional chargeable amount is more than £300 greater than the base liability:

$$2{,}408 - 0 = £2{,}408$$

Determine the maximum deduction to the chargeable amount

For each day in the financial year commencing 1 April 2023:

$$((2{,}408 - 0) \times 0.67) / 366 = £4.41$$

For each day in the financial year commencing 1 April 2024:

$$((2,408 - 0) \times 0.34) / 365 = £2.24$$

Determine the chargeable amount (liability) without transitional relief

As SBRR is the only other statutory relief applicable to the hereditament, the daily chargeable amount for the financial year commencing 1 April 2023 is:

$$(9,000 \times 0.535) / (366 \times 2) = £6.58$$

Determine the chargeable amount after transitional relief

For each day in the financial year commencing 1 April 2023, the daily chargeable amount is:

$$6.58 - 4.41 = £2.17$$

Example 2

A hereditament's rateable value increases from £10,000 to £13,000 and SBRR entitlement is removed.

Determine the base liability

On 31 March 2023, the hereditament is eligible for SBRR:

$$(10,000 \times 0.535) / 1.5 = £3,567$$

Determine the notional chargeable amount

On 1 April 2023, the hereditament is not eligible for SBRR:

$$13,000 \times 0.535 = £6,955$$

   Compare the base liability to the notional chargeable amount
Check whether the notional chargeable amount is more than £300 greater than the base liability:

$$6,955 - 3,567 = £3,388$$

Determine the maximum deduction to the chargeable amount

For each day in the financial year commencing 1 April 2023:

$$((6,955 - 3,567) \times 0.67) / 366 = £6.20$$

For each day in the financial year commencing 1 April 2024:

$$((6,955 - 3,567) \times 0.34) / 365 = £3.16$$

Determine the chargeable amount (liability) without transitional relief

As the hereditament is not eligible for any other statutory reliefs, the daily chargeable amount for the financial year commencing 1 April 2023 is:

$$(13,000 \times 0.535) / 366 = £19.00$$

Determine the chargeable amount after transitional relief

For each day in the financial year commencing 1 April 2023, the daily chargeable amount is:

$$19.00 - 6.20 = £12.80$$

Example 3

A hereditament's rateable value increases from £50,000 to £60,000.

Determine the base liability

On 31 March 2023, the hereditament is not eligible for SBRR:

$$50,000 \times 0.535 = £26,750$$

Determine the notional chargeable amount

On 1 April 2023, the hereditament is not eligible for SBRR:

$$60,000 \times 0.535 = £32,100$$

Compare the base liability to the notional chargeable amount

Check whether the notional chargeable amount is more than £300 greater than the base liability:

$$32,100 - 26,750 = £5,350$$

Determine the maximum deduction to the chargeable amount

For each day in the financial year commencing 1 April 2023:

$$((32,100 - 26,750) \times 0.67) / 366 = £9.79$$

For each day in the financial year commencing 1 April 2024:

$$((32,100 - 26,750) \times 0.34) / 365 = £4.98$$

Determine the chargeable amount (liability) without transitional relief

As the hereditament is not eligible for any other statutory reliefs, the daily chargeable amount for the financial year commencing 1 April 2023 is:

$$(60,000 \times 0.535) / 366 = £87.70$$

Determine the chargeable amount after transitional relief

For each day in the financial year commencing 1 April 2023, the daily chargeable amount is:

$$87.70 - 9.79 = £77.91$$

Example 4

A hereditament's rateable value increases from £15,000 to £16,000. The hereditament is eligible for charitable relief.

Determine the base liability

On 31 March 2023, the hereditament is not eligible for SBRR:

$$15,000 \times 0.535 = £8,025$$

Determine the notional chargeable amount

On 1 April 2023, the hereditament is not eligible for SBRR:

$$16,000 \times 0.535 = £8,560$$

Compare the base liability to the notional chargeable amount

Check whether the notional chargeable amount is more than £300 greater than the base liability:

$$8,560 - 8,025 = £535$$

Determine the maximum deduction to the chargeable amount

For each day in the financial year commencing 1 April 2023:

$$((8,560 - 8,025) \times 0.67) / 366 = £0.98$$

For each day in the financial year commencing 1 April 2024:

$$((8{,}560 - 8{,}025) \times 0.34) / 365 = £0.50$$

Determine the chargeable amount (liability) without transitional relief

As the hereditament receives 80% mandatory charitable relief, the daily chargeable amount for the financial year commencing 1 April 2023 is:

$$(16{,}000 \times 0.535) / (366 \times 5) = £4.68$$

Determine the chargeable amount after transitional relief

For each day in the financial year commencing 1 April 2023, the daily chargeable amount is:

$$4.68 - 0.98 = £3.70$$

This is the daily chargeable amount before any 'top-up' charitable relief (i.e. above the mandatory 80%) awarded at the discretion of the local authority is applied.

Example 5

A hereditament's rateable value increases from £15,000 to £16,000. The hereditament is unoccupied.

Determine the base liability

On 31 March 2023, the hereditament is not eligible for SBRR:

$$15{,}000 \times 0.535 = £8{,}025$$

Determine the notional chargeable amount

On 1 April 2023, the hereditament is not eligible for SBRR:

$$16{,}000 \times 0.535 = £8{,}560$$

Compare the base liability to the notional chargeable amount

Check whether the notional chargeable amount is more than £300 greater than the base liability:

$$8{,}560 - 8{,}025 = £535$$

Determine the maximum deduction to the chargeable amount

For each day in the financial year commencing 1 April 2023:

$$((8{,}560 - 8{,}025) \times 0.67) / 366 = £0.98$$

For each day in the financial year commencing 1 April 2024:

$$((8{,}560 - 8{,}025) \times 0.34) / 365 = £0.50$$

Determine the chargeable amount (liability) without transitional relief

For each day in the financial year commencing 1 April 2023 when empty property relief applies, the daily chargeable amount is £0.

For any other day in the financial year commencing 1 April 2023, the daily chargeable amount is:

$$(16{,}000 \times 0.535) / 366 = £23.39$$

Determine the chargeable amount after transitional relief

For each day in the financial year commencing 1 April 2023 when empty property relief applies, the daily chargeable amount is:

$$0 - 0.98 < £0$$

As the chargeable amount cannot be negative, it is set to zero.

For any other day in the financial year commencing 1 April 2023, the daily chargeable amount is:

$$23.39 - 0.98 = £22.41$$

Appendix 2

The whole area of non-domestic rating is in constant change whether brought about by changes to legislation or as a result of legal decisions.

The following is a short summary of the legislative changes that have been proposed but not yet enacted.

A2.1 Non-Domestic Rating Act 2023

The Non-Domestic Rating Act 2023 received royal assent on 26th October 2023. It makes a number of technical changes that had been planned for some time.

The various provisions include:

- Power to introduce 'improvement relief'. This is a twelve-month grace period after a property has been improved, during which any rise in liability due to set 'qualifying improvements' will not increase rates liability
- Introducing rate relief for low carbon heat networks
- Introducing improvement and charitable relief for central rating lists
- Providing for triennial not quinquennial revaluations. Previously (see section 15.1 in this book) the requirement was for 5 yearly revaluations. The period in the future for revaluations is reduced to every 3 years.
- Removing the legislative requirement for Transitional Relief schemes to be revenue-neutral (see sections 1.9 and 16.11.1 in this book)
- Permitting valuation officers to disclose analysis of rental information to ratepayers. This is in addition to the existing provision of 'summary valuation calculations' and is to assist ratepayers in understanding how their rateable values have been valued
- Creating a duty on ratepayers to provide information to their valuation officer, permitting financial penalties on ratepayers for non-compliance, and making it a new criminal offence where false information is provided to valuation officers
- Amending Sch 6 para 2(7) LFGA 1988 to provide that the effects of legislation, advice or guidance post AVDs are not to be taken into account in RVs

The last amendment is alluded to in chapter 6 of this book at 6.5.5. This makes some quite significant amendments to the existing legislation.

It renumbers the list of 'matters listed in Schedule 6, paragraph 2(7), (shown on page 129 at 6.5.3) that are to be taken as they are at the material day:

a)   **matters affecting the physical state of the hereditament,**
aa) **matters affecting the physical enjoyment of the hereditament,**
b)   **the mode of category of occupation of the hereditament,**
c)   **The quantity of minerals or other substances in or extracted from the hereditament,**
cc) **The quantity of refuse or waste material which is brought onto and permanently deposited on the hereditament,**
d)   **matters affecting the physical state of the locality in which the hereditament is situated,**
da) **matters which, though not affecting the physical state of the locality in which the hereditament is situated, are nonetheless physically manifest there, and**
e)   **the use or occupation of other premises situated in the locality of the hereditament.**

In succession to the coronovirus measures, it introduces a new paragraph 2ZA in Schedule 6 that applies for new lists and also any alteration to 2023 lists in England. It says that in determining RV no account to be taken of any change to matters that come under (aa), (da) and (e) in the new list above, if the change is directly or indirectly attributable to a relevant factor.

Relevant factors are:

• legislation
• provision made under, and given effect by, legislation
• advice or guidance given by a public authority
• anything done by a person with a view to compliance with any of those factors

'Legislation' and all the other factors apply to legislation etc of 'any country or territory' not just the UK.
'legislation' includes any provision of a legislative character
'public authority' includes any person exercising functions of a public nature.

This provision does not apply to the extent that it concerns whether a hereditament or some part of a hereditament is or is not:

• a domestic property
• is or is not exempt from local non-domestic rating.

The government's intention, like the original coronavirus regulations, is to prevent alterations due to post AVD events that are not dissimilar to economic or market changes. However it remains to be seen how wide-ranging the effect is given they refer to provisions 'made under, and given effect by, legislation of any country or territory'. Does this include demolition orders, planning permissions and the like?

A2.2 NDR Avoidance and Evasion: Consultation

The government announced a consultation to explore the causes of, and potential measures to combat, avoidance, evasion, and poor rating agent behaviour within the NDR system. Consultation closed on 28th September 2023. Its purpose is to:

- consult on specific measures to reform 'Empty Property Relief', to address known avoidance schemes;
- gather evidence on wider avoidance and evasion practises within NDR, and seek views on whether BAs have sufficient powers and information to combat them;
- gather evidence on "rogue" rating agent behaviour and seek views on how the government could address any problems.

Possible reforms include:

- Changing the present 6 week void period to 3 or 6 months
- Repeated short periods of re-occupation not allowing further 3 months periods before Empty Property Rates become payable
- Altering what constitutes occupation for Empty Property Rate purposes
- Altering the 'when next in use' charity exemption from Empty Property Rates

Index

Printed in the United States
by Baker & Taylor Publisher Services